Lecture Notes in Computer S

T0237887

Commenced Publication in 1973
Founding and Former Series Editors:
Gerhard Goos, Juris Hartmanis, and Jan van Leeuwen

André Gagalowicz   Wilfried Philips (Eds.)

Computer Vision/ Computer Graphics Collaboration Techniques

4th International Conference, MIRAGE 2009
Rocquencourt, France, May 4-6, 2009
Proceedings

 Springer

Volume Editors

André Gagalowicz
INRIA Rocquencourt
MIRAGES project
Domaine de Voluceau, 78153 Le Chesnay Cedex 1, France
E-mail: Andre.Gagalowicz@inria.fr

Wilfried Philips
Ghent University
Department of Telecommunication and Information Processing
St.-Pietersnieuwstraat 41, 9000 Gent, Belgium
E-mail: philips@telin.UGent.be

Library of Congress Control Number: Applied for

CR Subject Classification (1998): I.3, H.5.2, I.5, J.3, I.3.7

LNCS Sublibrary: SL 6 – Image Processing, Computer Vision, Pattern Recognition, and Graphics

ISSN   0302-9743
ISBN-10   3-642-01810-6 Springer Berlin Heidelberg New York
ISBN-13   978-3-642-01810-7 Springer Berlin Heidelberg New York

springer.com

© Springer-Verlag Berlin Heidelberg 2009
Printed in Germany

Typesetting: Camera-ready by author, data conversion by Scientific Publishing Services, Chennai, India
Printed on acid-free paper   SPIN: 12674831   06/3180   5 4 3 2 1 0

Preface

This volume collects the papers accepted for presentation at MIRAGE 2009.

The MIRAGE conference is continuing to receive international recognition, with this year's presentations coming from 25 countries despite the large worldwide financial crisis. This time Asia submitted far fewer papers than previously and fewer than Europe. France proved to be the most active scientifically with a total of 16 submitted papers. Germany came second (10 submitted papers) and China third (8 papers).

We received a total of 83 submissions and accepted 41 as oral presentations, over the three-day event. All papers were reviewed by three to four members of the Program Committee. The final selection was made by the Conference Chairs.

At this point, we wish to thank the Program Committee and additional referees for their timely and high-quality reviews. We also thank the invited speakers Luc Van Gool, Frank Multon and Raquel Urtasun for kindly accepting to present very interesting talks.

MIRAGE 2009 was organized by INRIA Rocquencourt and took place at INRIA, Rocquencourt, close to Versailles. We believe that the conference proved to be a stimulating experience for all.

March 2009

A. Gagalowicz
W. Philips

Organization

Mirage 2009 was organized by INRIA and Ghent University.

Conference Chair

André Gagalowicz   INRIA Rocquencourt, Le Chesnay, France

Conference Co-chairs

Jacques Blanc-Talon DGA, Bagneux, France
Frank Nielsen   LIX, Ecole Polytechnique, Palaiseau, France

Organizing Committee

André Gagalowicz   INRIA Rocquencourt, Le Chesnay, France
Nathalie Gaudechoux INRIA Rocquencourt, France
Chantal Girodon   INRIA Rocquencourt, Rocquencourt, France
Wilfried Philips   Ghent University - IBBT, Ghent, Belgium

Program Committee

Nina Amenta   University of California at Davis, Davis, USA
Moshe Ben-Ezra   Microsoft Asia, Beijing, P.R. China
Jacques Blanc-Talon DGA, Bagneux, France
Kadi Bouatouch   IRISA, Rennes, France
Edmond Boyer   INRIA Rhone-Alpes, Saint Ismier, France
José Braz   Polytechnic Institute of Setúbal, Setúbal,
         Portugal
Antonio Camurri   University of Genova, Genova, Italy
Leszek Chmielewski Institute of Fundamental Technological
         Research (IPPT PAN, Warsaw, Poland)
Adrian Clark   University of Essex, Colchester, UK
Peter Eisert   Heinrich-Hertz-Institut, Berlin, Germany
Alexandre Francois   Tufts University, Medford, USA
Bernd Froehlich   Bauhaus-Universität Weimar, Weimar,
         Germany
Andrea Fusiello   Università degli Studi di Verona, Verona, Italy
André Gagalowicz   INRIA Rocquencourt, Le Chesnay, France
Michael Gleicher   University of Wisconsin, Madison, USA
Oliver Grau   BBC, Tadworth, UK

Konrad Wojciechowski	Institute of Automation, Gliwice, Poland
Hau San Wong	City University of Hong Kong, Kowloon Hong Kong, China
Hyun S. Yang	Kaist (Korea Advanced Institute of Science and Technology), Daejeon, Korea
Cha Zhang	Microsoft Research, Redmond, USA
Jing Zhao	Peking University, Beijing, P.R. China
Tatjana Zrimec	University of New South Wales, Australia

Reviewers

Marie-Hélène Abel	Université de Compiègne, Compiègne, France
Alexis André	Sony Computer Science Laboratories, Tokyo, Japan
Marc Antonini	I3S, Sophia Antipolis, France
Jean-Yves Audibert	Université Paris Est 6, INRIA, Marne-la-Vallée, France
Abdessamad Ben Hamza	Concordia University, Montreal, Canada
Moshe Ben-Ezra	Microsoft Asia, Beijing, P.R. China
Jacques Blanc-Talon	DGA, Bagneux, France
Philippe Bolon	ESIA, Annecy, France
Sylvain Boltz	University of California, Los Angeles, USA
Kadi Bouatouch	IRISA, Rennes, France
Edmond Boyer	INRIA Rhone-Alpes, Saint Ismier, France
José Braz	Polytechnic Institute of Setúbal, Setúbal, Portugal
Antonio Camurri	University of Genova, Genova, Italy
Leszek Chmielewski	Institute of Fundamental Technological Research (IPPT PAN), Warsaw, Poland
Adrian Clark	University of Essex, Colchester, UK
Eric Debreuve	I3S, Sophia Antipolis, France
Peter Eisert	Heinrich-Hertz-Institut, Berlin, Germany
Alexandre Francois	Tufts University, Medford, USA
Bernd Froehlich	Bauhaus-Universität Weimar, Weimar, Germany
Andrea Fusiello	Università degli Studi di Verona, Verona, Italy
André Gagalowicz	INRIA Rocquencourt, Le Chesnay, France
Vincent Garcia	Ecole Polytechnique, Palaiseau, France
Michael Gleicher	University of Wisconsin, Madison, USA
Oliver Grau	BBC, Tadworth, UK
Radek Grzeszczuk	Nokia Research Lab, Palo Alto, USA
Cédric Guiard	THALES, Paris, France
Peter Hall	University of Bath, Bath, UK
Patrick Horain	TELECOM & Management SudParis, Evry, France

Raquel Urtasun	Berkeley, USA
Luc Van Gool	ETH -Zürich, Zürich, Switzerland
Thomas Vetter	Basel University, Basel, Switzerland
Qiang Wang	Microsoft Asia, Beijing, P.R. China
Harry Wechsler	George Mason University, Fairfax, USA
Konrad Wojciechowski	Institute of Automation, Gliwice, Poland
Hau San Wong	City University of Hong Kong, Kowloon Hong Kong, China
Hyun S. Yang	Kaist (Korea Advanced Institute of Science and Technology), Daejeon, Korea
Cha Zhang	Microsoft Research, Redmond, USA
Jing Zhao	Peking University, Beijing, P.R. China
Tatjana Zrimec	University of New South Wales, Australia

Table of Contents

Tracking Human Motion with Multiple Cameras Using an Articulated Model

Davide Moschini and Andrea Fusiello

Dipartimento di Informatica, Università di Verona,
Strada Le Grazie 15, 37134 Verona, Italy
davide.moschini@gmail.com
andrea.fusiello@univr.it

Abstract. This paper presents a markerless motion capture pipeline based on volumetric reconstruction, skeletonization and articulated ICP with hard constraints. The skeletonization produces a set of 3D points roughly distributed around the limbs' medial axes. Then, the ICP-based algorithm fits an articulated skeletal model (stick figure) of the human body. The algorithm fits each stick to a limb in a hierarchical fashion, traversing the body's kinematic chain, while preserving the connection of the sticks at the joints. Experimental results with real data demonstrate the performances of the algorithm.

1 Introduction

Tracking or capturing the motion of a human subject is a problem that has a long history in Computer Vision (see [1] for a survey) and several real-world applications, such as human-computer interfaces, motion transfer, animation of virtual characters, activity/gesture/gait recognition, biomechanical studies. Marker-based commercial systems are available that work at very high frame rates and very high precision. While it is out of doubt that such speed/accuracy combination is necessary in biomechanics, it is questionable whether it is needed when animating a virtual character in a videogame or building a user-interface. Therefore, there is a niche for less expensive markerless systems that work at a reduced speed. In this paper we present some preliminary results of an ongoing project aimed at building a system with those characteristics.

The literature on markerless body tracking in three dimensions can be broadly split into two groups: those using a stick model for the human body [2,3], roughly corresponding to its skeleton, and those using a full 3D model of the body's shape, in the form of a polygonal mesh or a volumetric model [4,5,6]. Since we aim at a real-time system, we are forced to work with a stick model. Indeed, a stick (or skeletal) model has fewer dependencies on anthropometric parameters than a shape model and can be tracked much faster because of its simplicity.

Our system bases on volumetric reconstruction from multiple cameras (*shape from silhouette* [7]) followed by skeletonization and model fitting. Proper skeletonization algorithms, like [8,9], are too computationally demanding to process

A. Gagalowicz and W. Philips (Eds.): MIRAGE 2009, LNCS 5496, pp. 1–12, 2009.

more than a few images per second, hence we are proposing here a novel strategy that produces a very coarse – but fast – approximation of the centerline of the human body.

The model fitting is based on the well-known Iterative Closest Point (ICP) algorithm [10]: the model is an articulated stick figure representing the body and it's kinematics, the data are 3D points roughly distributed around the centerline of the limbs. The data are registered to the model using a hierarchical approach that proceeds by traversing the kinematic chain.

Previous work on using ICP on articulated bodies include [11,6,12]. In [11] each segment is aligned independently to the data and articulated constraints are enforced *a-posteriori* by projection on the constraints surface. Likewise, [6] uses ICP to find a solution to a problem with relaxed joint constraints, and then forces hard constraints on that solution, thereby interfering with the result of ICP, which is optimal in the least-squares sense. Differently from these works we enforce joints constraint *during* the registration process. The only work with this feature is [12], that have been independently proposed. For the sake of clarity, the discussion on the differences is postponed to Section 5.

Other related approaches include those that optimize the same objective function as the articulated ICP (namely: the sum of squared distance between data and model with respect to the pose parameters of all the segments of the structure) with a different strategy, e.g. Expectation-Maximization [3] or Levenberg-Marquardt. The first is too computationally demanding for a tracking application, whereas the latter have been reported [12] to suffer from convergence to local minima more than articulated ICP.

2   Human Body Model

In this section we describe the articulated model representing the human body pose we used in the paper. It consists of a kinematic chain of ten sticks and nine joints, as depicted in Figure 1. The torso is at the root of tree, children represents

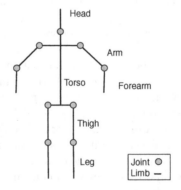

Fig. 1. The stick figure body model

limbs, each limb being described by a fixed-length stick and the corresponding rotation from its parent. Hence, the motion of one body segment can be described as the motion of the previous segment in the kinematic chain and an angular motion around a body joint. Only the torso contains a translation that accounts for the translation of the whole body. Rotations are represented with 3×3 matrices. For the sake of simplicity, all the joints are spherical (three d.o.f.) with no angle limits.

3   Shape from Silhouette

Shape from silhouette consist in recovering a volumetric approximate description of the human body (the *visual hull* [13]) from its silhouettes projected onto a number of cameras (three, in our case). Its main advantage over other

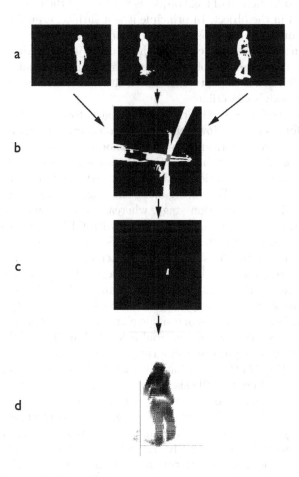

Fig. 2. a) Silhouettes; b) projection onto the sweeping plane; c) intersection (slice) d) final volumetric reconstruction

reconstruction techniques is that it seamlessly integrates the information from multiple cameras. Moreover, implementations have been demonstrated that achieves real-time performances [14] by exploiting the graphical hardware.

Silhouettes are obtained by background subtraction with the software distributed with the HumanEva dataset [15]. The reconstruction is accomplished using the technique described in [16,14], with a plane parallel to the floor sweeping the working volume (see Fig. 2). At each step the silhouettes are projected onto the current plane, using the projective texture mapping feature of OpenGL and the GPU acceleration, as described in [17]. The slice of the volume corresponding to the plane is reconstructed by doing the intersection of the projected silhouettes.

4   Skeletonization

The medial axis (or skeleton) of a 3D object is the locus of the centers of maximal spheres contained in the object. In principle it is a surface, even if it can degenerate to a curve or a point. A close relative is the *centerline* (or *curve-skeleton*) that is a curve in 3D space that captures the main object's symmetry axes and roughly runs along the middle of an object. This definition matches with the stick-figures model, hence the data onto which the model is to be registered will be points on the body's centerline.

There are many techniques in literature to find skeletons or centerlines of a 3D object (see [18] for a survey). However, they are too computationally demanding to fit our design, hence we introduce a new method based on slicing the volume along three axis-parallel directions (see Fig. 3). In each slice – which is a binary image – we compute the centroid of every connected component and add it to the set of centerline points. The slicing along the Z-axis comes for free from the previous volumetric reconstruction stage, whereas slicing along X and Y must be done expressly, but uses the same procedure with GPU acceleration.

Our method is similar to [16] which computes the centerline of a body by finding the centroids of the blobs produced by intersecting the body with planes orthogonal to Z-axes. Using a single sweep direction has some problems with some configurations of the body. Consider for example the "T" pose: using only the scan along the Z-axes we completely loose the arms because by cutting the body at the arms height produces one single elongated blob containing a slice of the torso and the two arms, whose centroid is located on the vertebral column.

Our method solves this problem using three sweeps, thus it can be considered as a refinement of [16]. On the other hand, it can also be regarded as a coarse approximation of [19], where first 2D skeletons are extracted for each axis-parallel 2D slice of the 3D volume and then they are intersected to obtain the 3D centerline of the object. When the centroid belongs to the centerline our method returns a subsampling of the centerline, and this is approximately the case for most configurations of the human body. Yet, when the 2D shape is strongly non convex and the centroid falls outside the shape itself the method yields spurious points. However, the subsequent fitting procedure, that will be described in the following section, has been designed to be robust with respect to outliers.

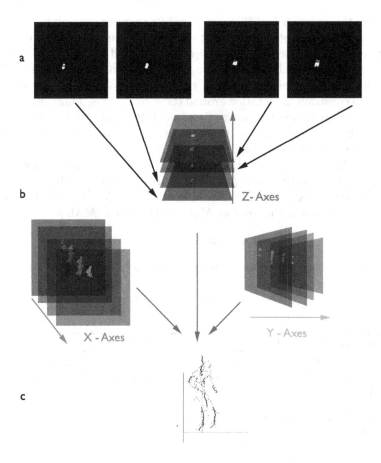

Fig. 3. a) slices along Z; b) slices along X and Y; c) centerline points with the stick figure overlaid

5   Hierarchical Articulated ICP

This section describes the Hierarchical Articulated ICP algorithm for registering an articulate stick model to a cloud of points. It is based on the well-known Iterative Closest Point (ICP) [20,10] that estimates the rigid motion between a given set of $3D$ data points and a set of $3D$ model points.

We assume that the data are 3D points distributed roughly around the centerline of the body's segments. The data are registered to the model using a hierarchical approach that starts from the torso and traverse the kinematic chain down to the extremities. At each step ICP computes the best rigid transformation of the current limb that fits the data while preserving the articulated structure.

The closest point search works from the data to the model, by computing for each data point its closest point on the body segments. Only the matches with the current segment are considered, all the other should be – in principle – discarded.

However, the rotation in 3D space of a line segment cannot be computed unambiguously, for the rotation around the axis is undetermined. In order to cope with this problem we formulate a *Weighted* Extended Orthogonal Procrustes Problem and give a small but non-zero weight also to points that match the descendants of the current segment in the kinematic chain. In this way they contribute to constrain the rotation around the segment axis. Think, for example, of the torso: by weighting the points that match the limbs as well, even if they cannot be aligned with single rigid transformation, the coronal (aka frontal) plane can be correctly recovered.

In order to improve ICP robustness against false matches and spurious points, following [21], we discard closest pairs whose distance is larger than a threshold computed using the X84 rejection rule. Let e_i be the closest-point distances, a robust scale estimator is the Median Absolute Deviation (MAD):

$$\sigma^* = 1.4826 \, \text{med}_i \, |e_i - \text{med}_j \, e_j|. \tag{1}$$

The X84 rejection rule prescribes to discard those pairs such that $|e_i - \text{med}_j \, e_j| > 3.5\sigma^*$.

The Hierarchical Articulate ICP is is described step by step in Algorithm 1.

Algorithm 1. HIERARCHICAL ARTICULATE ICP

Input: The model $\mathcal{S}$ composed by segments and the data set $\mathcal{A}$ of 3D points
Output: a set of rigid motions (referred to the kinematic chain) that brings the model onto the data

1. Traverse the body model tree structure using a level-order or a preorder traversal method.
2. Let $s_j \in \mathcal{S}$ be the current body segment.
3. Compute the closest points:
   (a) For each data point $\mathbf{a}_i \in \mathcal{A}$ and for each segment $s_\ell \in \mathcal{S}$ compute its projection $\mathbf{p}_{i\ell}$ onto the line containing s_ℓ ;
   (b) if $\mathbf{p}_{i\ell} \in s_\ell$ then add $\mathbf{p}_{i\ell}$ to $\mathcal{M}$ (the set of the closest-point candidates), otherwise add the endpoint of s_ℓ to $\mathcal{M}$.
   (c) Find $\mathbf{b}_i$, the closet point to $\mathbf{a}_i$ in $\mathcal{M}$.
4. Weight the points: If $\mathbf{b}_i$ belongs to s_j than its weight is 1, otherwise it is ε (chosen heuristically) for all the descendant and 0 for all the others.
5. If the distance of $\mathbf{b}_i$ to $\mathbf{a}_i$ is above the X84 threshold then the weight is set to 0.
6. Solve for the transformation of s_j.
7. Apply the transformation to s_j and its descendants.
8. Repeat from step 3 until the weighted average distance between closest points points is less than a given threshold.

Point 6, where a transformation is computed given some putative correspondences, deserves to be expanded, in order to make the paper self-contained. The problem to be solved is an instance of the *Extended Orthogonal Procrustes Problem* (EOPP) [22], which can be stated as follows: transform a given matrix A into a given matrix B by a similarity transformation (rotation, translation and scale) in such a way to minimize the sum of squares of the residual matrix. More precisely, since we introduced weights on the points, we shall consider instead the *Weighted Extended Orthogonal Procrustes Problem* (WEOPP) problem. In formulae:

$$\arg\min_{R} \left\| (cRA + \mathbf{t}\mathbf{u}^\top - B)W \right\|_F^2 \quad \text{subject to } R^T R = I \tag{2}$$

where matrices A and B are $(3 \times p)$ matrices containing p corresponding point in 3-D space, R is (3×3) orthogonal rotation matrix, $\mathbf{t}$ is a (3×1) translation vector, c is scale factor, $\mathbf{u}$ is a $p \times 1$ vector of ones, W is a $(p \times p)$ diagonal matrix weighting the p points, and $\|\cdot\|_F$ denotes the Frobenius norm.

The solution to the the problem (derived in [23]) is based on the Singular Value Decomposition (SVD). Let

$$UDV^\top = A_w \left(I_p - \frac{\mathbf{u}_w \mathbf{u}_w^\top}{\mathbf{u}_w^\top \mathbf{u}_w} \right) B_w^\top \tag{3}$$

be the SVD decomposition of the matrix on the right-hand side[1], where $A_w = AW$, $B_w = BW$, and $\mathbf{u}_w = W\mathbf{u}$. The sought transformation is given by (we omit the scale c that is not needed in our case):

$$R = V \begin{bmatrix} 1 & 0 & 0 \\ 0 & 1 & 0 \\ 0 & 0 & \det(VU^T) \end{bmatrix} U^\top \tag{4}$$

$$\mathbf{t} = (B_w - RA_w) \frac{\mathbf{u}_w}{\mathbf{u}_w^\top \mathbf{u}_w} \tag{5}$$

The diagonal matrix in (4) is needed to ensure that the resulting matrix is a rotation [24]

The *Weighted Orthogonal Procrustes Problem* (WOPP) problem is a special case of WEOPP and the solution can be derived straightforwardly by setting $\mathbf{u} = \mathbf{0}$. In our case we use WEOPP for the torso and WOPP (only rotation) for the limbs.

The hierarchical articulate ICP is deterministic, every limb is considered only once and brought into alignment with ICP. The transformation that aligns a limb s_j is determined mostly by the points the matches s_j and secondarily by the points that matches its descendants. The transformation is applied to s_j and its descendants, considered as a rigid structure. The output of the algorithm represents the pose of the body. In a tracking framework, the pose obtained at the previous time-step is used as the initial pose for the current frame.

A similar algorithm has been independently proposed in [12]. The main difference is in the way the basic ICP is applied to the articulated structure, which leads to different schema. In [12] at each step of the algorithm the subtree of the

---

[1] Please note that $A \frac{\mathbf{u}_w \mathbf{u}_w^\top}{\mathbf{u}_w^\top \mathbf{u}_w}$ is a matrix of the same size as A with identical columns, each of them equal to the centroid of the points contained in A.

selected joint is rigidly aligned using ICP with no weights, i.e., all the descendants of the joint plays the same role in the minimization. As a result, the same joint needs to be considered more than once to converge to the final solution. In this regard our approach is less computationally demanding. On the other hand one error in the alignment of a limb propagates downward without recovery, whereas in [12] a subsequent sweep may be able to correct the error, hence [12] seems to be more tolerant to a looser initialization.

6   Experimental Results

The body tracker has been tested on sequences taken from the HumanEva-I dataset [15]. All the sequences in HumanEva-I have been calibrated using the Vicon's proprietary software and the motion data saved in the common c3d file format. The dataset contains multiple subjects performing a variety of actions like walking, running, boxing, etc. In particular we used the sequences called *"S2 Jog"*, *"S2 Throwcatch"*. Figures 4 and 5 show some sample frames from these sequences together with the output of the silhouette extraction.

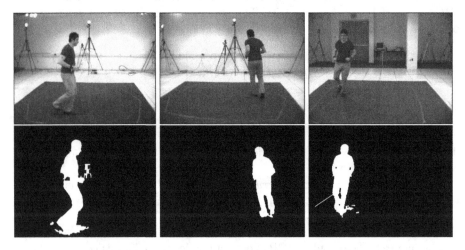

Fig. 4. Sample frames of *"S2 Jog"* and silhouettes

Validation of the algorithm is done by comparing the angles of the ground-truth with the angles of the computed model. Figure 6 reports the ground truth and estimated joint angles of the torso, right shoulder and right elbow in the two sequences. It can be seen that the estimated angles follows fairly closely the ground truth. There some spikes where the error grows but the tracker is able to recover in the subsequent frames. We expect that a Kalman filter will be able to smooth out significantly those spikes.

This results are remarkable if one considers the coarseness of the volumetric reconstruction, due to the small number of cameras (three) and the poor quality of image silhouettes.

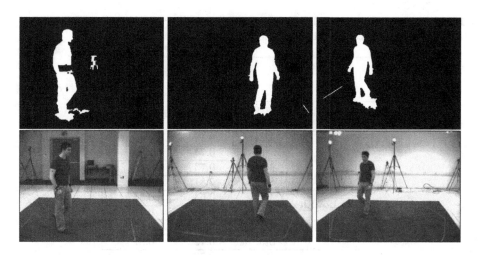

Fig. 5. Sample frames of *"S2 Throwcatch"* and silhouettes

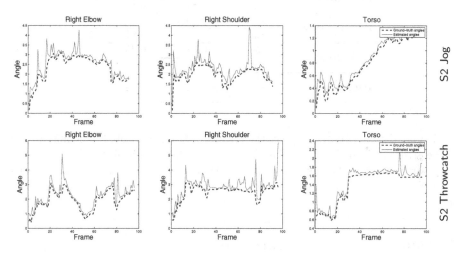

Fig. 6. Plots comparing ground truth and estimated joint angles of the torso, right shoulder and right elbow in the two sequences used for the experiments (the sequence name is on the right).

For a quantitative comparison we computed the following angular error for each joint, in each frame of the sequence:

$$e(R_1, R_2) = \angle(R_1 R_2^\top) \tag{6}$$

where $\angle(\cdot)$ denotes the angle of the axis-angle representation of the rotation, and can be computed with $\angle(R) = \arccos((\mathrm{tr}(R) - 1)/2)$.

Table 1. Mean and standard deviation of the errors (in radians) for each joint of the body for the sequences used in the experiments

		S2 Jog	S2 Throw-catch
Torso	*Mean*	0.29	0.21
	Std. dev.	0.40	0.57
Neck	*Mean*	0.59	0.82
	Std. dev.	0.74	1.09
Left shoulder	*Mean*	0.73	0.44
	Std. dev.	0.81	0.61
Right shoulder	*Mean*	0.63	0.53
	Std. dev.	0.79	0.53
Left hip	*Mean*	0.96	0.56
	Std. dev.	0.12	0.87
Right hip	*Mean*	0.76	1.66
	Std. dev.	1.08	1.67
Left elbow	*Mean*	0.68	0.55
	Std. dev.	0.81	0.80
Right elbow	*Mean*	0.48	0.56
	Std. dev.	0.65	0.81
Left knee	*Mean*	0.51	0.25
	Std. dev.	0.70	0.25
Right knee	*Mean*	0.33	0.70
	Std. dev.	0.38	0.77

Mean and standard deviation of the error are shown in Table 1. The magnitude of the error is still higher than the target standard, which is about three degrees, as reported in [25]. We expect, however, that a Kalman filter will be able to smooth out significantly the aforementioned spikes and thus reduce significantly the error.

7   Conclusions and Future Work

This paper has proposed a new ICP-based algorithm for tracking articulated skeletal model of a human body. The proposed algorithm takes as input multiple calibrated views of the subject, computes a volumetric reconstruction and the centerlines of the body and fits the skeletal body model in each frame using a hierarchic tree traversal version of the ICP algorithm that preserves the connection of the segments at the joints. The proposed approach uses only the kinematic constraints and no other assumptions are made on the position of the body. This implies that we can recognize potentially all the body configuration.

The results presented here demonstrate the feasibility of the approach, which is is intended to be used in complete system for vision-based markerless human body tracking.

The current Matlab implementation takes about 4 seconds to process a frame on a laptop with an Intel Core Duo Processor T2250. However, being the algorithm still in a prototypal stage, we are confident that a careful implementation in C/C++ could achieve nearly real-time performances. Indeed all the design choices focused on computational efficiency: the use of a simple stick model, the volumetric reconstruction on the GPU, the fast approximated skeletonization, the hierarchical ICP.

Future work will be aimed at optimizing the implementation and tackling the issue of pose initialization.

Acknowledgments

This paper was partially supported by PRIN 2006 project 3-SHIRT. Thanks to A. Giachetti for inspiring discussions.

References

1. Moeslund, T., Hilton, A., Kruger, V.: A survey of advances in vision-based human motion capture and analysis. Computer Vision and Image Understanding 103(2-3), 90–126 (2006)
2. Brostow, G.J., Essa, I., Steedly, D., Kwatra, V.: Novel skeletal representation for articulated creatures. In: Pajdla, T., Matas, J(G.) (eds.) ECCV 2004, part III. LNCS, vol. 3023, pp. 66–78. Springer, Heidelberg (2004)
3. Ménier, C., Boyer, E., Raffin, B.: 3d skeleton-based body pose recovery. In: Proceedings of the 3rd International Symposium on 3D Data Processing, Visualization and Transmission. Chapel Hill, USA (June 2006)
4. Anguelov, D., Koller, D., Pang, H.C., Srinivasan, P., Thrun, S.: Recovering articulated object models from 3D range data. In: Proc. of the 20th conference on Uncertainty in Artificial Intelligence, Arlington, Virginia, United States, pp. 18–26 (2004)
5. Knoop, S., Vacek, S., Dillmann, R.: Modeling joint constraints for an articulated 3D human body model with artificial correspondences in ICP. In: Proc. 5th IEEE-RAS International Conference on Humanoid Robots, pp. 74–79 (2005)
6. Mundermann, L., Corazza, S., Andriacchi, T.: Accurately measuring human movement using articulated ICP with soft-joint constraints and a repository of articulated models. In: Proc. IEEE Conference on Computer Vision and Pattern Recognition, pp. 1–6 (June 2007)
7. Martin, W.N., Aggarwal, J.K.: Volumetric descriptions of objects from multiple views. IEEE Transactions on Pattern Analysis and Machine Intelligence 5(2), 150–158 (1983)
8. Sharf, A., Lewiner, T., Shamir, A., Kobbelt, L.: On-the-fly curve-skeleton computation for 3d shapes. Comput. Graph. Forum 26(3), 323–328 (2007)
9. Dey, T.K., Sun, J.: Defining and computing curve-skeletons with medial geodesic function. In: SGP 2006: Proceedings of the fourth Eurographics symposium on Geometry processing, Aire-la-Ville, Switzerland, pp. 143–152 (2006)
10. Besl, P.J., McKay, N.D.: A method for registration of 3-d shapes. IEEE Transaction on Pattern Analysis and Machine Intelligence 14(2), 239–256 (1992)

11. Demirdjian, D., Ko, T., Darrell, T.: Constraining human body tracking. In: Proceedings of the Ninth IEEE International Conference on Computer Vision, Washington, DC, USA, p. 1071 (2003)
12. Pellegrini, S., Schindler, K., Nardi, D.: A generalisation of the icp algorithm for articulated bodies. In: Proceedings of the British Machine Vision Conference (2008)
13. Laurentini, A.: The visual hull concept for silhouette-based image understanding. IEEE Trans. Pattern Anal. Mach. Intell. 16(2), 150–162 (1994)
14. Li, M., Magnor, M., Seidel, H.P.: Hardware-accelerated visual hull reconstruction and rendering. In: Graphics Interface 2003, pp. 65–71 (2003)
15. Sigal, L., Black, M.: Humaneva: Synchronized video and motion capture dataset for evaluation of articulated human motion. Technical Report CS-06-08, Brown University, Department of Computer Science (2006)
16. Michoud, B., Guillou, E., Bouakaz, S.: Human model and pose Reconstruction from Multi-views. In: International Conference on Machine Intelligence (November 2005)
17. Everitt, C., Rege, A., Cebenoyan, C.: Hardware shadow mapping. In: ACM SIGGRAPH 2002 Tutorial Course no.31: Interactive Geometric Computations, pp. 38–51 (2002)
18. Cornea, N.D., Silver, D., Min, P.: Curve-skeleton applications. In: IEEE Visualization Conference, pp. 95–102 (October 2005)
19. Telea, A., van Wijk, J.J.: An augmented fast marching method for computing skeletons and centerlines. In: VISSYM 2002: Proceedings of the symposium on Data Visualisation 2002, Aire-la-Ville, Switzerland, p. 251 (2002)
20. Chen, Y., Medioni, G.: Object modelling by registration of multiple range images. Image and Vision Computing 10(3), 145–155 (1992)
21. Fusiello, A., Castellani, U., Ronchetti, L., Murino, V.: Model acquisition by registration of multiple acoustic range views. In: Proceedings of the European Conference on Computer Vision, pp. 805–819 (2002)
22. Schnemann, P., Carroll, R.: Fitting one matrix to another under choice of a central dilation and a rigid motion. Psychometrika 35(2), 245–255 (1970)
23. Akca, D.: Generalized procrustes analysis and its applications in photogrammetry. Technical Report, ETH, Swiss Federal Institute of Technology Zurich, Institute of Geodesy and Photogrammetry (2003)
24. Kanatani, K.: Geometric Computation for Machine Vision. Oxford University Press, Oxford (1993)
25. Rosenhahn, B., Brox, T., Kersting, U., Smith, A., Gurney, J., Klette, R.: A system for marker-less motion capture. Knstliche Intelligenz 20(1), 45–51 (2006)

Shape Recovery of Specular Surface Using Color Highlight Stripe and Light Source Coding

Yankui Sun[1], Chengkun Xue[1], Masatoshi Kimachi[2], and Masaki Suwa[2]

[1] Department of Computer Science and Technology, Tsinghua University,
Beijing 100084, P.R. China
syk@mail.tsinghua.edu.cn
[2] The Sensing & Control Technology Laboratory, OMRON Corporation, Kyoto, Japan

Abstract. Shape recovery of specular surface is a challenging task; camera images of these surfaces are difficult to interpret because they are often characterized by highlights. Structured Highlight approach is a classic and effective way for specular inspection, this paper suggests a new strategy to recover dense normals of a specular surface and reconstruct its shape by combining the ideas of Structured Highlight, color source coding, highlight stripe and its translations. Point sources with different colors are positioned on orbits to illuminate a specular object surface. These point sources are scanned, and highlights on the object surface resulting from each point source are used to derive local surface orientation. Dense normal information can be recovered by translating these orbits. Some experimental system configurations are given. The simulation results show that the new method is feasible and can be used to reconstruct shape of specular surface in a high precision.

1 Introduction

The problem of shape from shading has received great attention from the computer vision community. The brightness pattern of one or more images depends intricately on the prior knowledge of surface properties, imaging geometry, and lighting conditions. Most approaches to this problem have assumed that surfaces are Lambertian, that is, incident light is scattered by the surface so that the perceived brightness is independent of the direction of view. In order to reconstruct non-lambertian surfaces, many works have been done in the area of reconstructing or estimating surfaces from specularities[1~9]. In recent papers , Magda et al. [10] and Zickler et al. [11] have successfully made use of Helmholtz reciprocity in stereo reconstruction. This principle determines that the bidirectional reflectance distribution function (BRDF) of a surface is symmetric on the incoming and outgoing angles. [12] introduces a novel reconstruction algorithm for continuous surfaces that makes use of Helmholtz reciprocity without resorting to multiple image pairs.

Many practical tasks in robot vision and inspection require interpretation of images of specular, or shiny, surfaces where the perceived brightness becomes a very strong function of viewing direction due to highlights or reflections from the source.

A. Gagalowicz and W. Philips (Eds.): MIRAGE 2009, LNCS 5496, pp. 13–22, 2009.

For a purely specular surface, light is reflected such that the angle of incidence equals the angle of reflection. Therefore, illumination of a specular surface using a point source of light does not produce smooth shading on the surface. Camera images of such surfaces are difficult to interpret because they are characterized by bright points or highlights, and inspection and reconstruction of surface shape are challenging tasks. The structured highlight technique [13,14] uses a large number of point sources to illuminate the inspected object. The point sources are uniformly distributed around the object, and images of the object are obtained by using a camera. The binary-coded point sources are scanned, and highlights on the object surface resulting from each point source are used to derive local surface orientation. Some features of this approach include: 1) monochromatic point sources are distributed on a semi-sphere; 2) the recovered surface normals are sparse, which can be used in surface inspection but are not sufficient to reconstruct surface in a high precision. Zheng et al. proposes surface reconstruction of an object on a turn table from specularities [15], where highlight stripes are used to estimate surface normals. This method can be used to reconstruct specular objects such as metallic and plastic surfaces through object rotations by using a series of input images. However it is not an effective way for much curved objects such as sphere. Recently, shape from distortion [16] is introduced. It integrated the interference of patterns with different frequencies and the reconstruction of the surface from the environment matte to obtain the shape in a high accuracy.

In this paper, we propose a new approach to obtain 3D information of specular surface using color highlight stripe and point source coding. This method can estimate dense surface normals by using the binary coding scheme of point sources with different primitive colors, color highlight stripe and its translations. Not only does it adopt the algorithm based on Structured Highlight, but also combines with the idea of highlight stripe and its translation.

The rest of this paper is organized as follows. Sec.2 describes algorithm principle using color highlight stripe and its translations; Sec.3 describes two implementation schemes and the simulation experiments. Experimental results are given in Sec. 4 , discussions and conclusions are given in Sec. 5 and Sec. 6.

2   Algorithm Principle

The structured highlight method in [14] is difficult to obtain dense normal information of a specular surface since the limited point sources located on the semi-sphere in practice. In order to obtain dense normals of a specular surface by imaging the surface as few as possible, we propose new ways to estimate surface 3D information of a specular object by combining light source binary coding scheme, color highlight stripes and their translations.

In our initial design, the linear light consists of $2^N - 1$ point light sources, where $N=4$ or 5 is enough. When the point sources are scanned, the highlight stripe is generated by the linear light. According to [14], by using the binary coding scheme and taking N images of the specular surface, $2^N - 1$ normals can be estimated from the highlight stripe. Then, by translating the linear light at a distance and taking another N

images of the surface, the other $2^N - 1$ point normals can be obtained. Therefore, dense surface normals can be recovered by translating highlight stripe again and again.

One characteristics of specular surface is that the color of received reflection light is the same as the color of incident light source. By using this property, we can utilize color point sources to improve the above algorithm, which allows us to take fewer photos while keep the count of recovered point normals.

The system configuration is shown in Fig. 1, where the specular object is placed on a plane, and some point light sources are put above it in a line, we call this line "Light Stick", which can be moved from left to right. One camera is put just above the object. The point sources are distant from the object surface so that the size of the object is small compared with the distance between each point source and the object. Under this assumption, the angle of incidence of illumination is determined only by the position of the source and does not depend on the relative position of illumination on the object surface.

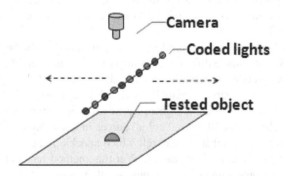

Fig. 1. System configuration illustration

The light sources on the light stick may have different colors, such as red, green and blue, and binary coding scheme for point sources is used for different colors separately. Our simulation test shows that the surface normal recovery algorithm with light translation is feasible and it is especially effective for less curved object. However, since the light stick is located above the object and it only translates on the plane parallel to the plane that object lies in, so the range of object lighted by the light sources is limited. To overcome the disadvantage, we further improve this scheme. That is, two orbits are used to fix the light sources (Fig.2a); each orbit is a bow which consists of three light sticks as depicted in Fig.2b. For each orbit, its height is half of its width.

In order to get well-distributed highlights on the specular surface, the light sources should be positioned in a particular way. Fig.2a shows the red light sources and green ones are fixed alternately. Fig.2b illustrates the simple principle we used to fix the light sources evenly on the orbit.

During the process of normal recovery, one orbit will translate along the X-axis (Fig.2a), and it will stop in several positions so that the camera can take photos. For convenience, we will use X-Orbit to indicate the orbit which moves along the X-Axis. Y-Orbit is named similarly. By moving the X-Orbit and the Y-Orbit in sequence, that

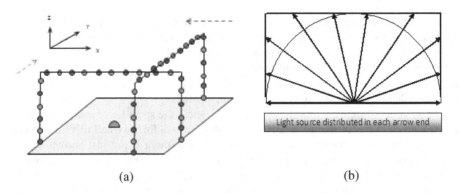

(a)   (b)

Fig. 2. (a) Two orbits used to fix the light sources (b) Distribution of light sources on an orbit

is, when the X-Orbit is moving, all the lights on the Y-Orbit are turned off, we can estimate the surface normals by using the binary coding scheme for different color sources on one orbit separately.

To reduce the photo number, a natural way is to translate the X-orbit and the Y-orbit simultaneously. But this made difficult to estimate surface normals if light sources on each orbit are binary-coded separately. For example, there will be two red specular speckles lighten by point source k (actually each orbit has a point source coded as k), and it can't be told which speckle is lighten by which point source. One solution is to change the color scheme: fix all the red color sources in one orbit and fix all the green sources in the other orbit so that we can tell which speckle is lighten by which point source by checking the color of the speckle. Yet this method has its drawback. First, when the number of point sources is fixed, putting all the same color lights in one orbit will reduce the estimation precision of surface normal because the distance between adjacent speckles is smaller, in the worst case they may mix up into a bigger speckle. Second, at most three orbits can be used because there are only three independent color components - red, green and blue. To solve this problem thoroughly, we introduce a new coding scheme, called united-coding, that is, we code point sources according to its color and ignore the orbit they are fixed on. With the united-coding scheme, light sources with different colors are placed alternately in each orbit, and the number of orbits can be set by our need. Another advantage is that the adjacent speckles will have different colors, so the center of them can be estimated more accurately.

To sum, we introduce the following new ideas to estimate normals and shape of a specular surface: 1) Replacing linear light with orbit so that highlight stripe is gener-ated by orbit which consists of point sources; 2) Different color light sources are used to reduce the number of imaging the specular surface while the amount of the recov-ered normals remain the same; 3) United-coding scheme is introduced so that we can take advantage of the color information while break the limitation of 3 independent color components; 4) Point sources with different colors are located alternately in each orbit so that center of speckles can be estimated more accurately.

3 Implementation

The Radiance software [17] is used to render the specular object and image it. In our simulation tests, a top-half metal sphere is created and the material of the sphere surface is listed in Table 1. The low diffuse, high specular and low coarseness parameters means the object is a high specular object.

Table 1. The material parameter for the specular object in Radiance

Color	Diffuse	Specular	Coarseness
(0.6 0.62 0.64)	0.05	0.95	0.02

3.1 United Binary Coding Scheme with Two Kinds of Color Sources

In the system setup shown in Fig.3, the 30 light sources with 15 reds and 15 greens are alternately fixed on each orbit. The X-orbit translates alone the X-axis, and pauses in predefined stops. The two X-orbits keep a distance of a fixed number of stops and translate toward the same direction simultaneously. The two Y-orbits are similar. In this implementation, the two X-orbits and the two Y-orbits are moving in sequence.

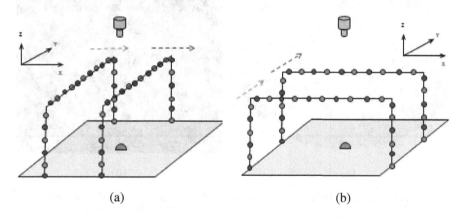

(a)   (b)

Fig. 3. System setup with two kinds of color sources

Applying the united-coding scheme, we code green point sources in two X-orbits (two Y-orbits) as a whole. For example, the red point sources in the X-orbit 1 are coded from 1 to 15, and the red point sources in the X-orbit 2 are coded from 16 to 30. According to the binary code scheme, to recover normals of 30 red speckles, 5 pictures are needed to be taken. Furthermore, these 5 pictures can also be used to recover normals of 30 green speckles. So we totally recover normals of 60 color speckles with 5 pictures. To recover 1440 normals, 120 pictures are taken and 24 translation times are done. Two sample input images are shown in Fig.4.

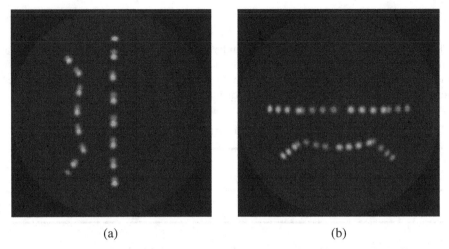

(a)   (b)

Fig. 4. Sample input images with red and green sources

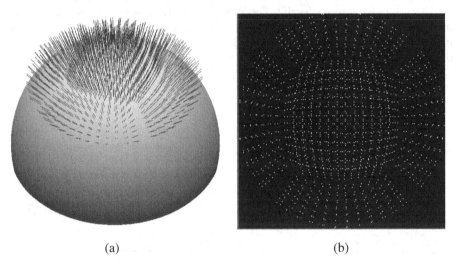

(a)   (b)

Fig. 5. (a) Illustration of the recovered normals (b) Recovered normals' distribution

The recovered 1440 normals are illustrated in Fig.5a, their distribution is dense and uniformly, as is shown in Fig.5b.

3.2 United Binary Coding Scheme with 3 Kinds of Color Sources

In the system setup shown in Fig.6a, the 45 light sources with 15 red, 15 green and 15 blue ones are fixed alternately in each orbit. Other configurations are the same as the setup of the Fig. 3.

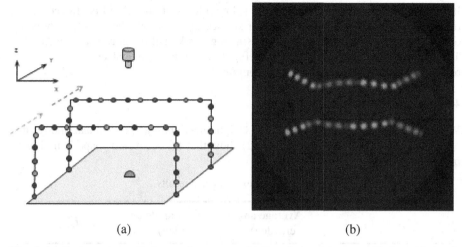

(a)   (b)

Fig. 6. (a) System setup with three color sources (b) Sample input images with three kinds of color sources

Applying the United-coding, to recover the 90 normals of colored speckles, only 5 pictures are needed too. One of the pictures taken is given in Fig.6b.

To recover 1440 surface normals, 80 pictures are taken, and 16 translation times are done. Since the speckles are much closer in this case, few normals of speckles were not able to be recovered because they are mixed up by more than one concolorous speckles. In our test, 1369 normals are recovered.

4 Result Evaluations

4.1 Evaluation Methods

Two different evaluation approaches are used to evaluate the accuracy of the result; they are Normal Error approach and Height error approach. In the Normal Error approach, the average dot product error and average angle error between real normal and the recovered normal on each speckle are measured.

Fig. 7. The reconstructed shape of the specular semi-sphere surface

In the Height error approach, firstly the spline interpolation method is utilized to estimate the other normals on the specular surface. And then the height of every point on the semi-sphere surface is obtained using a gauss kernel integration approach [18]. The reconstructed shape is depicted in Fig.7. The relative height error between the real object and the recovered object is computed.

4.2 Experiment Results

We use the Normal Error approach and Height error approach to evaluate the results of our experiments. The results (Table 2) show that the recovered shape is in a high precision.

Table 2. The experimental results

	Average angle err (deg)	Max angle err (deg)	E_{height}
Scheme in Sec. 3.1	0.6628	1.6206	0.500%
Scheme in Sec. 3.2	0.6917	1.8119	0.467%

5 Further Discussions

A more efficient solution is described here. Translate several orbits simultaneously both in X-axis and Y-axis can recover normals with even fewer images and less translation times. For example, we can place 45 light sources with 15 red, 15 green and 15 blue ones alternately in each orbit. Let the two parallel Y-orbits and the two parallel X-orbits translate simultaneously (Fig.8). In this case, taking 6 photos can recover 180 normals. If all the orbits translate 8 times, 48 photos are taken and up to 1440 normals can be recovered in theory.

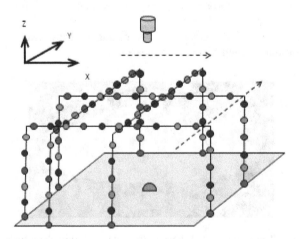

Fig. 8. System setup with the two X-orbits and the two Y-orbits translating simultaneously

6 Conclusions

Specular surface recovery is a challenging task. By combining the ideas of Structured Highlight, color source coding, highlight stripe and its translations, we propose a new strategy to recover dense specular surface normals and reconstruct its shape in a high precision. This method can be applied to both much curved and less curved specular objects in principle. Simulation test shows that this method is feasible and practical.

Acknowledgment

This research was supported in part by a grant from OMRON Corporation. 863 Project of China under Grand 2006AA02Z472 and 2008AA01Z419, the National Natural Science Foundation of China under Grant No. 60873249 and 60203003.

References

1. Blake, A., Brelstaff., G.: Geometry from specularities. In: Second International Conference on Computer Vision, pp. 394–403 (1988)
2. Zisserman, P.G., Blake, A.: The information available to a moving observer from specularities. Image and Vision Computing 7(1), 38–42 (1989)
3. Schultz, H.: Retrieving shape information from multiple images of a specular surface. IEEE Transactions on Pattern Analysis and Machine Intelligence 16(2), 195–201 (1994)
4. Zheng, J.Y., Murata, A.: Acquiring a complete 3d model from specular motion under the illumination of circular shaped light sources. IEEE Transactions on Pattern Analysis and Machine Intelligence 22(8), 913–920 (2000)
5. Bonfort, T., Sturm, P.: Voxel carving for specular surfaces. In: Int. Conf. Computer Vision, Nice, France, pp. 591–596 (2003)
6. Yang, R., Pollefeys, M., Welch, G.: Dealing with textureless regions and specular highlights—a progressive space carving scheme using a novel photo-consistency measure. In: Int. Conf. Computer Vision, Nice, France, pp. 576–584 (2003)
7. Solem, J.E., Aanæs, H., Heyden, A.: Pde based shape from specularities. In: Griffin, L.D., Lillholm, M. (eds.) Scale-Space 2003. LNCS, vol. 2695, Springer, Heidelberg (2003)
8. Solem, J.E., Heyden, A.: Estimating surface shape and extending known structure using specular reflections. In: International Conference on Pattern Recognition, Cambridge, UK (2004)
9. Solem, J.E., Aanæs, H., Heyden, A.: A variational analysis of shape from specularities using sparse data 3D Data Processing, Visualization and Transmission. In: Proceedings of 2nd International Symposium on 3DPVT 2004, pp. 26–33 (2004)
10. Magda, S., Kriegman, D.J., Zickler, T.E., Belhumeur, P.N.: Beyond lambert: Reconstructing surfaces with arbitrary BRDFs. In: Proc. 8th Int. Conf. on Computer Vision, Vancouver, Canada, vol. II, pp. 391–398 (2001)
11. Zickler, T., Belhumeur, P.N., Kriegman, D.J.: Helmholtz stereopsis: Exploiting reciprocity for surface reconstruction. In: Heyden, A., Nielsen, S.G.M., Johansen, P. (eds.) ECCV 2002. LNCS, vol. 2352, pp. 869–884. Springer, Heidelberg (2002)

12. Tu, P., Mendonca, P.R.S.: Surface Reconstruction via Helmholtz Reciprocity with a Single Image Pair. In: IEEE Computer Society Conference on Computer Vision and Pattern Recognition (2003)
13. Sanderson, A.C., Weiss, L.E., Nayar, S.K.: Structured Highlight Inspection of Specular Surfaces. IEEE Transactions on Pattern Analysis And Machine Intelligence 10(1), 44–55 (1988)
14. Nayar, S.K., Sanderson, A.C., Weiss, L.E., et al.: Specular surface inspection using structured highlight and Gaussian images. IEEE Transactions on Pattern Analysis And Machine Intelligence 10(1), 208–218 (1990)
15. Zheng, J.Y., Fukagawa, Y., Abe, N.: 3D surface estimation and model construction from specular motion in image sequences. IEEE Transactions on Pattern Analysis and Machine Intelligence 19(5), 513–520 (1997)
16. Tarini, M., Lensch, H.P.A., Goesele, M., et al.: 3D Acquisition of Mirroring Objects using Striped Patterns. Graphical Models 67, 233–259 (2005)
17. Debevec, P.: Image-Based Lighting. IEEE Computer Graphics and Applications 22(2), 26–34 (2002)
18. Tan, W., Wang, Y.: Surface reconstruction by a gauss kernel integration approach. In: ICSP 2004 Proceedings, pp. 1252–1255 (2004)

Geometric Mesh Denoising via Multivariate Kernel Diffusion

Khaled Tarmissi and A. Ben Hamza

Concordia Institute for Information Systems Engineering
Concordia University, Montréal, QC, Canada
{k_tarmis,hamza}@encs.concordia.ca

Abstract. We present a 3D mesh denoising method based on kernel density estimation. The proposed approach is able to reduce the oversmoothing effect and effectively remove undesirable noise while preserving prominent geometric features of a 3D mesh such as curved surface regions, sharp edges, and fine details. The experimental results demonstrate the effectiveness of the proposed approach in comparison to existing mesh denoising techniques.

Keywords: Mesh denoising; kernel density; anisotropic diffusion.

1 Introduction

Recent advances in computer and information technology have increased the use of 3D models in many fields including medicine, the media, art and entertainment. With the increasing use of 3D scanners to create 3D models, which are usually represented as triangle meshes, there is a rising need for robust mesh denoising techniques to remove inevitable noise in the measurements. Even with high-fidelity scanners, the acquired 3D models are usually contaminated by noise, and therefore a reliable mesh denoising technique is often required.

In recent years, a variety of techniques have been proposed to tackle the 3D mesh denoising problem [1–5]. The most commonly used mesh denoising method is the so-called Laplacian flow which repeatedly and simultaneously adjusts the location of each mesh vertex to the geometric center of its neighboring vertices [1]. Although the Laplacian smoothing flow is simple and fast, it produces, however, the shrinking effect and an oversmoothing result. The most recent mesh denoising techniques include the mean, median, and bilateral filters [6–8] which are all adopted from the image processing literature. Also, a number of anisotropic diffusion methods for triangle meshes and implicit surfaces have been proposed recently. Desbrun *et al.* [9, 10] introduce a weighted Laplacian smoothing technique by choosing new edge weights based on curvature flow operators. This mesh denoising method avoids the undesirable edge equalization from Laplacian flow and helps preserve curvature for constant curvature areas. However, re-computing new edge weights after each iteration results in a more expensive computational cost. Clarenz *et al.* [11] propose a multiscale surface smoothing method based on

A. Gagalowicz and W. Philips (Eds.): MIRAGE 2009, LNCS 5496, pp. 23–33, 2009.

the anisotropic curvature evolution problem. By discretizing nonlinear partial differential equations, this method aims to detect and preserve sharp edges by two user defined parameters which are a regularization parameter for filtering out high frequency noisy and a threshold for edge detection. This multiscale method was also extended to the texture mapped surfaces [12] in order to enhance edge type features of the texture maps. Different regularization parameters and edge detection threshold values, however, need to be defined by users onto noisy surfaces and textures respectively before the smoothing process. Bajaj et al. [13] present a unified anisotropic diffusion for 3D mesh smoothing by treating discrete surface data as a discretized version of a 2D Riemannian manifold and establishing a partial differential equation (PDE) diffusion model for such a manifold. This method helps enhance sharp features while filtering out noise by considering 3-ring neighbors of each vertex to achieve a nonlinear approach of the smoothing process. Tasdizen et al. [14, 15] introduce a two-step surface smoothing method by solving a set of coupled second-order PDEs on level set surface models. Instead of filtering the positions of points on a mesh, this method operates on the normal map of a surface and manipulates the surface to fit the processed normals. All the surface normals are processed by solving second-order equations using implicit surfaces. In [16], Hildebrandt et al. present a mesh smoothing method by using a prescribed mean curvature flow for simplicial surfaces. This method develops an improved anisotropic diffusion algorithm by defining a discrete shape operator and principal curvatures of simplicial surfaces.

Roughly speaking, mesh denoising techniques can be defined as the requirement to adjust vertex positions without changing the connectivity of the 3D mesh, and may be classified into two main categories: one-step and two-step approaches. The one-step approaches directly update vertex positions using the original vertex coordinates and a neighborhood around the current vertex, and sometimes face normals too. On the other hand, the two-step approaches first adjust face normals and then update vertex positions using some error minimization criterion based on the adjusted normals. In many cases, a single pass of a one-step or two-step approach does not yield a satisfactory result, and therefore iterated operations are performed. In this paper, we present a 3D mesh denoising method based on kernel density estimation. The proposed technique falls into the category of one-step approaches. The main idea is to use Laplacian smoothing algorithm combined with Gaussian kernel density estimators in order to reduce the over-smoothing problem and remove the noise effectively while preserving the nonlinear features of the 3D mesh such as curved surface regions, sharp edges, and fine details.

The rest of this paper is organized as follows. In the next section, we briefly recall some basic concepts of 3D mesh data, and then a general formulation of the 3D mesh denoising problem is stated. In Section 3, a kernel-based nonlinear diffusion is introduced. In Section 4, we provide experimental results to demonstrate a much improved performance of the proposed method in 3D mesh denoising. Finally, some conclusions are included in Section 5.

2   Problem Formulation

In computer graphics and geometric-aided design, triangle meshes have become the de facto standard representation of 3D objects. A triangle mesh $\mathbb{M}$ may be defined as $\mathbb{M} = (\mathcal{V}, \mathcal{E})$ or $\mathbb{M} = (\mathcal{V}, \mathcal{T})$, where $\mathcal{V} = \{v_1, \ldots, v_m\}$ is the set of vertices, $\mathcal{E} = \{e_{ij}\}$ is the set of edges, and $\mathcal{T} = \{t_1, \ldots, t_n\}$ is the set of triangles. Each edge $e_{ij} = [v_i, v_j]$ connects a pair of vertices $\{v_i, v_j\}$. Two distinct vertices $v_i, v_j \in \mathcal{V}$ are adjacent (denoted by $v_i \sim v_j$ or simply $i \sim j$) if they are connected by an edge, i.e. $e_{ij} \in \mathcal{E}$. The neighborhood (also referred to as a ring) of a vertex v_i is the set $v_i^\star = \{v_j \in \mathcal{V} : v_i \sim v_j\}$. The degree d_i of a vertex v_i is simply the cardinality of $v_i^\star$. Fig. 1 depicts an example of a neighborhood $v_i^\star$, where the degree d_i of the vertex v_i is equal to 6.

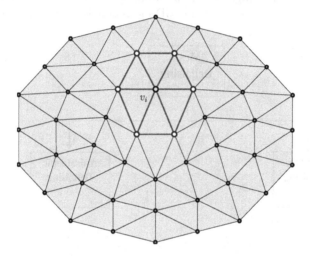

Fig. 1. Illustration of vertex neighborhood $v_i^\star$

The mean edge length $\bar{\ell}$ of the mesh $\mathbb{M}$ is given by

$$\bar{\ell} = \frac{1}{|\mathcal{E}|} \sum_{e_{ij} \in \mathcal{E}} \|e_{ij}\|, \tag{1}$$

where $\|e_{ij}\| = \|v_i - v_j\|$ if $v_i \sim v_j$, and $\|e_{ij}\| = 0$ otherwise.

2.1   Mesh Denoising Model

In all real applications, measurements are perturbed by noise. In the course of acquiring, transmitting or processing a 3D model for example, the noise-induced degradation often yields a resulting vertex observation model, and the most commonly used is the additive one,

$$v = u + \eta, \tag{2}$$

where the observed vertex v includes the original vertex u, and the random noise process η which is usually assumed to be Gaussian with zero mean and standard deviation σ.

Mesh smoothing refers to the process of recovering a 3D model contaminated by noise. The challenge of the problem of interest lies in recovering the vertex u from the observed vertex v, and furthering the estimation by making use of any prior knowledge/assumptions about the noise process η.

Generally, 3D mesh denoising methods may be classified into two major categories: isotropic and anisotropic. The former techniques filter the noisy data independently of direction, while the latter methods modify the diffusion equation to make it nonlinear or anisotropic in order to preserve the sharp features of a 3D mesh. Most of these nonlinear methods were inspired by anisotropic-type diffusions in the image processing literature. The diagram shown in Fig. 2 summarizes the main classification of 3D mesh denoising approaches.

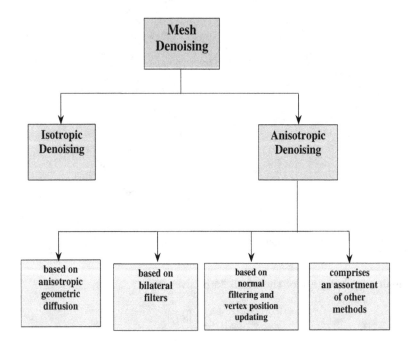

Fig. 2. Classification of 3D mesh denoising techniques

3   Proposed Method

Kernel density estimates are output as smooth curves with the amount of smoothing governed by a bandwidth value used during calculation [18]. Densities are calculated by placing kernels over the distribution of data points [18]. Kernels that overlap one another increase density values in shared areas of the distribution. For univariate and multivariate data, the Gaussian kernel is the most commonly used one. In particular, for 3D data, the Gaussian kernel is given by

$$K(\boldsymbol{x}) = \frac{1}{(2\pi)^{\frac{3}{2}}} \exp\left(-\frac{\|\boldsymbol{x}\|^2}{2}\right), \quad \forall \boldsymbol{x} \in \mathbb{R}^3. \tag{3}$$

The proposed approach updates iteratively each mesh vertex according to the following rule

$$\boldsymbol{v}_i \leftarrow \boldsymbol{v}_i + \sum_{\boldsymbol{v}_j \in \boldsymbol{v}_i^*} \mathcal{K}_{H_i}(\boldsymbol{v}_i - \boldsymbol{v}_j) \frac{1}{\sqrt{d_i}} \left(\frac{\boldsymbol{v}_j}{\sqrt{d_j}} - \frac{\boldsymbol{v}_i}{\sqrt{d_i}}\right) \tag{4}$$

where

$$\mathcal{K}_{H_i}(\boldsymbol{v}_i - \boldsymbol{v}_j) = \frac{1}{\det(H_i)} K\left(H_i^{-1}(\boldsymbol{v}_i - \boldsymbol{v}_j)\right) \tag{5}$$

and H_i is a symmetric positive semi-definite matrix, which defines the covariance matrix around the neighborhood of vertex v_i, and it is given by

$$H_i = \sum_{j \sim i} (\boldsymbol{v}_j - \boldsymbol{c}_i)(\boldsymbol{v}_j - \boldsymbol{c}_i)^T, \quad \text{where } \boldsymbol{c}_i = \frac{1}{d_i} \sum_{j \sim i} \boldsymbol{v}_j. \tag{6}$$

It is worth pointing out that H_i is also called the bandwidth matrix in the context of kernel smoothing and it measures the amount of smoothing.

4   Experimental Results

This section presents experimental results where the mean filtering [6], angle median filtering [6], Laplacian flow [1], weighted Laplacian flow [9, 10], geometric diffusion [11], bilateral filtering [7], and the proposed method are applied to noisy

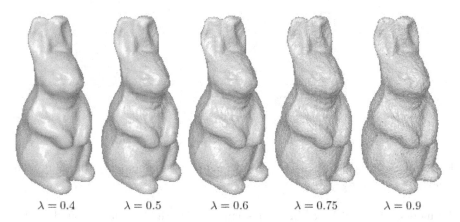

$\lambda = 0.4$ $\qquad$ $\lambda = 0.5$ $\qquad$ $\lambda = 0.6$ $\qquad$ $\lambda = 0.75$ $\qquad$ $\lambda = 0.9$

Fig. 3. Output results of our proposed mesh denoising approach for different values of the regularization parameter. The number of iterations is set to 5.

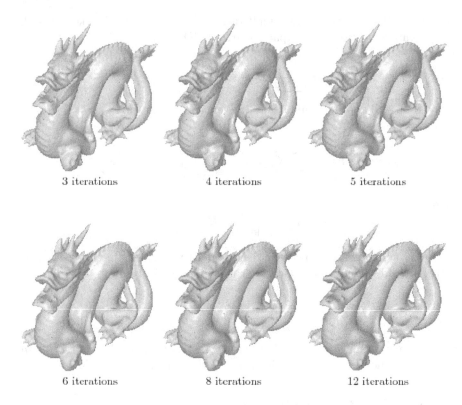

3 iterations   4 iterations   5 iterations

6 iterations   8 iterations   12 iterations

Fig. 4. Output results of our proposed mesh denoising approach at different iteration numbers. The regularization parameter is set to $\lambda = 0.8$.

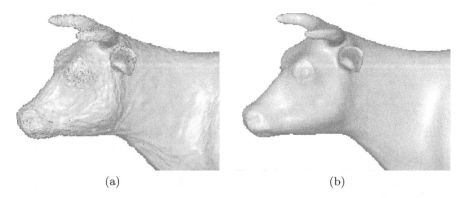

(a)   (b)

Fig. 5. (a) Noisy model; (b) output result of our proposed approach with $\lambda = 0.8$. The number of iterations is set to 6.

3D models obtained by adding Gaussian noise to the original 3D models. The standard deviation of the noise was set to 2% of the mean edge length, that is $\sigma = 0.02\,\bar{\ell}$, where $\bar{\ell}$ is given by Eq. (1).

In practical applications, the covariance matrix H_i given by Eq. (6) may become singular. To circumvent this singularity problem and also to ensure the stability of the proposed algorithm, we use a regularized covariance matrix as follows

$$\widetilde{H}_i = H_i + \lambda I \tag{7}$$

where I is a 3×3 identity matrix and λ is a positive regularization parameter.

Fig. 3 displays the mesh denoising results obtained by our proposed method for different values of the regularization parameter λ, where the number of iterations was to set to 5. As can be seen in Fig. 3, the value $\lambda = 0.4$ gives the best denoising result for the 3D rabbit model. And as a the value of λ increases, the rabbit model becomes more noisier. Also, we noticed through extensive experimentation that a smaller value of λ often tends to produce a distorted shape of the 3D object. Therefore, the regularization parameter should be tuned to be small enough to capture the intrinsic shape of a 3D object and large enough not to recapture noise.

Fig. 4 depicts the output results of the proposed approach at different iteration numbers. These results show that using the proposed approach, the noise can be removed with just a small number of iterations and that the sharp features are well preserved when the regularization parameter is appropriately chosen.

4.1   Qualitative Evaluation of the Proposed Method

Fig. 5(b) depicts the output result of the proposed algorithm on an enlarged view of the noisy 3D cow model's head shown in Fig. 5(a). Note that the geometric structures and the fine details around the eye and the ear of the denoised cow model are very well preserved. Fig. 6(c) through Fig. 6(h) show the denoising results obtained via Laplacian flow, weighted Laplacian flow, mean filtering, angle median filtering, bilateral filtering, and the proposed method respectively. These results clearly show that our method outperforms all the mesh filtering techniques used for comparison. Moreover, the proposed method is simple and easy to implement. One main advantage of the proposed algorithm is that it requires only a few iterations to smooth out the noise, whereas the weighted Laplacian flow, the mean and the angle median filters require substantial computational time. This better performance is in fact consistent with a variety of 3D models used for experimentation. Fig. 7 shows the denoising results for the 3D Igea model.

4.2   Quantitative Evaluation of the Proposed Method

Let $\mathbb{M}$ and $\widehat{\mathbb{M}}$ be the original model and the denoised model, with vertex sets $\mathcal{V} = \{v_i\}_{i=1}^m$ and $\widehat{\mathcal{V}} = \{\hat{v}_i\}_{i=1}^m$ respectively. To quantify the performance of the proposed approach, we compute the visual error metric [19] given by

$$E = \frac{1}{2m} \left(\sum_{i=1}^m \|v_i - \hat{v}_i\|^2 + \sum_{i=1}^m \|\mathcal{I}(v_i) - \mathcal{I}(\hat{v}_i)\|^2 \right), \tag{8}$$

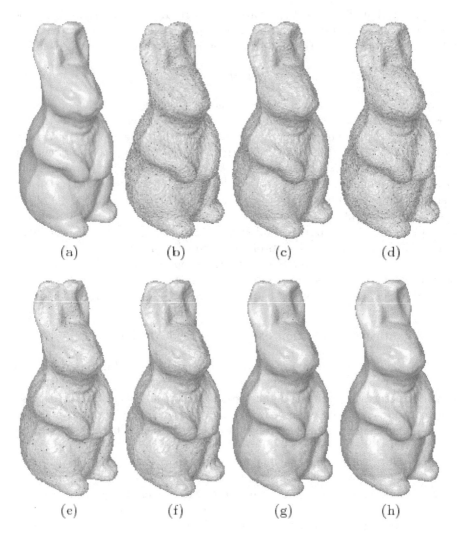

Fig. 6. Denoising results for the 3D rabbit model: (a) original model; (b) noisy model; (c) Laplacian flow; (d) weighted Laplacian flow; (e) mean filtering; (f) angle median filtering; (g) bilateral mesh flow; (h) our proposed approach. The number of iterations is set to 3.

where $\mathcal{I}$ is the geometric Laplacian operator defined as

$$\mathcal{I}(\boldsymbol{v}_i) = \boldsymbol{v}_i - \frac{1}{d_i} \sum_{\boldsymbol{v}_j \in \boldsymbol{v}_i^*} \boldsymbol{v}_j.$$

The values of visual error metric for some denoising experiments are depicted in Fig. 8(a) and Fig. 8(b) which clearly show that the proposed method gives the best results, indicating the consistency with the subjective comparison.

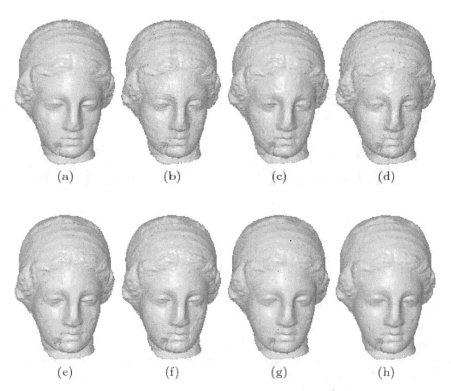

Fig. 7. Denoising results for the 3D Igea model: (a) original model; (b) noisy model; (c) Laplacian flow; (d) weighted Laplacian flow; (e) mean filtering; (f) angle median filtering; (g) bilateral mesh flow; (h) our proposed approach. The number of iterations is set to 6.

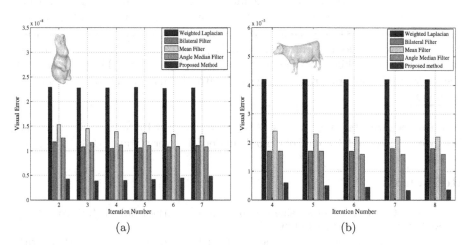

Fig. 8. Visual error comparison results between the proposed approach and other methods for the (a) rabbit and (b) cow models

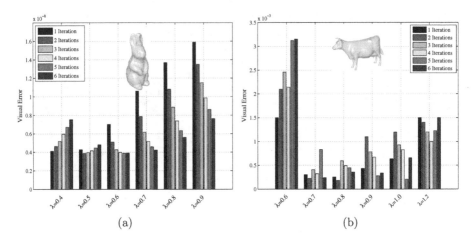

Fig. 9. Visual error vs. regularization parameter with different number of iterations for the (a) rabbit and (b) cow models

4.3   Choice of the Regularization Parameter

The regularization parameter should be tuned to be small enough to capture the intrinsic shape of a 3D object and large enough not to recapture noise. This parameter may be estimated experimentally using the visual error as shown in Fig. 9(a) and Fig. 9(b), which display the plots of the visual error vs. the regularization parameter for different iteration numbers of the proposed approach.

5   Conclusions

In this paper, we introduced a simple and fast 3D mesh denoising method using the concept of multivariate kernel density estimation. The main idea behind our proposed approach is to use a regularized bandwidth matrix of the kernel density in order to avoid over-smoothing and to fully preserve the geometric structure of the 3D mesh data, while effectively removing undesirable noise. The experimental results showed that our proposed method outperforms existing mesh denoising techniques.

Acknowledgments. This work was supported in part by Natural Sciences and Engineering Research Council of Canada under Discovery Grant no. 311656-2008.

References

1. Taubin, G.: A signal processing approach to fair surface design. In: Proc. ACM SIGGRAPH, pp. 351–358 (1995)
2. Ohtake, Y., Belyaev, A.G., Bogaevski, I.A.: Polyhedral surface smoothing with simultaneous mesh regularization. In: Proc. Geom. Modeling and Processing, pp. 229–237 (2000)

3. Taubin, G.: Linear anisotropic mesh filtering, IBM Research Report, RC2213 (2001)
4. Petitjean, S.: A survey of methods for recovering quadrics in triangle meshes. ACM Computing Surveys 34(2), 211–262 (2002)
5. Shen, Y., Barner, K.E.: Fuzzy Vector Median-Based Surface Smoothing. IEEE Trans. Visualization and Computer Graphics 10(3), 252–265 (2004)
6. Yagou, H., Ohtake, Y., Belyaev, A.: Mesh smoothing via mean and median filtering applied to face normals. In: Proc. Geometric Modeling and Processing - Theory and Applications, pp. 124–131 (2002)
7. Fleishman, S., Drori, I., Cohen-Or, D.: Bilateral mesh denoising. In: Proc. ACM SIGGRAPH, pp. 950–953 (2003)
8. Jones, T., Durand, F., Desbrun, M.: Non-iterative, feature preserving mesh smoothing. In: Proc. ACM SIGGRAPH, pp. 943–949 (2003)
9. Desbrun, M., Meyer, M., Schröder, P., Barr, A.: Implicit fairing of irregular meshes using diffusion and curvature flow. In: Proc. ACM SIGGRAPH, pp. 317–324 (1999)
10. Desbrun, M., Meyer, M., Schröder, P., Bar, A.: Anisotropic feature-preserving denoising of height fields and bivariate data. In: Graphics Interface, pp. 145–152 (2000)
11. Clarenz, U., Diewald, U., Rumpf, M.: Anisotropic geometric diffusion in surface processing. In: Proc. IEEE Visualization, pp. 397–405 (2000)
12. Clarenz, U., Diewald, U., Rumpf, M.: Processing textured surfaces via anisotropic geometric diffusion. IEEE Trans. Image Processing 13(2), 248–261 (2004)
13. Bajaj, C.L., Xu, G.: Anisotropic diffusion of subdivision surfaces and functions on surfaces. ACM Trans. Graphics 22(1), 4–32 (2003)
14. Tasdizen, T., Whitaker, R., Burchard, P., Osher, S.: Geometric surface smoothing via anisotropic diffusion of normals. In: Proc. IEEE Visualization, pp. 125–132 (2002)
15. Tasdizen, T., Whitaker, R., Burchard, P., Osher, S.: Geometric surface processing via normal maps. ACM Transactions on Graphics 22(4), 1012–1033 (2003)
16. Hildebrandt, K., Polthier, K.: Anisotropic filtering of non-linear surface features. Computer Graphics Forum 23(3), 391–400 (2004)
17. Chung, F.R.: Spectral Graph Theory. American Mathematical Society, Providence (1997)
18. Silverman, B.W.: Density estimation for statistics and data analysis. Chapman and Hall, London (1986)
19. Karni, Z., Gotsman, C.: Spectral compression of mesh geometry. In: Proc. ACM SIGGRAPH, pp. 279–286 (2000)

Automatic Segmentation of Scanned Human Body Using Curve Skeleton Analysis

Christian Lovato[1], Umberto Castellani[2], and Andrea Giachetti[2]

[1] Dipartimento di Scienze Morfologico-Biomediche, Università di Verona
[2] VIPS lab, Dipartimento di Informatica, Università di Verona
andrea.giachetti@univr.it
http://www.univr.it

Abstract. In this paper we present a method for the automatic processing of scanned human body data consisting of an algorithm for the extraction of curve skeletons of the 3D models acquired and a procedure for the automatic segmentation of skeleton branches. Models used in our experiments are obtained with a whole-body scanner based on structured light (Breuckmann bodySCAN, owned by the Faculty of Exercise and Sport Science of the University of Verona), providing triangulated meshes that are then preprocessed in order to remove holes and create clean watertight surfaces. Curve skeletons are then extracted with a novel technique based on voxel coding and active contours driven by a distance map and vector flow. The skeleton-based segmentation is based on a hierarchical search of feature points along the skeleton tree.

Our method is able to obtain on the curve skeleton a pose-independent subdivision of the main parts of the human body (trunk, head-neck region and partitioned limbs) that can be extended to the mesh surface and internal volume and can be exploited to estimate the pose and to locate more easily anthropometric features.

The curve skeleton algorithm applied allows control on the number of branches extracted and on the resolution of the volume discretization, so the procedure could be then repeated on subparts in order to refine the segmentation and build more complex hierarchical models.

Keywords: Whole-body scanner, Curve skeleton, segmentation.

1   Introduction

Recent advances on scanning techniques make possible to acquire high resolution models of the human body that can be extremely useful for anthropometric studies and for other applications like medical diagnosis, clothing design, computer animation and entertainment. Most of these applications could benefit of an automatic processing of the scanner data able to segment and recognize the different parts of the body and to locate reference points useful, for example, to perform anthropometric measurements. In this paper we present a processing pipeline based on the analysis of the curve skeletons of acquired models that can be used for this task.

A. Gagalowicz and W. Philips (Eds.): MIRAGE 2009, LNCS 5496, pp. 34–45, 2009.

The skeleton is extracted as a tree structure with curve segments joined at the extrema and the segmentation is performed by detecting features on the extracted branches using a priori information on the body structure. The result is a labelling of the curve skeleton and of the original mesh that can be used for several applications: feature points can be located on the basis of the functional decomposition obtained and a stick figure representing the pose of an articulated human model can be easily computed and used to control mesh animation or to fit an articulated deformable model over the data.

The paper is structured as follows: section 2 presents the state of the art of whole body scanner technology, section 3 a short literature review on scanned body data segmentation, section 4 deals with the curve skeleton extraction problem and describes the method we developed for this task. In section 5 the segmentation procedure is finally described and experimental results are presented in section 6.

2   Whole-Body Scanner Technology

Whole-body scanners are used to build models applied in a wide range of applications such as ergonomic design, creation of sizing charts for clothing manufacture, creation of avatars for computer games and animation industry, anthropometric surveys, medical diagnosis.

Roughly speaking there are two main categories of technologies employed in whole-body scanning [16]: laser-based and Moiré-fringing-based technologies. In the former a laser stripe is projected onto the body surface. Then, the laser stripe is detected from several cameras and the set of 3D points representing the body shape are recovered by triangulation. Examples of this kind are the scanners developed by Cyberware (www.cyberware.com), Hamamatsu (www.hamamatsu.com) and Vitronic (www.vitronic.de).

In the latter, a white light source projects contour patterns, e.g. sinusoidal fringes, on the body surface. Therefore, a 3D cloud of points is estimated by observing the pattern deformations on the body surface, again from a set of cameras. Both technologies avoid direct contact with the body and fall into the category of shape from multiple views approach. The performance of different body scanners differ by resolution, accuracy, acquisition time, and organization of cameras position [16]. Examples of this kind are the scanners developed by Textile and Clothing Technology Corporation (www.dh.aist.go.jp), inSpeck (http://www.inspeck.com), and the product used for our experiments, i.e. the bodySCAN developed by Breuckmann GmbH (www.breuckmann.com). A bodySCAN typical acquisition creates a triangulated mesh with about 400.000 nodes and a resolution varying from 0.2 mm to 1.4 mm. The acquisition time ranges between 2.5 and 5.5 seconds. We found in our experiment that this time is sufficient to obtain an acceptable mesh quality (limited motion artifacts). Meshes, however, present various types of defects like holes, non manifold edges, bad shaped triangles and outliers, that should be corrected during the processing. The scanner also provides grayscale information. We do not use this information in our pipeline, but it may be extremely important to acquire, for example,

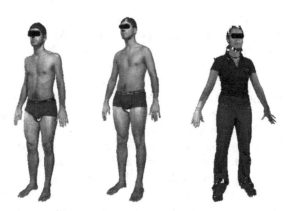

Fig. 1. Three meshes generated by the Breuckmann bodySCAN. It is possible to observe missing parts in shadowed regions and in correspondence with clothing and hair.

markers positions useful to validate automatic measurements. Fig. 1, shows examples of acquired textured models. It is possible to see holes and inaccuracies caused by occlusions and reflective materials.

3   Related Work

A huge literature is available on mesh segmentation. However, not so many paper deal with the reliable partitioning of a human body model into semantically consistent parts. A recent detailed review on scanned human body processing methods [16], presents and compare only few methods applied in literature to perform this task, most of them limited to standard postures, except for those developed by the authors, based on Reeb Graphs [17,15].

Mortara et al. [7] proposed the use of a surface point classification called *plumber* in order to identify tubular region and extract body parts, performing also anthropometric measurements. Yu et al. [19] proposed a method able to find automatically joints by computing specific measurements on volume sections. The method, however, requires a previous detection of body landmarks and limbs direction.

Our approach also uses similar ideas, but starts its automatic processing by first extracting and segmenting the curve skeleton of the model. In this way it is possible to process the curve skeleton branches corresponding, for example to the limbs and to perform local measurements useful to locate landmarks or joints without slicing in pre-defined directions the surface. Recently, we discovered that this segmentation approach is close to that proposed by Reniers and Telea [2], who implemented a generic method for skeleton-based mesh segmentation. Our method differs from theirs because it exploits a priori information on human body structure in order to simplify the curve skeleton and to recognize its parts and for the use a feature point search on skeleton branches in order to define cut points for the skeleton.

4   Curve Skeleton Extraction

The literature on curve skeleton extraction is also huge and research on this topic is still active. Despite its apparent simplicity, in fact, the extraction of a 1D connected curve skeleton from volume data presents several problems, and even the definition of a curve skeleton is not easy, as pointed out by Cornea et al. [10].

The first methods successfully applied were those based on topological thinning [11], i.e. the iterative removal of external voxels preserving the topology or on the computation of distance maps from the border as in the voxel coding method [18]. They gave useful results, especially in the medical field, where the estimation of a centerline path in vessels is fundamental their characterization and measurement. These techniques, however, usually required interaction to place seed points or extremal points of the skeleton to be preserved. Furthermore results obtained were usually not reliable for non tubular objects.

A variety of approaches has then been proposed to overcome these problems. Telea and van Wijk [1] used the intersection of 2d skeletons for a fast 3D skeleton extraction, Cornea et al. [9] used a fast marching method, Sharf et al. [13] obtained the skeleton "on the fly" while reconstructing the mesh with a surface growth. Shapira et al. [12] used a function defined on the surface (Shape Diameter Function) in order to find approximate skeletal points then fitted into curves. Dey and Sun [4] removed ambiguities in curve-skeleton definition by considering it as the subset of the medial axis where a function called Medial Geodesic Function can be defined and is singular. A similar approach, but defined on voxelized volumes has been used by Reniers et al. [3]. Drawbacks of these approaches are the complexity of the discretization steps and the computational weight of the geodesic path evaluation.

For our mesh processing pipeline we adopted for the curve skeleton extraction a novel method based on voxel coding and active contours that strongly improves and extends to general objects the one used in [5] for vascular reconstruction. This method is good for our task because it allows a fast extraction of a connected structure, thin and smooth, well centered in tubular regions and with no closed loops by construction. Furthermore, it is possible to limit a priori the number of branches to extract neglecting shorter ones and to define the resolution of the curve skeleton, an useful feature for the hierarchical processing of parts and subparts of the human body we plan to develop.

Let us describe the method in more detail. The process starts with a raw extraction of a tree structure with a voxel coding method, like in [18]. The mesh is discretized in a 3D grid of given dimension (the resolution can be adapted to the level of detail, for human body we start with 5 mm of voxel size). Border voxels are then extracted and a distance map computing the distance of internal voxels from the border (DFB map) is generated. The classical voxel coding method by Zhou and Toga then extracts iteratively branches by giving a seed point, computing the "distance from seeds" (DFS) map and then using it to compute the shortest voxel path joining the farthest point to the seed. The path is then centered by replacing each voxel of the chain with the voxel obtained by first finding the cluster of the connected voxels with the same distance from the seed,

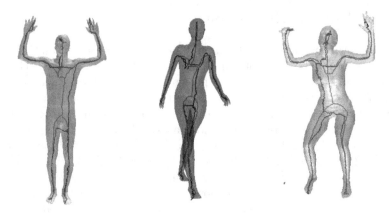

Fig. 2. Examples of curve skeletons extracted on human body models, meshes are represented as transparent surfaces, non-centered paths are represented in black, centered and cleaned trees are represented in gray. Examples are chosen in order to underline that the characterization obtained is almost pose-independent.

and then taking the voxel with the highest distance from border. It has been shown in [5] that this centering procedure can give very bad results in case of non tubular shapes because clusters can be large and the curve may result not continuous. We therefore do not use this method to center the branches (we just perform optionally this step as pre processing), but exploit an active contour approach instead.

Our method consists of first creating a tree structure joining several branches. After the extraction of the first branch as the shortest path joining a seed placed near the mesh border and the farthest voxel of the volume, and its rough centering, a recursive shortest path extraction is done in the same way updating the distance from seed at each iteration as the distance from the previously extracted skeleton point. The procedure can be iterated until branches are shorter than a threshold or a sufficient number if branches has been found (this is our case). The skeleton, consisting of nodes with floating point coordinates and links is then centered moving iteratively the nodes according to external forces driving the contour and other constraints. In detail:

- two image based forces are applied to the skeleton points: the first is directed as the gradient of the interpolated DFB map, the other is generated with a fast propagation of the internal normal vector at the boundaries. This last force is used not to have ambiguous stationary points in the medial surface.
- The internal forces are generated by a mass-spring model, with masses at the node positions and zero length springs connecting them.
- During the curve evolution, terminal nodes are kept fixed, links are preserved and new nodes are inserted when the distance between neighbors is higher than a threshold and removed when it is lower than another threshold.

– If two curve segments are partially superimposed, the duplicated part is removed and new links created.
– The iterative procedure is stopped when the global displacement is lower than a threshold.

In our implementation, after the mesh discretization, the seed point for the extraction of the first branch is taken as the surface voxel with the highest z coordinate. Rules chosen for the curve evolution makes, in any case, the results substantially independent on this choice. Weights for the snake forces have been set by trial and error.

Results obtained are satisfactory: the algorithm is fast and robust and curve skeletons obtained are well centered (see Fig. 2). The active contour approach presents the big advantage of keeping the curves smooth, avoiding the discontinuity problems of classical methods based on local geometrical properties and clustering.

5   Curve-Skeleton Based Segmentation

The curve skeleton is then subdivided in segments, i.e. chains with no links except at the extremal nodes. Segments can be divided in leaves, i.e. chains with links at only one of the extremal nodes, and internal segments, i.e. chains with links at two extremal nodes (Figure 3 A).

The processing pipeline consists then in two steps: first the candidate skeleton branches containing the main body structures are isolated, then they are processed in order to find feature points used to segment them correctly. In order to find the human body components we process the skeleton as follows: first we remove the shortest leaves as follows:
-leaves are put in a list ordered by length
-the shortest leaf is removed and if its endpoint is linked to two segments, these segments are merged. The procedure is ended when the list includes only 5 leaves. Each leaf should include the complete skeleton of one structure among the limbs and to the head/neck (Fig 3 B). Finally, a simple decision tree based

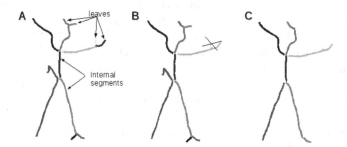

Fig. 3. The procedure used to purge the skeleton tree preserving the five main leaves including the parts to be segmented. A: original skeleton. B Shortest leaf is removed and the connected branches merged. C: The procedure is stopped with five leaves left.

on length and distance functions of the leaves is able to label them according to the structure included (arm, leg, head).

5.1   Feature Extraction on the Curve-Skeleton

Leaves, i.e. candidate skeleton branches that should include limbs and head/neck, are then processed in order to find the optimal position of feature points. These points are the approximate location of limbs an neck attachments to the trunk, and the approximate position of knee and elbow joints, wrist and ankle joints. In order to approximately locate these points on the curve skeleton we considered the behavior of four selected scalar values along the leaves extracted: the local interpolated distance from border (obtained from the map), estimates of local average diameter and eccentricity of the section perpendicular to the curve (obtained from a number of sample rays traced on that plane), and an estimate of the local principla curvature. Fig. 4 shows examples of the typical behavior of the first three values on leaves including arm, leg and head/neck. Curvature is not reported being pose-dependent, so its behavior is not characteristic. It is possible to see that the interpolated DFB is a slow-varying measure and can capture the general trend of the diameter of the mesh around the evaluated branch. Mean diameter is more sensitive to the local structure (and to noise) and can be used used to capture subtle features. Particularly interesting is the eccentricity estimate, well-suitable for joint feature extraction as also reported in [19].

Our segmentation method is hierarchical: first the attachments of the limb or of the neck to the trunk are found starting from the middle of the branches and moving towards the trunk until the eccentricity value presents a big step edge. Currently we place the starting point of limbs in the location where the value becomes stationary. Wrists and ankles are then searched by finding in the lower part of the branches the local minimum of the section diameter following the large local maximum corresponding to forearm/calf muscles. Feature points are then located in the subsequent edge in the section diameter (location of maximum of the diameter increase). More difficult is to locate accurately elbows and knee. They are currently identified by the maximum in the curvature if well characterized, otherwise they are located at the extrema of the deviation of the diameter and of the distance value from the linear growth. If the feature point is not clearly visible in this way we locate first a different feature point, i.e. the location where the forearm/calf muscles starts, characterized by a minimum in the local diameter, and the position is then refined by shifting the detected point on the path of a distance proportional to the average distance on the two feature points on test cases and to the leg length. The position of the detected points is then finally refined with an iterative procedure adding the constraints of symmetry between left-right legs and arms. Finally, we locate the feature point used to separate the head from the trunk by extracting from the eccentricity plot not only the edge corresponding to the beginning of the neck but also the one corresponding to the neck attachment to the head and put the feature point in the middle of the neck.

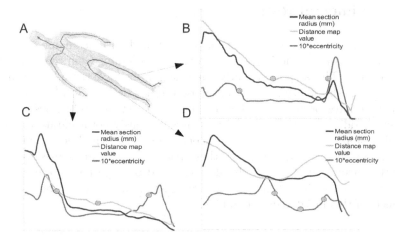

Fig. 4. Example of the behavior of selected scalar measures computed on curve skeleton points on the different skeleton leaves. A: The original mesh, with the curve skeleton extracted. Leaves are represented with different intensities. B,C,D: Plots of DFB, average section radius and eccentricity along the leaf including an arm, a leg and head and neck. Dots indicate the approximate extracted position of the feature points.

When a reasonable number of scans will be available, we plan to build statistical models of the behavior of the computed values near the interested points and to locate them fitting the models to the data.

5.2   Mesh Segmentation

Mesh points can be easily labelled according to the skeleton partitioning. We label the discretized space computing a distance map of the volume voxels from the final skeleton. Triangles of the mesh are then labelled according to the label of the voxel including its center. Being the space partitioning not depending on the mesh triangulation, the mesh labelling can be performed also on the processed mesh as well as on the original one.

5.3   Pose Estimation

Human pose estimation can be achieved by finding the best fit an articulated stick figure over the acquired data. This is extremely simple in our case, having a rough estimation of the positions of the nodes of the articulated model from the skeleton analysis. It is therefore immediate to calculate a rough pose estimate by connecting the detected feature points with segments. Methods based on hierarchical articulated model fitting (like, for example, the method described in [8]) could be applied as well.

6   Experimental Results

We tested our pipeline on 8 meshes acquired with the Breuckmann scanner. A pre-processing is performed by generating a closed volume from the original incomplete mesh. This step is currently done using Polymender, a fully automated software for mesh repairing based on the procedures described in [6]. Algorithm and package have been chosen due to their simplicity and robustness, it must however considered that the reconstruction is obviously not accurate where the original mesh presents large holes due to occlusions (i.e. at limb joints, under the shoulders and between the legs). In these regions, the repaired meshes do not follow the natural curvature of the skin as would be expected. However, being our processing pipeline based on the curve skeleton and not on surface features, the effects of this inaccuracy are limited.

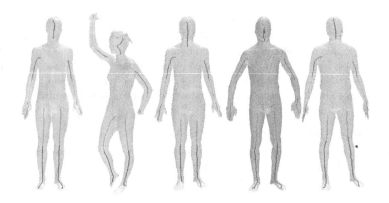

Fig. 5. Examples of curve skeleton partitioning on data obtained from the Breuckmann body-scanner

On all the acquired mesh the curve skeletons automatically extracted appear correctly located, smooth and well centered in the tubular parts.

Fig. 5 shows example partitioned curve skeletons extracted on the Breuckmann scanner models. The labelling of the body parts is correct for all the datasets tested. The position of the feature points is, of course, rough and also more influenced by holes and inaccuracies of the original meshes. Only in a few cases, however, points appears a bit shifted from the expected position. Despite these limits in the accuracy, the body partitioning obtained can be extremely useful for several practical tasks, like fitting hierarchical models or stick figures to the data or reducing the search space for anatomical features on the mesh surface.

In order to evaluate the robustness of the curve-skeleton extraction and labeling methods against pose variations, we also tested the body partitioning on five watertight meshes representing human body in different poses, available from the Aim@Shape Watertight dataset[14]. Results obtained (see Fig. 6) show that

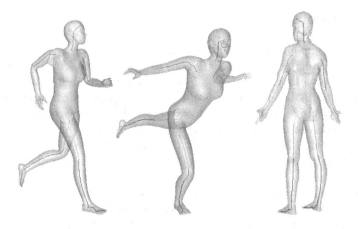

Fig. 6. Results obtained on human body meshes from the aim@shape database

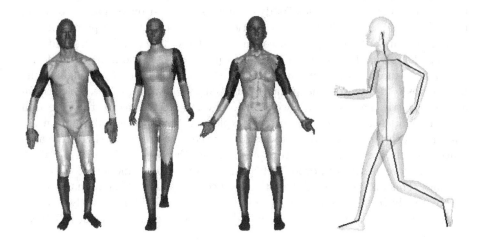

Fig. 7. Left: Examples of mesh labelling derived from the skeleton partitioning. Right: A stick figure obtained from the feature points located with the described technique.

the recognition of the body parts and the location of feature points is possible independently on the pose (but obviously requiring the sphere-like topology).

The left part of Fig. 7 shows examples of mesh labelling induced by the skeleton segmentation. The accuracy of the surface boundaries is obviously limited, we plan, however, to refine the segmentation using surface information. The right part of the figure shows an example of stick figure obtained from the processed skeletons superimposed to the original meshes. Results may be improved by considering that the feature point at the beginning of the limbs are not exactly corresponding to the articulation, and should be slightly displaced. The articulated models obtained seem, however, suitable for articulated model fitting and computer animation.

7   Conclusions

We presented a pipeline for the fully automated segmentation of volumetric data acquired with a whole body scanner. This pipeline is intended as a first step on order to have an automatic extraction of useful information for anthropometric studies. The original contribution proposed in this paper consists mainly of

-A novel algorithm for curve-skeleton extraction, based on voxel coding and active contours driven by a distance map and vector flow.

-A curve-skeleton based body segmentation algorithm, based on a priori information and the search of feature points on scalar functions computed on skeleton points.

An interesting aspect of the procedure proposed is that it can be performed iteratively on the segmented parts, increasing the resolution of the curve skeleton extraction in order to have more detailed models after the first rough classification of the parts. We plan to segment automatically in this way hands and feet. Results obtained are promising even if preliminary and a lot of further work is planned to carry on the whole body scanner segmentation pipeline. We plan, in fact, to characterize better the position of the characteristic points on the curve-skeleton by building statistical models of the skeleton features and registering them with the acquired data for a precise localization. We should also improve the mesh segmentation derived by the skeleton labeling by making the boundaries computed on the surface attracted by curvature edges. To perform anthropometric measurements, we plan to develop algorithms for the detection of meaningful features on the surface, exploiting the a priori information acquired with the mesh partitioning to simplify the problem narrowing the search space. Finally, we want to improve and exploit the pose estimation obtained from the curve skeleton in order to fit generic human body models to the scanner data. This procedure could greatly enhance the quality of the human models automatically reconstructed from the scanner data, providing directly realistic watertight meshes also when the acquired data are incomplete or not reliable.

Acknowledgements

This work is partially supported by the University of Verona (Joint Project AnthropoScan) and by the Italian Ministry of Research and Education (project 3-ShIRT, 3D SHape Indexing and Retrival Techniques). Special thanks to prof. Carlo Zancanaro for making the Breuckmann scanner available for this study.

References

1. van Wijk, J.J., Telea, A.: An augmented fast. marching method for computing skeletons and centerlines. In: Proc. IEEE VisSym 2002, pp. 251–260. ACM Press, New York (2002)
2. Reniers, D., Telea, A.: Skeleton-based hierarchical shape segmentation. In: Proc. of the IEEE Int. Conf. on Shape Modeling and Applications (SMI), pp. 179–188 (2007)

3. Reniers, D., Van Wijk, J.J., Telea, A.: Computing multiscale curve and surface skeletons of genus 0 shapes using a global importance measure. IEEE Transactions on Visualization and Computer Graphics 14(2), 355–368 (2008)
4. Dey, T.K., Sun, J.: Defining and computing curve-skeletons with medial geodesic function. In: SGP 2006: Proceedings of the fourth Eurographics symposium on Geometry processing, Aire-la-Ville, Switzerland, pp. 143–152. Eurographics Association (2006)
5. Giachetti, A., Zanetti, G.: Aquatics reconstruction software: The design of a diagnostic tool based on computer vision algorithms. In: Sonka, M., Kakadiaris, I.A., Kybic, J. (eds.) CVAMIA/MMBIA 2004. LNCS, vol. 3117, pp. 48–63. Springer, Heidelberg (2004)
6. Ju, T.: Robust repair of polygonal models. In: Proc. of ACM SIGGRAPH 2004, pp. 888–895. ACM Press, New York (2004)
7. Mortara, M., Patané, G., Spagnuolo, M.: From geometric to semantic human body models. Computers & Graphics, 185–196 (2006)
8. Moschini, D., Fusiello, A.: Tracking stick figures with hierarchical articulated icp. In: Proceedings THEMIS 2008, pp. 61–68 (2008)
9. Cornea, N., Silver, D., Yuan, X., Balasubramanian, R.: Computing hierarchical curve-skeleton of 3d objects. The Visual Computer (11), 945–955 (2005)
10. Cornea, N., Silver, D., Yuan, X., Balasubramanian, R.: Curve-skeleton applications. In: IEEE Visualization, pp. 95–102 (2005)
11. Gagvani, N., Silver, D.: Parameter-controlled volume thinning. Graph. Models and Image Proc. (3), 149–164 (1999)
12. Shapira, L., Shamir, A., Cohen-Or, D.: Consistent mesh partitioning and skeletonisation using the shape diameter function. Vis. Comput. 24(4), 249–259 (2008)
13. Sharf, A., Lewiner, T., Shamir, A., Kobbelt, L.: On-the-fly curve-skeleton computation for 3d shapes. The Visual Computer 26(3) (2007)
14. Veltkamp, R.C., ter Haar, F.B.: Shrec 2007 3d retrieval contest. Technical Report UU-CS-2007-015, Department of Information and Computing Sciences (2007)
15. Werghi, N.: A robust approach for constructing a graph representation of articulated and tubular-like objects from 3d scattered data. Pattern Recognition Letters 27, 643–651 (2007)
16. Werghi, N.: Segmentation and modeling of full human body shape from 3-d scan data: A survey. IEEE Transactions on Systems, Man, and Cybernetics, Part C 37(6), 1122–1136 (2007)
17. Xiao, Y., Siebert, P., Werghi, N.: A discrete reeb graph approach for the segmentation of human body scans. In: Proceedings of Fourth International Conference on 3-D Digital Imaging and Modeling, 2003. 3DIM 2003, pp. 378–385 (2003)
18. Zhou, Y., Toga, A.: Efficient skeletonization of volumetric objects. TVCG (3), 196–209 (1999)
19. Yu, Y., Wang, Z., Xia, S., Mao, T.: Automatic joints extraction of scanned human body. HCI (12), 286–293 (2007)

Multi-view Player Action Recognition in Soccer Games

Marco Leo, Tiziana D'Orazio, Paolo Spagnolo, Pier Luigi Mazzeo,
and Arcangelo Distante

Institute of Intelligent Systems for Automation, via Amendola 122/d, Bari , Italy
leo@ba.issia.cnr.it

Abstract. Human action recognition is an important research area in
the field of computer vision having a great number of real-world applications. This paper presents a multi-view action recognition framework
able to extract human silhouette clues from different synchronized static
cameras and then to validate them by analyzing scene dynamics. Two
different algorithmic procedures were introduced: the first one performs,
in each acquired image, the neural recognition of the human body configuration by using a novel mathematical tool called Contourlet transform.
The second procedure performs, instead, 3D ball and player motion analysis. The outcomes of both procedures are then merged to accomplish
the final player action recognition task. Experiments were carried out
on several image sequences acquired during some matches of the Italian
"Serie A" soccer championship.

Keywords: Human Pose Estimation, Contourlet Transform, Neural
Networks, Soccer Player Action Recognition.

1 Introduction

Human action recognition aims at automatically ascertaining the activity of a
person, i.e. to identify if someone is walking, dancing, or performing other types
of actions. It is an important area of research in the field of computer vision and
the ever growing interest in it is fueled, in part, by the great number of real-
world applications such as surveillance scenarios, content-based image retrieval,
human-robot interaction, sport video analysis, smart rooms etc.

Human action recognition has been widely studied (for an extensive review
see [2]). In general human activity recognition approaches are categorized on the
basis of the representation of the human body: representation can be extracted
either from a still image or a dynamic video sequence.

In [1] three dimensional space-time shapes such as local space-time saliency,
action dynamics, shape structure, and orientation are extracted from silhouette
images of humans and they are used to classify actions. In [3], shape and motion
cues are used for the action recognition of two different actions in the movie
"Coffee and Cigarettes". [4] introduces instead a biologically inspired action
recognition approach which uses hierarchically ordered spatio-temporal feature

A. Gagalowicz and W. Philips (Eds.): MIRAGE 2009, LNCS 5496, pp. 46–57, 2009.

detectors. In [5] an unsupervised learning method for human action categories is presented. A video sequence is represented as a collection of spatial-temporal words by extracting space-time interest points. In [6] two types of features are fused treating different features as nodes in a graph, where weighted edges between the nodes represent the strength of the relationship between entities.

In general action recognition extracted from video sequences requires complex models to understand the dynamics. On the other hand recent studies demonstrated that a static human pose encapsulates many useful clues for recognizing the ongoing activity.

In [7], a bag-of-rectangles method is used for action recognition, effectively modeling human poses for individual frames and thereby recognizing various action categories. [8] use Principal Component Analysis (PCA) to extract eigenshapes from silhouette images for behavior classification. Other static pose descriptors were suggested in [9] [10] and [11]. Unfortunately, due to possible large variations in body appearance, both static and dynamic representations have a considerable failure rate unless specific databases are used.

In this paper we introduce a new multi-view action recognition framework that extracts human silhouette clues from different synchronized static cameras and then validates them by analyzing scene dynamics. In particular this paper deals with the challenging problem of player action recognition in soccer games from images acquired by 6 fixed cameras placed around the playing field. The proposed framework combines human body representation and the motion information of the scene. It consists of two different algorithmic procedures: the first one runs on the processing unit dedicated to each camera and it is an evolution of the classical action recognition methods based on a static representation of the human body. In particular its main novelty is the use of the Contourlet transform to obtain a more suitable body representation to be given as input to a neural classifier. The second procedure runs on a unit with supervisor function and it performs ball and player motion analysis: in particular it reconstructs the 3D ball trajectories and projects them onto a virtual playing-field where players are also localized. The outcomes of both procedures are then merged to accomplish the player action recognition task. Experiments were carried out on several image sequences acquired during some matches of the Italian "Serie A" soccer championship.

2   System Overview

Six high resolution cameras (labeled as FGi, where i indicates the $i - th$ camera) were placed on the two sides of the pitch assuring double coverage of almost all the areas by either adjacent or opposite cameras. In figure 1 the location of the cameras is shown. The acquired images are transferred to six processing nodes by fiber optic cables. The acquisition process is guided by a central trigger generator that guarantees synchronized acquisition between all the cameras. Each node, using two hyper-threading processors, records all the images of the match on its internal storage unit, displays the acquired images and, simultaneously, processes them with parallel threads, in an asynchronous way with respect to the other nodes.

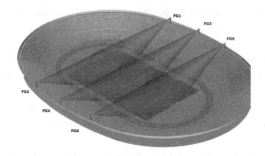

Fig. 1. The location of the cameras around the pitch

The six processing nodes, are connected to a central node, which has having the supervisor function. It synchronizes the data coming from nodes and performs high level processing.

Each node uses a background subtraction algorithm for motion detection. It is based on a modified version of a well known approach for background creation and maintenance [12]. Information relative to moving objects is then sent to two parallel processing threads: the first one performs human blob detection, classification [13] and tracking [14] as well as neural player action recognition using static representation by Contourlet transform; the second one performs ball detection by means of a correlation based approach using six reference sets containing some ball examples acquired in different positions with respect to the camera (near, far, very far) and in different lighting conditions (sunny days, evening or cloudy days).

At each trigger pulse, the outcomes of the algorithmic procedures running on each processing node, are sent to the central node which analyzes them in order to localize the ball and the players on a virtual play-field, to compute their trajectories and to validate player action by using motion information.

3   Player Body Posture Estimation

The first step in the proposed framework deals with the recognition of the player body configuration *(postures)* in each of the different cameras. For this purpose, in this paper, a learning based approach is used: first of all the player silhouettes are binarized, mirrored (only those coming from the three cameras FG2, FG4 and FG6) in order to use the same left-right labeling system for body configuration, re-sized to avoid scaling effects, and described by using Contourlet coefficients that are extracted via a double iterated filter bank structure providing a flexible multi-resolution, local and directional image expansion [15].

The new Contourlet representation is then provided as input to a back propagation neural network able to recognize (after a training phase with about 20 positive examples for each class) seven different human configurations associated with seven player actions: walking, running left, running right, running front-back, still, shooting left, shooting right.

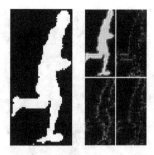

Fig. 2. The binarized player silhouette (on the left) and its Contourlet representation (on the right)

The neural network architecture (experimentally set) consists of three processing layers: 30 hidden neurons with sigmoidal activation functions and 6 output neurons with *softmax* activation functions [16] are used. The neural output values are managed as follows: the greatest output value is considered and if it is greater than $th = 0,5$ the input patch is labeled by the corresponding action, otherwise it is labeled as undetermined. In figure 2, on the left the initial binary silhouette of a running player is reported whereas on the right the relative Contourlet representation is shown.

4   Ball and Player Trajectory Computation

The second algorithmic procedure runs on the processing unit with the supervisor function. The supervisor makes use of a virtual play-field (having the same dimension as the real play-field) to project the extracted information: in particular the player and referee data are projected onto the virtual play-field by homographic transformation assuming that their feet are always in contact with the ground. However, the projection of the same player using data relative to different cameras are not coincident due to the different segmentations into the image planes caused by different appearances of the same player (different position with respect the camera, different lighting conditions, shadows and so on). To overcome this drawback, the mid-point of the line connecting the different projections of the player in the virtual play-field is considered for further processing.

The projection of the ball position onto the virtual play-field requires, instead, a different procedure which considers that the ball is not always in contact with the ground. The 3D ball position then has to be firstly recovered by triangulation (if ball information coming from two opposite or adjacent views is available) and then its projection onto the virtual play-field can be performed. Assuming that the ball is observed from two cameras c_1 and c_2, we obtain the corresponding projections b_1 and b_2 on the ground plane using homography. Let l_1 and l_2 be the two lines from c_1 to b_1 and c_2 to b_2, respectively. Theoretically, the intersection point between l_1 and l_2 should correspond to the 3D ball position but, in practice l_1 and l_2 do not intersect due to errors caused by camera calibration and

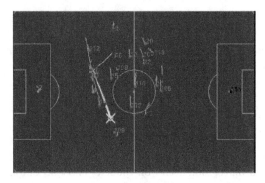

Fig. 3. The virtual play-field

object detection. Thus, the 3D ball position, can be estimated by assuming the camera error of both cameras is of the same order, and fixing the ball position $P(x, y, z)$ as the mid-point of the minimum distance segment between l_1 and l_2. The projection of $P(x, y, z)$ into the virtual play-field is finally obtained setting to 0 the z-component i.e. $P(x, y, 0)$.

In figure 3 the virtual play-field is reported. The red and cyan rectangles indicate the player positions computed by merging data coming from two opposite or adjacent views (relative IDs assigned from nodes are also reported) whereas the ball position is indicated by the yellow cross. The white lines behind each object indicate recent ball and player displacements.

Starting from the estimated ball and player positions their temporal trajectories can be computed. The player trajectories in the virtual play-field cannot be mathematically modeled by straight lines or curves; they vary continuously in an unpredictable way and so they can be represented only by collecting the player positions into the play-field.

For the ball trajectories in the virtual play-field can be approximated by straight lines: this allows the system to predict the successive positions, to recover missed intermediate ones, and to introduce high level reasonings useful for understanding the soccer game developing. For the sake of precision we have to explain that we dispose of both 3D and 2D ball trajectories. In order to detect shots, as abrupt changes of trajectories, we consider in this paper only 2D trajectories, obtained by projecting the 3D ball positions onto the virtual play-field. This simplification allows the system to avoid false shot detections when there are ball rebounds on the field.

5   Multi-view Player Action Recognition

Finally the outcomes of the algorithmic procedures described in sections 3 and 4 are given as input to a higher level functional step running on the supervisor unit that performs a multi-view player action recognition. To do that, first of all, the supervisor processing unit merges body posture information coming from different views of the same player: the M available estimation scores extracted by the single view procedure (one for each camera acquiring the considered player) for the $i-th$ player are averaged to obtain k values of MPV (Multi-view Probability Value) :

$$(MPV)_k = P(X_k|z_1...z_M) = \frac{1}{M}\sum_{i=1}^{M}p(X_k|z_i) \quad k=1,...N$$

where X is the player posture class, z_i are the single view estimated configurations and N indicates the maximum number of body configuration classes to be recognized. In this way a global estimation score is obtained for each of the k configuration classes. The problem now becomes how to decide which body configuration class has to be associated to the $i-th$ player on the basis of the relative available $(MPV)_k$ with $k=1,...N$.

To solve this problem a preliminary statistical evaluation of the neural outcomes is done: these values are evaluated for both correct and incorrect occurrences during a preliminary experimental phase and they are then used to estimate the relative gaussian probability distributions (respectively in red and blue in figure 4). Notice that the most probable values in the case of correct body configuration estimations in a single view are close to 0.85, whereas in the case of wrong estimations they are close to 0.5. Starting from this statistical consideration a multi-view decision rule, based on available MPVs, is introduced: the player's body configuration K is associated to the $i-th$ player if

$$\{(K = \arg\max_{k=1,...N} P_k) \wedge (P_K > th) \wedge (P_i < th, \ i = \{1...N, i \neq K\})$$

where th is the intersection point of the estimated $pdfs$ in figure 4. If these conditions are not simultaneously satisfied the considered player body configuration is labeled as undetermined.

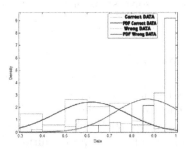

Fig. 4. Probability distribution of the winning neural output in case of correct (red) and uncorrect (blue) action recognition from binarized player silhouette

5.1   Integration of Motion Information

After that, a static multi-view player action estimation is available for each player in the scene and it can be validated by using motion information. In particular 'running', 'walking' and 'still' actions are validated by using the motion information of the relative player, whereas 'shooting' action is validated taking into consideration also the 3D ball trajectory. In fact for an estimated 'still' player the system checks his motion: if, considering the last three frames, his position in the virtual playing field does not significatively change, the player action is definitively labeled as 'still'. The same approach, based on the analysis of position changes in the play-field, is also used for validating estimated 'running' and 'walking' players: in this case the estimated player body action is validated if player position (considering the last three frames) changes according to the common running and walking velocity values for a human being whereas running directions (left, right, front-back) are validated considering the recovered player trajectory in the play-field. In particular a player is validated as 'walking' if this velocity in the virtual play-field varies between 3 km/h and 6 km/h; running action is instead validated if the velocity of the relative player is greater than 6 km/h.

A quite different approach is finally used for validating estimated 'shooting' players: in this case both ball and player motion information are used. In fact, 'shooting' player is validated if : 1) the 3D ball trajectory indicates that the ball is really going away from the considered player (as expected for the kicking player) 2) the player is near (at lest 2 meters) to $P(x_s, y_s)$, i.e. the intersection point of two consecutive validated ball trajectories.

Four different situations can happen: 1)classified actions are validated by motion and definitively labeled 2)undetermined actions (two or more multi-view scores greater than th) are solved if only one of them is validated by motion information; 3)Undetermined actions (No multi-view score greater than th) are solved if the maximum score (independently from th) is validated by motion information 4)classifications are not validated and become undetermined;

In this way motion information helps to improve player action recognition but, at the same time, could solve some of the previously unclassified action occurrences.

6   Experimental Results

The proposed multi-steps method was applied to several image sequences acquired during some matches of the Italian "Serie A" championship. Experiments were carried out on a set of 5000 different pairs of synchronized patches (10000 total patches) relative to players acquired during different soccer matches. These patches were preliminarily labeled by a human operator who assigned to each pair one of the seven considered actions: *Running right side; Running left side; Running front-back; Walking; Still; Shooting left side; Shooting right side.*

The ground truth relative to the 10000 patches is reported in table 1.

Table 1. The ground truth relative to the 10000 considered binary patches used in the experimental phase

Running right side	Running left side	Running front-back	Walking	Still	Shooting left side	Shooting right side
1930	**2304**	**622**	**2590**	**1358**	**674**	**522**

The first step of the experimental phase concentrates on the recognition of the human action on each single image by using Contourlet representation and a neural classifier as described in section 3.

The results of this first experiment are reported in table 2.

Table 2. The scatter matrix relative to the first experiment regarding the recognition of the player action recognition in a single image

Running right side	Running left side	Running front-back	Walking	Still	Shooting left side	Shooting right side	Und.
1459	14	228	164	0	0	0	65
0	**1999**	0	179	120	0	0	6
0	0	**461**	41	45	47	0	28
0	0	25	**2396**	42	68	0	59
0	33	69	0	**1193**	0	0	63
12	0	24	0	0	**613**	0	25
18	11	9	0	0	0	**462**	22

The experimental results reported in table 2 were very encouraging: almost 86% of the testing patches were automatically labeled by the system in the same way as the human operator. Some miss-classifications happened due to the similarity of appearance, under certain conditions, of the body silhouettes relative to players performing different actions. For example, in figure 5, three wrongly classified patches are reported: the player on the left was classified as running towards the camera by the human operator whereas the automatic system consider him as running right. The player in the center was instead classified as running towards the camera by the human operator whereas the

Fig. 5. Three different cases where the proposed system missrecognized player body configurations

Table 3. Action recognition performance integrating information coming from different camera views. The test set consists of 5000 pairs of binary patches.

Running right side	Running left side	Running front-back	Walking	Still	Shooting left side	Shooting right side	Und.
929	7	5	8	0	0	7	9
0	**1108**	0	16	12	8	0	8
0	0	**271**	8	12	12	0	8
0	0	14	**1226**	16	18	0	21
0	17	11	0	**632**	0	0	19
0	0	9	0	0	**311**	0	17
7	0	13	0	0	0	**230**	11

automatic system classified him as still. Finally, the player on the right was labeled as kicking by the human operator and running right by the automatic system. As you can see in this cases it is not easy to definitively decide real player configurations and then you have to consider that experimental results strongly depend, on the operator that generated the ground truth.

The action data coming from this first experimental step were then merged, for each pair of opposite cameras, by using the procedure described in section 5. In table 3 the multi-view action estimation results on the 5000 pairs of binary patches are reported.

Fig. 6. Different pairs of patches containing the same player acquired from different cameras

More than 94% of the 5000 tested pairs were correctly recognized. Less than 2% of the tested patches were not classified due to the ambiguities in the probability values provided by the neural algorithms running on the images coming from each camera.

In figure 6, the two pictures on the left report the player's silhouettes acquired by two opposite cameras at a shot instant. In this case the analysis of human body configuration performed on each view agreed and they indicated that the player was shooting the ball with probability values respectively of 0.97 and 0.89. The two pictures on the right show, instead, a case in which the procedure running on each single view disagreed: one camera recognized the player as

Table 4. Action recognition performance after validation by motion information

Running right side	Running left side	Running front-back	Walking	Still	Shooting left side	Shooting right side	Und.
954	0	0	0	0	0	3	8
0	**1135**	0	4	2	0	0	11
0	0	**295**	0	4	0	0	12
0	0	4	**1274**	3	0	0	14
0	3	3	0	**654**	0	0	19
0	0	0	0	0	**324**	0	13
0	0	0	0	0	0	**250**	11

Fig. 7. Two examples in which motion information overcome drawback of the neural approach for player action recognition

Fig. 8. An example in which motion information did not solve action miss-classification

walking (probability value 0.87) and the other one as still (probability value 0.54). The multi-view approach solved this ambiguity and labeled the player "walking" as the human operator also did. Finally, the remaining 3% of the tested patches were miss-classified.

Finally the validation procedure based on motion information described in section 5.1 was tested in order to verify its capability to improve player action recognition. The final results of the proposed player action recognition approach are then reported in table 4.

Introducing action validation by motion information drastically reduces both uncertainty occurrences and increases correct classification. In figure 7 two examples pointing out the benefits of using motion information are reported. On the left two patches (acquired from FG3 and FG4) are relative to a player kicking the ball. The neural approach did not classify player action because, for the patch on the left, it rightly recognizes the player as shooting left but, for the patch on the right, it erroneously classifies the player as running left. Introducing reasonings about motion and ball proximity the system verified that both the ball was close to the player and the distance of the player from the intersection point of the ball trajectory was very small. For these reasons the player was correctly classified as shooting the ball. On the right the two patches are relative to a player running right (acquired from FG5 and FG6). Unfortunately both patches were classified as "walking" by the neural approach based on Contourlet representation (most probably due to perspective distortions). The motion validation procedure did not validate the player as walking due to his velocity on the pitch (9 km/h) and then his action was considered as "undermined" avoiding a wrong classification.

Finally in figure 8 an example in which motion information did not solve miss-classification is reported. The two players were, actually, both classified as shooting right by the multi-view approach described in section 5.

7 Conclusion

In this paper we present a multi-view action recognition framework able to extract human silhouette clues from different synchronized static cameras and then to validate them by analyzing scene dynamics. A number of experiments were carried out on several image sequences acquired during some matches of the Italian "Serie A" soccer championship and they demonstrated that action recognition performance increased by both merging player posture information coming from different overlapped camera views and introducing motion constraints. Future works will deal with the analysis of the action relationship between a player and his neighbors, as well as on the temporal analysis of the player postures.

References

1. Gorelick, L., Blank, M., Shechtman, E., Irani, M., Basri, R.: Actions as Space-Time Shapes. Transactions on Pattern Analysis and Machine Intelligence 29(12), 2247–2253 (2007)
2. Turaga, P., Chellappa, R., Subrahmanian, V.S., Udrea, O.: Machine Recognition of Human Activities: A Survey. IEEE Transactions on Circuits and Systems for Video Technology 18(11), 1473–1488 (2008)
3. Laptev, I., Marszałek, M., Schmid, C., Rozenfeld, B.: Learning Realistic Human Actions from Movies. In: IEEE Conference on Computer Vision & Pattern Recognition (2008)
4. Jhuang, H., Serre, T., Wolf, L., Poggio, T.: A Biologically Inspired System for Action Recognition. In: Proceedings of the Eleventh IEEE International Conference on Computer Vision, ICCV (2007)

5. Niebles, C., Wang, H., Fei-Fei, L.: Unsupervised learning of human action categories using spatial-temporal words. International Journal of Computer Vision, 2247–2253 (December 2008)
6. Liu, J., Ali, S., Shah, M.: Recognizing Human Actions Using Multiple Features. In: International Conference on Computer Vision and Pattern Recognition CVPR 2008, Anchorage, Alaska (2008)
7. Ikizler, N., Duygulu, P.: Human Action Recognition Using Distribution of Oriented Rectangular Patches. In: Elgammal, A., Rosenhahn, B., Klette, R. (eds.) Human Motion 2007. LNCS, vol. 4814, pp. 271–284. Springer, Heidelberg (2007)
8. Goldenberg, R., Kimmel, R., Rivlin, E., Rudzsky, M.: Behavior Classification by Eigen-decomposition of Periodic Motions. IEEE transactions on systems, man and cybernetics. Part C, Applications and reviews 38(7), 1033–1043 (2005)
9. Lu, W.L., Little, J.J.: Simultaneous Tracking and Action Recognition using the PCA-HOG Descriptor. In: Proceedings of the The 3rd Canadian Conference on Computer and Robot Vision, CRV 2006, Quebec, Canada (2006)
10. Zhang, L., Wu, B., Nevatia, R.: Detection and Tracking of Multiple Humans with Extensive Pose Articulation. In: Proceedings of the Eleventh IEEE International Conference on Computer Vision, ICCV 2007, Rio de Janeiro, Brazil, pp. 1–8 (2007)
11. Thurau, C.: Behavior Histograms for Action Recognition and Human Detection. In: Elgammal, A., Rosenhahn, B., Klette, R. (eds.) Human Motion 2007. LNCS, vol. 4814, pp. 299–312. Springer, Heidelberg (2007)
12. Kanade, T., Collins, R., Lipton, A., Burt, P., Wixson, L.: Advances in cooperative multi-sensor video surveillance. In: Proceedings of DARPA Image Understanding Workshop, November 1998, vol. 1, pp. 3–24 (1998)
13. Spagnolo, P., Mosca, N., Nitti, M., Distante, A.: An Unsupervised Approach for Segmentation and Clustering of Soccer Players. In: IMVIP 2007: Proceedings of the International Machine Vision and Image Processing Conference, NUI Maynooth, Ireland (2007)
14. D'Orazio, T., Leo, M., Spagnolo, P., Mazzeo, P.L., Mosca, N., Nitti, M.: A Visual Tracking Algorithm for Real Time People Detection. In: Wiamis 2007: Proceedings of the International Workshop on Image Analysis for Multimedia Interactive Services, Santorini, Greece (2007)
15. Do, M.N., Vetterli, M.: The Contourlet Transform: An Efficient Directional Multiresolution Image Representation. IEEE Transactions on Image Processing 12(14), 2091–2106 (2005)
16. Bishop, C.M.: Neural Networks for Pattern Recognition. Oxford University Press, Oxford (1995)

Heart Cavity Segmentation in Ultrasound Images Based on Supervised Neural Networks

Marco Mora[1], Julio Leiva[2], and Mauricio Olivares[1]

[1] Department of Computer Science
Catholic University of Maule
Talca, Chile
mora@spock.ucm.cl
http://ganimides.ucm.cl/mmora/
[2] Department of Mathematics
Catholic University of Maule
Talca, Chile
jleiva@ucm.cl

Abstract. This paper proposes a segmentation method of heart cavities based on neural networks. Firstly, the ultrasound image is simplified with a homogeneity measure based on the variance. Secondly, the simplified image is classified using a multilayer perceptron trained to produce an adequate generalization. Thirdly, results from classification are improved by using simple image processing techniques. The method makes it possible to detect the edges of cavities in an image sequence, selecting data for network training from a single image of the sequence. Besides, our proposal permits detection of cavity contours with techniques of a low computational cost, in a robust and accurate way, with a high degree of autonomy.

1 Introduction

Ultrasound images are characterized by low contrast and high level of *speckle* noise. Such characteristics make it difficult to process and to analyze the images. Cavity edge detection is a complex task, the noise present in ultrasound images causes errors when detecting contours. Besides, the low contrast makes cavity contours imperceptible in certain zones.

There are approaches of the most varied nature proposed in ultrasound image segmentation literature [1], particularly artificial intelligence techniques such as neural networks have been reported. Ultrasound image segmentation methods based on back-propagation neural networks are proposed in [2,3,4], Kohonen networks in [5,6], radial basis neural networks in [7], models of incremental networks in [8], hybrid neural networks trained with genetic algorithms are proposed in [9,10], cellular neural networks in [11], and support vector machines are adopted in [12].

In this paper we propose a method that combine supervised neural networks with different image processing techniques. Our approach is made up by three main stages. During the first stage, an original representation of the ultrasound

A. Gagalowicz and W. Philips (Eds.): MIRAGE 2009, LNCS 5496, pp. 58–68, 2009.

image is proposed, which is obtained after applying to the image a neighborhood homogeneity operator based on the variance. We adopted such homogeneity operator mainly for simplifying the images and differentiating its zones.

In the second stage, ultrasound image segmentation is formulated as a classification problem. A Multilayer Perceptron (MLP) is adopted since it corresponds to a universal function estimator [13]. It is also considered a more efficient version than the classical back-propagation algorithm for its training. In fact, the Levenberg-Marquardt optimization algorithm is adopted [14] so as to improve the training convergence speed. The computation of network parameters is carried out by adding to the traditional cost function (the sum of squared errors) a term which corresponds to the sum of squares of the network weight [15]. The cost function parameters are automatically computed by using a Bayesian approximation [16], and for the estimation of hidden layer neurons the effective parameters of the network are considered [17]. To solve the classification problem, we formulated the hypothesis that after the representation stage, the image remains constituted by three kinds of pixels; those exterior to the cavity, of contour, and inside the cavity. The MLP is also trained to recognize neighborhoods and not pixels, in order to propose a robust solution.

The final stage consists in improving classification through a simple and fast sequence of image processing techniques such as erosion, dilation, median filtering, and size segmentation [18], establishing an order in the application of operators which has a general validity. All in all, our method allows to detect contours in a robust and accurate way, with techniques of a low computational cost.

This article is structured as follows. In part 2, the stages of our method are detailed. Part 3 shows results considering the classification of original intensity image, and the classification of the proposed for the ultrasound image representation. Both approximations are compared through a FOM indicator. Finally, part 4 presents the conclusions of the paper.

2   Cavity Segmentation Method Using a MLP

This section presents the detail of all three stages of our proposal, that is to say, image representation, classification and improvements.

2.1   Ultrasound Image Representation

Our method proposes to represent the ultrasound image by using a homogeneity measure based on the variance in pixel neighborhoods [18]. The image based on the variance corresponds to:

$$Ivar(I_{i,j}) = \frac{1}{1 + var(V_{i,j})} \tag{1}$$

where var corresponds to the variance operator, and $V_{i,j}$ corresponds to a pixel neighborhood (i, j) in the image $I_{i,j}$. The values of $Ivar(I_{i,j})$ belong to the interval $(0, 1]$.

We have proposed such representation for the ultrasound image since it permits to simplify the image without going through a noise iterative filtering process. Besides, the fact that pixels of an image based on the variance are between 0 and 1, allows limiting the image value variability, which facilitates the classification.

2.2   Classification of Ultrasound Image Representation

For the problem, we formulated the hypothesis that in image representation we can find three kinds of pixels: the class of those outside the cavity, those on the edge, and others inside the cavity. In experimentations done during the research, this hypothesis was verified adequate for the problem. The patterns to be classified correspond to vectors representing the neighborhood of a pixel (i, j). This means that the classifier has as many inputs as elements has the neighborhood, and as many outputs as classes of neighborhoods defined for the problem. Figure 1 shows the adopted neural classifier, which considers a 3×3 neighborhood of the image, meaning 9 inputs. Besides, the classifier has 3 outputs; s1, s2 and s3, representing the exterior, edge, and interior classes respectively.

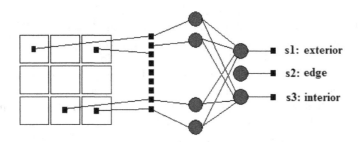

Fig. 1. MLP based Classifier

The network operation is represented in table 1, in which it is possible to observe the values designated to outputs depending on the pattern type which is being classified.

Table 1. Network Operation

Pattern	s1	s2	s3
Exterior class	1	0	0
Edge class	0	1	0
Interior class	0	0	1

For training the network, a procedure based on effective network parameters is considered, which permits establishing the number of neurons in the hidden layer [17], and also, to regularize the network interpolation in order to improve the MLP generalization. The procedure consists in training the network with more neurons in the hidden layer each time, starting with 1 neuron. Getting to a certain number of neurons in such layer, the effective parameters will stabilize and will not increase even when the number of neurons does. The smallest value permitting stabilization of effective parameters is considered to be an adequate number of neurons.

2.3   Classification Improvements

For improving results of the classification stage we propose the following sequence of image processing operations [18]:

- Erosion of the classified image, which permits disconnecting objects from the main object's body.
- Size segmentation, which allows eliminating small bodies.
- A median filtering, to round up object contours.
- Dilation of the filtered image, to recuperate original size before erosion.

After the mentioned operation sequence, object contours are detected with the Sobel operator [18]. The classification improvement sequence can be generalized; this means it can be applied to several images without altering the order of the operations. In this sense, we can talk about an automatic improvement.

3   Results

Results in contour detection on the intensity image and on the variance-based image are presented. For each representation of the ultrasound image, a training set is built, selecting neighborhoods outside, on the edge, and inside the cavity. The MLP is trained, and with the obtained parameters the image is classified. Also, an image sequence whose neighborhoods haven't been considered in the training set is classified using the same network.

To numerically evaluate the results of our method, the Pratts merit figure (FOM) is used [19]. The FOM indicator is a similarity measure between two curves whose values are between 0 and 1. If the curves are similar, the FOM is close to 1, and if the curves are way different, their value is closer to 0. The FOM expression is next:

$$FOM = \frac{1}{max(I_I, I_E)} \sum_{i=1}^{I_A} \frac{1}{(1 + a * d^2(i))} \tag{2}$$

where I_I and I_E represent the number of pixels of the real contour (traced by the specialist) and that of the found contour respectively, and a is a scale factor (usually $1/9$). Finally, $d(i)$ is the distance which separates a pixel i of the found contour from its nearest over the real contour. It is important to notice that this indicator can only be used if the real contour is available.

3.1   Results in Training Set Image

Figure 2 shows images from which training set have been obtained. Figure 2(a) shows the intensity image, and figure 2(b) shows the image based on the variance. Figure 2(c) shows a manually traced contour considered as reference contour. The resulting contour will be compared respecting this last curve.

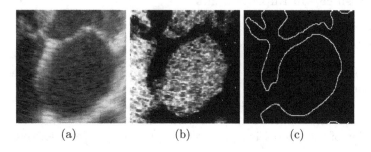

(a)   (b)   (c)

Fig. 2. Initial images: (a) original image (b) representation based on variance (c) reference contour

Tables 2 and 3 show the training sequences that allow determining the number of neurons of the hidden layer, for the intensity image and for the variance based image respectively. In such table *Epochs* (first column), correspond to the number of epochs required for training; S (second column), corresponds to the number of neurons in the hidden layer; E_D (third column), corresponds to the sum of squared errors; E_W (fourth column), is the sum of squares of the network weights; γ (fifth column), the effective parameters; and N (sixth column), the number of total parameters. For the case of intensity image, table 2 shows that effective network parameters stabilize starting from 4 neurons in the hidden layer. For the case of variance based image, table 3 shows that effective parameter stabilization starts from 3 neurons. Then, for intensity image, the

Table 2. Hidden layer neuron selection for intensity image

Epochs	S	ED	EW	γ	N
162/5000	1	80.8333	3962.29	12.471	16
508/5000	2	16.1596	25381.9	15.287	29
526/5000	3	16.1898	23810.8	17.316	42
4744/5000	**4**	**16.2309**	**22849.3**	**19.971**	**55**
5000/5000	5	16.2314	22560.8	19.993	68
5000/5000	6	16.284	22038.6	20.535	81
5000/5000	7	16.2319	22308.5	20.112	94
5000/5000	8	16.2597	22309.7	21.318	107
5000/5000	9	16.2602	22310.3	21.708	120
5000/5000	10	30.1469	22.2403	22.240	133

Table 3. Hidden layer neuron selection for variance based image

Epocas	S	ED	EW	γ	N
45/5000	1	120.834	15.0006	7.506	16
2337/5000	2	120.984	16.3345	7.951	29
400/5000	**3**	**118.026**	**17.682**	**9.610**	**42**
828/5000	4	118.119	16.8845	9.503	55
895/5000	5	118.134	16.5869	9.442	68
2914/5000	6	118.134	16.4341	9.450	81
5000/5000	7	118.258	15.8473	9.331	94
1568/5000	8	118.323	15.5152	9.379	107
2042/5000	9	118.318	15.4438	9.358	120
3241/5000	10	118.353	15.2545	9.343	133

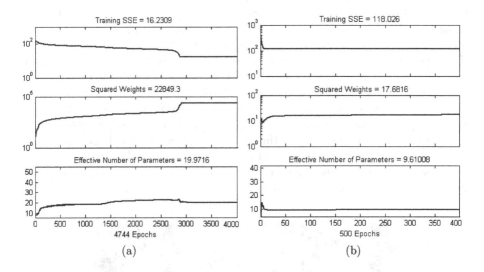

Fig. 3. Training curves: (a) original image (b) variance based image

selected architecture is 9-4-3 neurons, and for variance based image the selected architecture is 9-3-3 neurons.

The training curves considering the amount of neurons in the hidden layer established thanks to tables 2 and 3 are shown in figure 3. Figure 3(a) corresponds to training for intensity image, and figure 3(b) to training for variance based image. The curve on the upper part corresponds to the evolution of the sum of squared errors (SSE), the center curve corresponds to the the sum of squares of the network weights (SSW), and the lower curve corresponds to the evolution of effective parameters. For both images it is observed that all curves have stabilized in the last iterations of the training. This means that the network has been correctly trained.

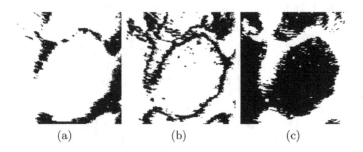

Fig. 4. Intensity image classification: (a) exterior class (b) edge class (c) interior class

After training, we proceeded to classify image neighborhoods. Figure 4 shows the direct classification of intensity image from figure 2(a). Figure 4(a) corresponds to the exterior class; figure 4(b) corresponds to edge class; and figure 4(c) corresponds to interior class. It is observed we accomplish the hypothesis that heart image can be split up in three kinds of neighborhoods. Besides, the figure shows that interior class allows a better visualization of the cavity.

Figure 5 shows variance based image classification from figure 2(b). Figure 5(a) corresponds to exterior class; figure 5(b) corresponds to edge class; and figure 5(c) corresponds to interior class. We confirm again the hypothesis that in the image we are able to find three types of pixel neighborhoods. It is also observed that the exterior class results in low noise, but in the interior class the cavity size is closer to that determined by the reference edge. Because of the previously exposed reasons, we adopted the interior class image to be improved.

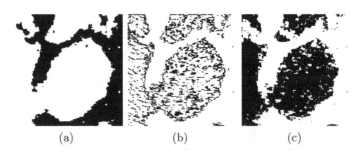

Fig. 5. Variance based image classification: (a) exterior class (b) edge class (c) interior class

Figures 6 and 7 show the sequence of improvements made to the classification of intensity image and variance based image respectively. For both figures the first image corresponds to interior class; second image corresponds to erosion; the third one corresponds to size segmentation; the fourth corresponds to dilatation; the fifth corresponds to a median filtering; and the last image corresponds to the final contour.

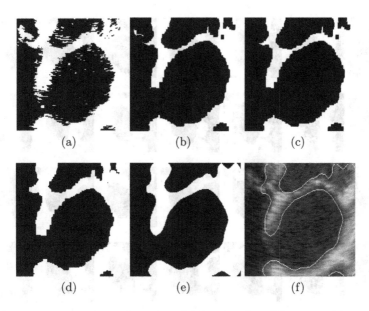

Fig. 6. Improvements to intensity image: (a) interior class (b) erosion (c) size segmentation (d) dilation (e) median filtering (f) resulting contour (g) final contour

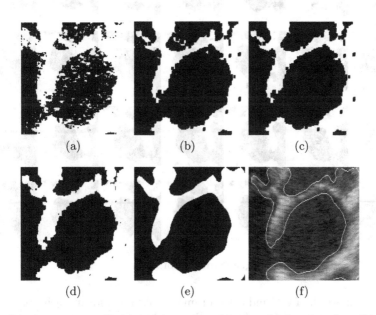

Fig. 7. Improvements applied to variance based image: (a) interior class (b) erosion (c) size segmentation (d) dilation (e) average filter (f) final contour

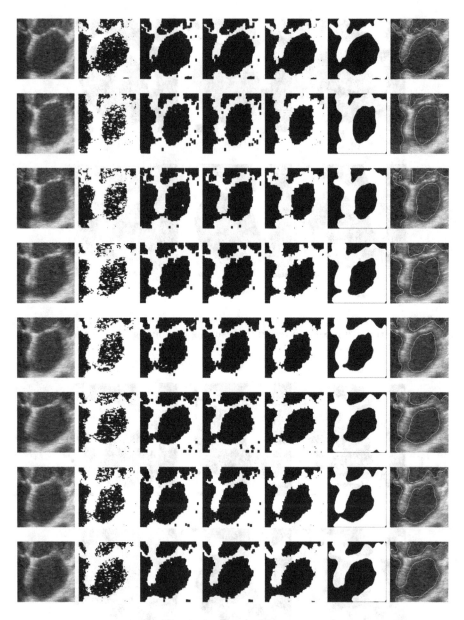

Fig. 8. Sequence of images

Table 4 shows the FOM indicator of images obtained for intensity image, and for variance based image. It is observed that the FOM for the variance based image is closer to the unit, which reflects that the resulting contour is closer to the desired contour. What was said before shows that variance based image yields better results in contrast with intensity image when it comes to contour detection.

Table 4. FOM for intensity and variance based image contour

Class	Intensity image	Variance based image
Interior	0.6475	0.7677

3.2   Results over a Sequence of Images

From the previous section, it has been determined that the variance based image permits better results in cavity contour detection, and that the neural network which allows classifying this kind of patterns has three neurons in the hidden layer. Using this neuronal network, we have classified an ultrasound image sequences. It is important to notice that the neighborhoods of such images do not belong to the training set. Figure 8 shows the sequence of 8 images of a heart cavity in motion. Each line shows the sequence of improvements applied to the interior class of the variance based representation, for each of the eight images. The first image of the sequence corresponds to the original image, and the rest to the interior class of the variance based image, erosion, size segmentation, dilation, median filtering and final contour, respectively. It is observed that our method has granted us to obtain the contours of all the images correctly. The method behaving in the same manner for these multiple images and for the training set image as well.

4   Conclusions

This paper has presented an original ultrasound image segmentation method based on a MLP. The proposed method presents different aspects of interest. A variance based ultrasound image representation has been proposed, which has led all the images to a same value range, permitting a simplification of the segmentation process by classification. The adopted training method allows a correct generalization and to establish the amount of neurons in the hidden layer, being determined the network architecture by using an objective criterion. Finally, an image processing operations sequence has been established to improve the result from classification. This sequence can be generalized to other images.

Another interesting aspect of the proposed method is that it does not consider a noise model of the image, nor the geometrical shape of the cavities, or any kind of a priori knowledge, which gives it higher degree of autonomy and simplicity to the solution.

At last, a method has been conceived that allows classifying multiple images, obtaining data for the training of only one of them. This is possible thanks to the standardization given by variance based representation, and by the technique adopted for the MLP training.

All in all, the characteristics of the proposed approximation allow solving the ultrasound image cavity segmentation problem in a robust and accurate way, with techniques of a low computational cost.

References

1. Noble, J., Boukerroui, D.: Ultrasound image segmentation: A survey. IEEE Transaction on Medical Imaging 25(8), 197–1009 (2006)
2. Piccoli, L., Dahmer, A., Scharcanski, J., Navaux, P.: Fetal echocardiographic image segmentation using neural networks. In: Proceedings of the 7th International Conference on Image Processing and its Applications, vol. 2, pp. 507–511 (1999)
3. Binder, T., Sussner, M., Moertl, D., Strohmer, H., Baumgartner, T., Maurer, G., Porenta, G.: Artificial neural networks and spatial temporal contour linking for automated endocardial contour detection on echocardiograms: A novel approach to determine left ventricular contractile function. Ultrasound in Medicine and Biology 25, 1069–1076 (1999)
4. Herng, E., de Andrade, M., Nicolisi, D., Pontes, S.: Artificial neural network: border detection in echocardiography. Medical and Biological Engineering and Computing 46, 841–848 (2008)
5. Siqueira, M., Scharcanski, J., Navaux, P.: Echocardiographic image sequence segmentation and analysis using self-organizing maps. Journal of VLSI Signal Processing System 32, 135–145 (2002)
6. Iscan, Z., Kurnaz, M., Dokur, Z., lmez, T.: Ultrasound image segmentation by using wavelet transform and self-organizing neural network. Neural Information Processing-Letters and Reviews 10, 183–191 (2006)
7. Kotropoulos, C., Nikolaidis, N., Bors, A., Pitas, I.: Robust and adaptive techniques in self-organizing neural networks. International Journal of Computer Mathematics 67, 183–200 (1998)
8. Kurnaz, M., Dokur, Z., lmez, T.: An incremental neural network for tissue segmentation in ultrasound images. Computer Methods and Programs in Biomedicine 85, 187–195 (2007)
9. Dokur, Z., lmez, T.: Segmentation of ultrasound images by using a hybrid neural network. Pattern Recognition Letters 23, 1825–1836 (2002)
10. Dokur, Z., Kurnaz, M.N., Olmez, T.: Segmentation of ultrasound images by using quantizer neural network. In: Proceedings of the 15th IEEE Symposium on Computer-Based Medical Systems (CBMS 2002), p. 257 (2002)
11. Rekeczky, C., Tahy, A., Vegh, Z., Roska, T.: Cnn-based spatio-temporal nonlinear filtering and endocardial boundary detection in echocardiography. International Journal of Circuit Theory and Aplications 27, 171–207 (1999)
12. Kotropoulos, C., Pitas, I.: Segmentation of ultrasonic images using support vector machines. Pattern Recognition Letters 24, 715–727 (2003)
13. Funahashi, K.: On the approximate realization of continuous mappings by neural networks. Neural Networks 2, 183–192 (1989)
14. Luenberger, D.: Linear and Nonlinear Programming, 2nd edn. Addison Wesley, Reading (1984)
15. Hinton, G.: Connectionist learning procedures. Artificial Intelligence 40, 185–234 (1989)
16. Mackay, D.: Bayesian interpolation. Neural Computation 4, 415–447 (1992)
17. Foresee, D., Hagan, M.: Gauss-Newton approximation to bayesian learning. In: Proceedings of the 1997 International Joint Conference on Neural Networks (IJCNN 1997), vol. 3, pp. 1930–1935 (1997)
18. Gonzalez, R., Woods, R.: Digital image processing, 2nd edn. Addison-Wesley, Reading (2001)
19. Pratt, W.: Digital image processing, 2nd edn. John Wiley and Sons, Chichester (1991)

Automatic Fitting of a Deformable Face Mask Using a Single Image

Annika Kuhl, Tele Tan, and Svetha Venkatesh

Department of Computing
Curtin University of Technology
GPO Box U1987, Perth, 6845, Western Australia
(a.kuhl,t.tan,s.venkatesh)@curtin.edu.au

Abstract. We propose an automatic method for person-independent fitting of a deformable 3D face mask model under varying illumination conditions. Principle Component Analysis is utilised to build a face model which is then used within a particle filter based approach to fit the mask to the image. By subdividing a coarse mask and using a novel texture mapping technique, we further apply the 3D face model to fit into lower resolution images. The illumination invariance is achieved by representing each face as a combination of harmonic images within the weighting function of the particle filter. We demonstrate the performance of our approach on the IMM Face Database and the Extended Yale Face Database B and show that it outperforms the Active Shape Models approach [6].

Keywords: Deformable Face Mask, Eigenfaces, PCA, ASM.

1 Introduction

Accurately detecting faces and geometrically aligning key features of the face is an important pre-requisite for many face recognition algorithms. The appearances of faces can be affected by the rotation of the face (both in-plane and out-of-plane rotations), image resolution and lighting conditions which all make the face detection problem even more challenging. Although some of these factors may not be an issue in a controlled indoor environment (e.g. access control system), it become profoundly important when dealing with outdoor environment and conditions where CCTV cameras are becoming more prevalent. In this paper, we propose a solution to accurately localise a face by automatically aligning the key landmarks of the face with a 3D face model. The proposed technique is designed to cope with low image resolution and varying illumination conditions.

The problem of fitting a deformable mask to an image has previously been addressed by finding facial features first and fitting the face mask to these points afterwards [16]. Active Shape Models (ASM) are commonly used for detecting facial feature points. ASM is a statistical model of the shape of a class of object. During the model matching process the position of landmark points along the contour lines of the shape are optimised. First proposed by [6], the ASM has

A. Gagalowicz and W. Philips (Eds.): MIRAGE 2009, LNCS 5496, pp. 69–81, 2009.

been extended by [19] using an hierarchical CONDENSATION (particle filter) framework and by [24], where the shape estimation problem is formulated in a Bayesian framework.

While the ASM only optimises around the contour lines of the object, Active Appearance Models (AAM) [11] represent both shape and texture of a given class of object. During the training process a generative model is built such that both shape and texture are controlled by a set of parameters. AAM have been extended to fit faces in low resolution images [7]. The authors therefore incorporated the image formation process into the fitting criterion. In [22], AAMs were combined with 3D Morphable Models to allow for a three-dimensional fitting. Furthermore in [2], AAMs have been extended to incorporate lighting changes using the work of [3].

Our approach differs from AAMs in that the possible shape deformations are already given by the 3D mask mesh we are utilising. We do not attempt to learn them from a set of training images. By using a 3D face mask for fitting instead of a 2D AAM we also incorporate three-dimensional information about the shape and the pose of the face. Furthermore we do not aim to generate a 3D model of the face like in [23,4,14], since the utilised face mask is not flexible enough to adapt to slight differences in bone structure.

This work addresses the problem of automatically fitting a deformable face mask to a previously unseen near frontal image of a face and thus implicitly finding facial feature. We use Principle Component Analysis (PCA) and a novel texture mapping technique to build a face model which is then used by a particle filter. Once the model is build offline it can be applied to input image resolutions that differ from the resolution of the training images. We achieve lighting invariance by incorporating the work of [3] into the weighting function of the particle filter.

Accurate detection of facial feature points is a pre-processing step for a number of applications, like fitting a deformable model for tracking [18] or 3D face modelling and recognition [4]. While feature point detection is feasible in high resolution images of faces under neutral lighting conditions most approaches struggle in low resolution and changing lighting. We therefore propose a method that estimates the pose parameters as well as person-specific shape parameters to fit a deformable 3D face model to a previously unseen image, instead of detecting facial feature points and then fitting a mask. This approach is applicable in low resolution and in arbitrary lighting conditions.

The significance of the proposed method is (1) automatic fitting, (2) light invariance, (3) automatic face shape fitting. By automatically fitting a deformable face mask model to a near frontal face, important facial feature points are located implicitly. Thus avoiding laborious manual feature point selection or manual mask alignment. The work of [3] is incorporated into the weighting function of the particle filter in order to compensate for arbitrary illumination, which makes our approach more robust and increases its application. The particle filter is used to estimate the person specific shape parameters to fit a deformable face mask to previously unseen faces. There are no information about the person required, which again increases the robustness and applicability of our approach.

2   Method

Our proposed method first builds a face model by fitting a deformable mask to a set of different face images. Using PCA a set of principle components, the so called Eigenfaces, are then calculated from these textured masks. During runtime a particle filter is used to estimate the pose and the shape parameters of the mask given a single image of a new face. Illumination invariance is achieved by incorporating the work of [3] into the weighting function of the particle filter.

2.1   The Face Model

The offline step includes building a face model from a set of training images using principle component analysis (PCA) [20]. Instead of using vectorised 2D face images we fit a deformable mask to near frontal faces and apply PCA on a vector comprising the grey values of each mask triangle.

We utilise the CANDIDE-3 face model [1]. This 3D deformable face model consists of 104 vertices $\mathbf{P}_i, i = 1, .., n$, where $\mathbf{P}_i$ are the 3D coordinates of the ith vertex. These vertices form the complete mesh that consists of 184 triangles which are described by the vector $\mathbf{g}$

$$\mathbf{g} = \overline{\mathbf{g}} + \mathbf{S}\gamma \tag{1}$$

where $\overline{\mathbf{g}}$ is the neutral face and the columns of $\mathbf{S}$ are the shape parameters that are controlled by $\gamma_k \in [-1, .., 1]$. Each shape parameter is a list of vertices and their displacement that result in the corresponding shape deformation. The original model allows for 14 different shape parameters; we reduce this number to 7, namely *'eyebrows vertical position'*, *'eyes vertical position'*, *'eyes width'*, *'eye separation distance'*, *'nose vertical position'*, *'mouth vertical position'* and *'mouth width'*. This is done because we are mainly interested in the actual position of the facial features with respect to the face rather than the exact shape of the overall face. The number of parameters and the deformability of the mask do not allow for a precise estimation of a person-independent face shape.

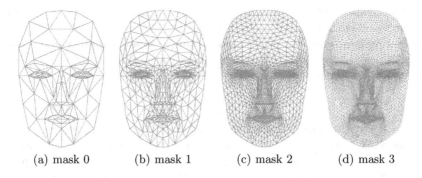

(a) mask 0   (b) mask 1   (c) mask 2   (d) mask 3

Fig. 1. (a) Original CANDIDE-3 face mask, (b), (c) and (d) are subdivided masks after 1, 2 and 3 subdivision steps respectively. Source: [15] © 2008 IEEE.

This face mask is subdivided using the Modified Butterfly algorithm [26] three times to finally produce 5984 vertices and 11776 triangles. This subdivision of the mask allows for different mask resolutions varying from only a few triangles to a very fine mask mesh when projected into the image. Fig. 1 shows the result.

After the mask is fitted to a particular face by adjusting the elements of γ, the middle of each mask triangle is projected into the image

$$\mathbf{I} = \mathcal{P}(\mathbf{g}, T) \text{ with } T = [\mathrm{T_{int}}, \mathrm{T_{ext}}] \tag{2}$$

where T_{int} are the intrinsic camera parameters determined using camera calibration techniques as in [25], T_{ext} are the extrinsic camera parameters, ie translation and rotation and $\mathcal{P}$ the projection of the face mask model $\mathbf{g}$ into the image using T. The resulting vector $\mathbf{I}$ contains one grey value for each mask triangle. This differs from standard texture mapping techniques that usually require warping and interpolating the texture to fit the 3D model. This model has been applied previously for super-resolving faces in low resolution images [15]. By using a face mask model and a vector $\mathbf{I}$ of concatenated grey values no geometric normalisation is required, the face masks of different persons are aligned implicitly.

These texture vectors of concatenated grey values $\mathbf{I}$ are then used to calculate the average face $\overline{\mathbf{x}}$ and a set of principle components $\mathbf{X}$, the so called Eigenfaces, using PCA [20]. A new face $\mathbf{I}_{new}$ can then be represented as $\hat{\mathbf{I}}$, reconstructed from a combination of principle components $\mathbf{X}$

$$\hat{\mathbf{I}} = \overline{\mathbf{x}} + \mathbf{X}\mathbf{X}^T(\mathbf{I}_{new} - \overline{\mathbf{x}}) \tag{3}$$

2.2   Light-Invariant Automatic Head Pose and Shape Initialisation

After the face model is built offline it is used to fit the mask to new unseen faces. We therefore utilise the algorithm developed by [21] to detect near frontal faces, the first step of our approach. The image is then cropped according to the detection result. We initialise the face mask to fit this window by centralising the neutral face mask $\overline{\mathbf{g}}$ such that it fills out the whole window.

Since every face might be cropped differently this first rough initialisation is further refined by a fast grid-search, to get a better approximation of the z-value, that controls the depth, ie size of the face mask. Starting from the initial z-value, we assign z with 14 different values in the range of $\pm 5\%$ of the inital mask size and solve for the best x and y using a locally exhaustive and direct search as proposed by [9]. The error function e, that guides this search is the reconstruction error in face space

$$e(T_{ext}, \gamma) = ||\mathbf{I} - \hat{\mathbf{I}}||_2 \tag{4}$$

with $\gamma = 0$ and the rotation parameters within T_{ext} set to 0, since we assume a near frontal face. The set of pose parameters (x, y, z) that results in the smallest reconstruction error is then used for initialisation. To refine this initial pose a particle filter [10] is then used to converge to the correct pose (T_{ext}) and person-specific shape parameters γ by 'tracking' repeatedly on a single image, essentially performing an incremental randomised search for the global maximum.

The reconstruction error in face space (Eq. 4) serves as the weighting function for the particle filter. Illumination invariance is achieved by incorporating the work of [3]. They have proven that an image J can be expressed as a linear combination of so called harmonic images b_{nm}

$$J_i = \sum_{n=0}^{\infty} \sum_{m=-n}^{n} \alpha_{nm} b_{nm}(p_i) \tag{5}$$

where α_{nm} are the linear coefficients. Given a 3D model of the object, each harmonic image is a function of the albedo and the surface normal of each surface point p_i. Throughout this paper we use the first nine harmonic images [3]. Equation 5 is rewritten for simplicity as

$$J = \sum_{j=1}^{9} \beta_j V_j(n, \rho) \tag{6}$$

where β_j is the jth coefficient and the function V_j returns the ith harmonic image given the surface normals n and the albedo ρ of all surface points of the 3D model.

2.3   Particle Filter Refinement

Particle filters are a statistical model commonly used in tracking [10]. They are based on the idea of Monte Carlo sampling to approximate the posterior probability distribution of a state – in our case the pose and shape parameters of the human face. This posterior is factorised as follows

$$P(x_t|y_{1:t}) \propto P(x_1) \prod_{t=1}^{T} P(y_t|x_t) \prod_{t=2}^{T} P(x_t|x_{t-1}) \tag{7}$$

where x_t is the state at time t (consisting of $\{T_{ext}, \gamma_{1:K}\}$) and y_t is the observation at time t. For the bootstrap particle filter, Monte Carlo samples $x_t^{(i)}$ can be drawn as follows [10]

$$x_t^{(i)} \sim P(x_t|x_{t-1}) \tag{8}$$

$$\widetilde{w}_t^{(i)} = P(y_t|x_t), \qquad w_t^{(i)} = \frac{\widetilde{w}_t^{(i)}}{\sum_i \widetilde{w}_t^{(i)}} \tag{9}$$

where $\sim$ means "sample from" and $w_t^{(i)}$ is the weight associated with particle i.

For this paper we set $P(x_t|x_{t-1}) = \mathcal{N}(x_t; x_{t-1}, \Sigma)$, that is we are sampling in the neighbourhood of the previous state. Σ is a diagonal covariance, ie the dimensions are sampled independently, with values set heuristically. In terms of standard deviation, 3D position is set to equal a shift of about one sixth of the face size, 3D rotation is 15° and the shape parameters γ_k are all set to 0.3.

These parameters are then used to deform the neutral face mask $\bar{g}$ according to Eq. 1 and project it into the image using Eq. 2. The face image vector $\mathbf{I}$

is extracted according to Eq. 3 and the distance in feature space is then defined as the norm of the difference between the retrieved image $\mathbf{I}$ and the illumination-adjusted, reconstructed image $\hat{\mathbf{I}}$.

$$d(x_t) = \min_{\beta} ||\mathbf{I} - \beta V(n, \hat{\mathbf{I}})||_2 \qquad (10)$$

where n are the surface normals of the face mask that has been deformed by γ and the reconstructed face $\hat{\mathbf{I}}$ is used as the reference albedo. Using this distance d, the weighting function $P(y_t|x_t)$ is defined as a normalised vector

$$\widetilde{w}_t^{(i)} = [\eta - d(x_t^{(i)})]^\lambda \qquad (11)$$

where $\eta = \max_i(d(x_t^{(i)}))$ and λ is an annealing factor to increase the spread of the particle weights [8], empirically set to $\lambda=4$. Note that we provide the particle filter with the *same* image at every t, which essentially performs an incremental refinement of the face mask fit rather than a tracking task. This utilises the genetic-algorithm-like nature of particle filters [8]. In order to ensure a convergence, we adjust Σ so that it declines as per $\Sigma(t) = 0.8 \cdot \Sigma(t-1)$.

3   Experiments

We use two face databases to perform our experiments, namely the the IMM Face Database [17] and the Extended Yale Face Database B [12]. The IMM Face Database consists of 240 images of 40 individuals, 6 images per individual, each exhibiting variations in pose, expression and lighting. We only use the first image (*'full frontal face, neutral expression, diffuse light'*) and the fifth image (*'full frontal face, neutral expression, spot light added at the person's left side'*) of each person and we do not incorporate any prior knowledge on the lighting conditions in our model.

Each image of the IMM Face Database is labelled with 58 landmarks at various facial feature points, which can be used to automatically fit the CANDIDE-3

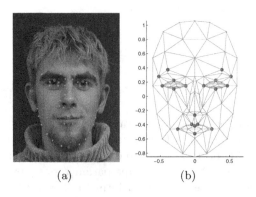

(a)   (b)

Fig. 2. (a) Sample image of the IMM face database with annotated facial landmarks, (b) CANDIDE-3 mask, red dots indicate point-to-point correspondences

mask as a ground truth. We therefore assigned point-to-point-correspondences between these landmarks and the corresponding mask vertices as shown in Fig. 2. The pose T_{ext} and shape parameters γ of the deformable face mask are estimated by utilising the Levenberg-Marquardt algorithm to minimise the Euclidean distance between the landmarks and the mask vertices. This mask fitting is performed on the first image of each of the 40 persons in the IMM Face Database which are used for training purpose as well as creating the ground truth.

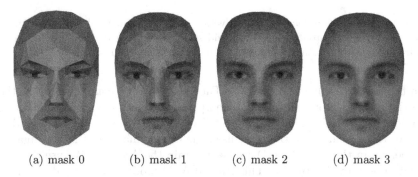

(a) mask 0   (b) mask 1   (c) mask 2   (d) mask 3

Fig. 3. Mean face for each mask sized used after applying PCA to the first image of each of the 40 individuals of the IMM Face Database

After the fitting process each mask triangle is projected onto the image and its grey value is extracted to form the face image vector **I** for each of the 40 person. These vectors of concatenated grey values are then used to calculate the Eigenfaces as described in Section 2.1. We calculate the Eigenfaces and the mean face $\bar{x}$ for each mask size, varying from coarse to fine as shown in Fig. 3. We keep the top 70% of all eigenvectors, using the Matlab code provided by [5].

Using this face model we conducted the following experiments on IMM Face Database: (1) testing on training, (2) leave one out, (3) testing on different lighting condition and (4) comparison with ASM. The progressive results of the proposed algorithm are shown in Fig. 4.

For experiment number one we used the first image of each person, the same that was used for calculating the face model, for fitting each of the four different mask sizes, from coarse (mask 0) to fine (mask 3). Additionally we divided the initial image resolution of 640×480 in half three times resulting in images of size 320×240, 160×120 and 80×60 pixels, generating corresponding average face sizes of 250×160, 125×80, 60×40 and 30×20 pixels.

In the second experiment we calculated the mean face and the Eigenfaces using only 39 persons, the first image of each person. Using this model the mask is fit the image of the person that was left out. Experiment number three uses the face model, that is trained on the first image of each person, to fit the mask to the fifth image of each of the 40 persons, which are illuminated with a spotlight. Again we use all 4 different masks as well as dividing the original image resolution of 640×480 pixels in half three times.

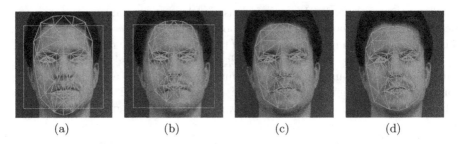

| (a) | (b) | (c) | (d) |

Fig. 4. (a) Result after face detection [white rectangle] and mask initialisation, (b) Result after refined initialisation using grid-search, (c) Final result of our approach, (c) Ground truth mask that was fitted to the labelled landmarks

In the last experiment on the IMM Face Database we compared our method with the method proposed in [16]. The approach finds significant facial feature points using an Active Shape Model (ASM) first and then a 3D mesh model is fitted to these points. We therefore used the 58 landmarks of the first image of each person in the IMM Face Database to train the ASM [6]. We use the ASM Matlab code provided by [13]. This trained ASM model is then used to retrieve these 58 landmarks in each of the 40 images again. The ASM face model is initialised to fill the cropped image window found by the frontal face detector [21]. After the landmarks are retrieved by ASM, the CANDIDE-3 mask is fitted to these landmarks as described earlier.

Table 1. Mean Euclidean vertex point differences in pixel on images of size 640×480, using mask 2 and 10,000 particles. The mean error of the ASM is calculated for only 26 acceptable mask fits and for the subset 2 of the Yale Face Database only 433 mask fits were reasonable. The face model that was used to fit the mask to the Yale Face Database was trained entirely from the IMM Face Database.

	Mean Vertex Difference
IMM Face Database	
(1) Testing on training	2.6 pixels
(2) Leave one out	4.2 pixels
(3) Lighting variety	4.8 pixels
(4) ASM comparison	6.2 pixels (26/40)
Yale Face Database	
Subset 1	6.5 pixels
Subset 2	6.2 pixels (433/456)

We further use the Yale Face Database B and its extension to conduct more experiments. This database contains images of 10 persons under 9 different poses and 64 different illumination conditions; the extended Yale Face Database B contains additional 28 persons under the same pose and lighting conditions. We use both sets. Furthermore the images for each person are divided into subsets according to the angle between the light source and the camera axis. We use subsets 1 which contains 7 images of each person with spot lights at up to 12° and subset 2 which contains 12 images with angles up to 25°. The greater the angle the longer the shadows on the person's face.

We used the face model that was built from 40 images of 40 persons in the IMM Face Database to fit the mask to all 266 images of the first subset. Again the resolution of the images (640×480 pixels), it is cut in half three times to result in images of size 320×240, 160×120 and 80×60 pixels, with corresponding average face sizes of 250×160, 125×80, 60×40 and 30×20 pixels. We also fitted the mask to the second subset, consisting of 456 images, 12 for each person using the original image resolution of 640×480 pixels.

As a ground truth we fitted the mask to all 38 individuals of the Yale Face Database B and its extension manually by adjusting the pose and shape parameters. We thereby assume that the pose and shape of each face remained constant across different lighting settings. Again we use the mean Euclidean distance between the manually fitted and the automatically detected mask vertices as a quality measurement.

Our results are summarized in table 1. The error is the mean Euclidean distance in pixels between the manually fitted and the automatically detected mask vertices. Images are 640×480 pixels, we use mask 2 and 10,000 particles.

We tested the number of particles versus the processing time needed to achieve best results and found that 10,000 particles is the best trade off between accuracy and speed. The processing time depends on the size of the mask, ie the number of mask triangles. The evaluation of 10,000 particles in Matlab on a Intel Core 2 Quad 2.33GHz PC (using only a single core) takes about 9 s, 20 s, 66 s and 315 s for mask 0, 1, 2 and 3 respectively. The face detection algorithm in Matlab takes 0.40 s and the fast grid-search takes about 0.15 s per image.

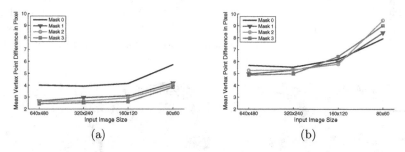

(a)   (b)

Fig. 5. Mean vertex point difference for different input resolutions and different mask sizes used for (a) testing on training and (b) testing on different lighting condition

Fig. 5 shows our results for the testing on training experiment and the testing on different lighting condition in more detail. The graph shows the mean vertex point differences averaging over 40 persons for different image resolutions and mask sizes. It shows that the original mask 0 is too coarse and results in the least accurate fitting. Since the mean vertex point difference depends on the image resolution we calculated it with respect to a image resolution of 640×480 in order to make the results comparable. Therefore the recovered pose and shape parameters for resolutions 320×240 and below are used to project the mask into the image of size 640×480. The graph shows more clearly that the error increases with decreasing image resolution as well as with decreasing mask size. The finer the mask and the higher the resolution of the input image the better the estimation of the mask fit.

We use seven different shape parameters from which some are estimated better than others. Shape parameters controlling positions, namely *'eyes vertical position'*, *'eye separation distance'*, *'nose vertical position'*, *'mouth vertical position'*, are estimated more precisely than shape parameters that only apply to a few number of triangles, like *'eyebrows vertical position'*, *'eyes width'*, *'mouth width'*, which are more difficult to estimate.

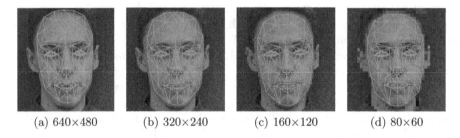

(a) 640×480   (b) 320×240   (c) 160×120   (d) 80×60

Fig. 6. Result of the testing on training for different image resolutions

Figure 6 shows the result of testing on training experiment for one person of the IMM Face Database using different image resolutions. The mean vertex point differences are 1.86, 0.54, 0.77 and 0.31 for image resolutions of 640×480, 320×240, 160×120 and 80×60 pixels respectively. Note that the face model is only trained with images of resolution 640×480 but by using different mask sizes it can also be applied to image resolutions different from the training resolution.

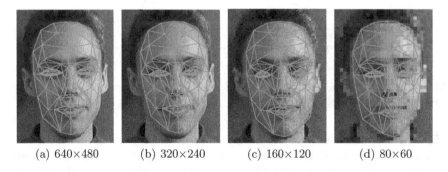

(a) 640×480   (b) 320×240   (c) 160×120   (d) 80×60

Fig. 7. Result of the proposed automatic fitting for different image resolutions

In comparison Fig. 5(b) shows the results for the testing on different lighting condition, using the fifths image of each person. Again masks 1, 2 and 3 achieve best results for images resolutions of 640×480 and 320×240 pixels. But modelling the lighting condition is more difficult in low resolution, so the performance decreases for images of size 160×120 and below. On average the mean vertex difference increases by about two pixel when fitting the mask to images that exhibit spot lights compared to images with diffuse lighting only. Fig. 7 shows the result for one person of the IMM Face Database for all 4 different resolutions.

When comparing our results with the ASM, we found that the ASM is very sensitive to the initialisation and the parameters used. We found that is it was only able to find the facial feature points of 26 persons correctly, in 14 out of 40 images the ASM drifted completely off the face region using the original resolution of 640×480. After fitting the face mask the mean vertex distance for these 26 persons amounted to 6.2 pixels compared to 2.7 pixels achieved by our approach. As it must be trained on the original image resolution, the ASM was unable to find the facial feature points in resolutions that differ from the training resolution, unlike our approach.

The generalisation of our approach is shown in experiment number 2 and the experiments on the Yale Face Database. The mean vertex point difference for the leave one out experiment is 4.2 pixels as shown in table reftab:results. This is only an increase of 1.6 pixels compared to testing on training. The results on the subset 1 of the Yale Face Database show another increase of two pixels. But non of the 38 persons in the Yale Face Database B are included in the face model and all images are recorded under different non-diffuse lighting conditions.

When using the second subset of the Yale Face Database, consisting of 456 images, 12 for each person, our approach was able to correctly fit the mask to 433 of these images, using the original resolution and mask size 2. The mean vertex difference amounts to 6.2 pixels, compared to 6.5 pixels for full resolution images of the first subset. Results will decrease with decreasing resolution and more extreme lighting conditions.

4   Conclusion

We proposed a new method for automatic fitting of a deformable face mask to a previous unseen face. By subdividing a coarse mask mesh to produce different mask sizes we are able to fit this mask to different input image resolutions. The face model is built with the highest image resolution and can then be applied to any smaller resolution unlike ASM. Using different mask sizes we are able to fit a deformable mask to faces with average face sizes of down to 30×20.

Traditional approaches usually find facial features first before fitting a face mask to these points whereas our method combines these two steps. The fitting of a deformable face mask implies the detection of facial feature points. By incorporating the work of [3] the proposed method can handle different lighting conditions as shown in experiments on the IMM and the extended Yale Face

Database. The approach is shown to generalise well even when testing on a completely different data set to the training data. Future work involves extending the approach to handle facial expressions and partial occlusions of the face.

References

1. Ahlberg, J.: CANDIDE-3 - an updated parameterized face. Technical Report LiTH-ISY-R-2326, Dept. of Elec. Eng., Linkoping University, Sweden (2001)
2. Ayala Raggi, S., Altamirano Robles, L., Cruz Enriquez, J.: Towards an illumination-based 3D active appearance model for fast face alignment. In: Ruiz-Shulcloper, J., Kropatsch, W.G. (eds.) CIARP 2008. LNCS, vol. 5197, pp. 568–575. Springer, Heidelberg (2008)
3. Basri, R., Jacobs, D.W.: Lambertian reflectance and linear subspaces. IEEE Transactions on Pattern Analysis and Machine Intelligence 25, 218–233 (2003)
4. Blanz, V., Vetter, T.: Face recognition based on fitting a 3D morphable model. IEEE Trans. Pattern Anal. Mach. Intell. 25(9), 1063–1074 (2003)
5. Cai, D.: Codes and datasets for subspace learning (2007), http://www.cs.uiuc.edu/homes/dengcai2/Data/data.html (retrieved, October 2007)
6. Cootes, T.F., Taylor, C.J., Cooper, D.H., Graham, J.: Active shape models—their training and application. Comp. Vis. Image Underst. 61(1), 38–59 (1995)
7. Dedeoglu, G., Baker, S., Kanade, T.: Resolution-aware fitting of active appearance models to low-resolution images. In: Leonardis, A., Bischof, H., Pinz, A. (eds.) ECCV 2006. LNCS, vol. 3952, pp. 83–97. Springer, Heidelberg (2006)
8. Deutscher, J., Reid, I.: Articulated body motion capture by stochastic search. International Journal of Computer Vision 61(2), 185–205 (2005)
9. Dornaika, F., Ahlberg, J.: Fitting 3D face models for tracking and Active Appearance Model training. Image and Vision Computing 24(9), 1010–1024 (2006)
10. Doucet, A., Godsill, S., Andrieu, C.: On sequential Monte Carlo sampling methods for Bayesian filtering. Statistics and Computing 10(3), 197–208 (2000)
11. Edwards, G.J., Taylor, C.J., Cootes, T.F.: Interpreting face images using active appearance models. In: Face & Gesture Recognition, p. 300. IEEE, Los Alamitos (1998)
12. Georghiades, A., Belhumeur, P., Kriegman, D.: From few to many: Illumination cone models for face recognition under variable lighting and pose. IEEE Transactions on Pattern Analysis and Machine Intelligence 23(6), 643–660 (2001)
13. Hamarneh, G.: Multi-resolution active shape models (2008), http://www.cs.sfu.ca/~hamarneh/software/asm/index.html (retrieved, February 2008)
14. Kalinkina, D., Gagalowicz, A., Roussel, R.: 3d reconstruction of a human face from images using morphological adaptation. In: Gagalowicz, A., Philips, W. (eds.) MIRAGE 2007. LNCS, vol. 4418, pp. 212–224. Springer, Heidelberg (2007)
15. Kuhl, A., Tan, T., Venkatesh, S.: Model-based combined tracking and resolution enhancement. In: Intl. Conf. on Pattern Recognition. IEEE, Los Alamitos (2008)
16. Lu, L., Zhang, Z., Shum, H.-Y., Liu, Z., Chen, H.: Model- and exemplar-based robust head pose tracking under occlusion and varying expression. In: IEEE Workshop on Models versus Exemplars in Computer Vision (2001)
17. Nordstrøm, M.M., Larsen, M., Sierakowski, J., Stegmann, M.B.: The IMM face database - an annotated dataset of 240 face images. Technical report, Technical University of Denmark, DTU (May 2004)

18. Tang, H., Huang, T.S.: MPEG4 performance-driven avatar via robust facial motion tracking. In: IEEE International Conference on Image Processing (2008)
19. Tu, J., Zhang, Z., Zeng, Z., Huang, T.: Face localization via hierarchical condensation with fisher boosting feature selection. In: CVPR, pp. II: 719–724 (2004)
20. Turk, M., Pentland, A.: Face recognition using eigenfaces. In: IEEE Conference on Computer Vision and Pattern Recognition (CVPR), pp. 586–591 (1991)
21. Viola, P., Jones, M.: Rapid object detection using a boosted cascade of simple features. In: IEEE Conference on Computer Vision and Pattern Recognition (2001)
22. Xiao, J., Baker, S., Matthews, I., Kanade, T.: Real-time combined 2D+3D active appearance models. In: IEEE Comp. Vis. and Pattern Recog., pp. 535–542 (2004)
23. Xin, L., Wang, Q., Tao, J., Tang, X., Tan, T., Shum, H.: Automatic 3D face modeling from video. In: International Conference on Computer Vision, October 2005, vol. 2, pp. 1193–1199. IEEE, Los Alamitos (2005)
24. Yan, S., Li, M., Zhang, H., Cheng, Q.: Ranking prior likelihood distributions for bayesian shape localization framework. In: Intl. Conf. on Comp. Vis. IEEE, Los Alamitos (2003)
25. Zhang, Z.: A flexible new technique for camera calibration. IEEE Transactions on Pattern Analysis and Machine Intelligence 22(11), 1330–1334 (2000)
26. Zorin, D., Schroder, P.: Subdivision for modeling and animation. In: Subdivision for modeling and animation, Computer Graphics and Interactive Techniques. ACM Siggraph, vol. 36 (July 2000)

Re-projective Pose Estimation of a Planar Prototype

Georg Pisinger and Georg Maier

University of Passau,
Innstr. 43, 94032 Passau, Germany
{georg.pisinger,georg.maier}@uni-passau.de

Abstract. We present an approach for robust pose estimation of a planar prototype. In fact, there are many applications in computer graphics in which camera pose tracking from planar targets is necessary. Unlike many other approaches our method minimizes the Euclidean error to re-projected image points. There is a number of recent pose estimation methods, but all of these algorithms suffer from pose ambiguities. If we know the positions of some points on the plane we can describe the 3D position of the planar prototype as a solution of an optimization problem over two parameters. Based on this formulation we develop a new algorithm for pose estimation of a planar prototype. Its robustness is illustrated by simulations and experiments with real images.

Keywords: Pose estimation, absolute orientation.

1 Introduction

Computation of the position and orientation of a camera from a single image with respect to a known object is crucial for many computer and robot vision tasks. Camera calibration, object recognition and self-localization of mobile robots are typical examples for the use of pose estimation ([1,2]).

At least three point correspondences are needed to solve the pose estimation problem. For every 3D object point expressed in world coordinates its 2D projection expressed in image coordinates must be given. For three or four non-collinear 3D object points an analytical solution can be found ([3,4]). Therefore a fourth or fifth-degree polynomial system can be formulated and the problem can be solved by finding roots of the polynomial system. However, this method can be only applied to three or four 3D object points and the roots of the polynomial systems are susceptible to noise. There is no closed form solution for more than four 3D object points. The classical approach is to formulate the pose estimation problem as a nonlinear least-squares problem. A local minimum of this problem can be computed by using nonlinear optimizations algorithm, e.g. the Gauss-Newton method proposed by Lowe [5] or the globally convergent iterative method of Lu [6]. But the solution of such optimization methods does not necessarily correspond to the global minimum or correct pose. They always find only a local minimum which depends on the initial values.

A. Gagalowicz and W. Philips (Eds.): MIRAGE 2009, LNCS 5496, pp. 82–93, 2009.

This paper focuses on the special case in which the object points are coplanar. A main area of application of vision based tracking systems which are based on planar targets is Augmented Reality ([2]).

For this case there are also closed form solutions for three and four coplanar object points ([3,4]). These methods can be applied to more points by taking subsets and finding common solutions to several polynomial systems. But the results are highly sensitive to noise. There are iterative methods based on minimizing the error, either on the image (*image space error*) or in $\mathbb{R}^3$ (*object space error*). The magnitudes and the positions of local minima are very similar for both error functions ([7]). In this paper we consider the object space error since this is easier to parameterize. There are some iterative methods ([6,7,8,9]) but most of them have problems to determine the correct pose since the error function typically has multiple local minima. In [7] it is shown by tests that, in general, there are two local minima of the object space error function. To determine such a minimum the roots of a multivariate polynomial of order five in three variables must be computed.

However, before we start describing the particular approaches to get the correct minimum of the object space error function, we will first define our camera model and its parametrization (section 2). In section 3 we show that the object space error function with 6 degrees of freedom can be formulated as a function with only two variables. Therefore the graph of this function can be easily visualized. Based on this simplification, we develop our new robust pose estimation algorithm. Section 4 presents experimental results and comparison with state-of-the-art pose algorithms.

2   The Camera Model

2.1   The Projective Camera Mapping

In our context the term "camera" means an image capturing device including the lens, a CCD camera, a frame grabber and the displayed image. The camera mapping $\mathcal{K} : \mathbb{R}^3 \to \mathbb{R}^2$ defines the way by which an object point $p \in \mathbb{R}^3$ will be transformed into the image coordinate system. The common way to model the camera mapping $\mathcal{K} : \mathbb{R}^3 \to \mathbb{R}^2$ in computer vision is to use a rigid motion $T : \mathbb{R}^3 \to \mathbb{R}^3$ followed by a dimension reducing mapping $\Pi : \mathbb{R}^3 \to \mathbb{R}^2$, resulting in $\mathcal{K} = \Pi \circ T$.

The so called *extrinsic parameters* of the rigid motion T define the transformation from a given world coordinate system to the camera coordinate system. Since we are interested in the pose and orientation of a planar prototype with respect to the camera only we assume $T = \mathrm{id}$.

Π is a central projection followed by a coordinate system transformation. For theoretical analysis many authors consider only a pure pinhole camera model. But for a realistic camera modelling a distortion mapping in the image plane has to be considered. This leads to a pinhole camera model with distortion ([10]).

The first part of Π describes the projection of 3D points on the camera plane. Let $P_3 : \mathbb{R}^2 \times (\mathbb{R} \setminus \{0\}) \to \mathbb{R}^2$, $(x, y, z) \mapsto (\frac{x}{z}, \frac{y}{z})$ denote the central projection with respect to the third coordinate. The distortion mapping $\delta : \mathbb{R}^2 \to \mathbb{R}^2$ is defined in the camera plane with respect to the camera coordinate system. The most common distortion model uses only radial distortions

$$\delta\left(\begin{pmatrix} u \\ v \end{pmatrix}\right) = \begin{pmatrix} u + u \cdot r \\ v + v \cdot r \end{pmatrix}, \; r = \sum_{i=1}^{D} k_i(u^2 + v^2)^i \tag{1}$$

with parameters $k_1, \ldots, k_D \in \mathbb{R}$ (usually $D = 2$).

The last step accomplishes the change of the camera coordinate system to the image coordinate system. This is also a change of units: from metric to pixel geometry. We set

$$P : \mathbb{R}^2 \to \mathbb{R}^2 : \begin{pmatrix} u \\ v \end{pmatrix} \mapsto \begin{pmatrix} \alpha & \gamma \\ 0 & \beta \end{pmatrix} \begin{pmatrix} u \\ v \end{pmatrix} + \begin{pmatrix} u_0 \\ v_0 \end{pmatrix}, \tag{2}$$

where (u_0, v_0) is the principal point. If f is the focal length of the camera and $d_u \times d_v$ is the dimension of a single sensor element, then the parameters α and β can be interpreted as $\alpha = \frac{f}{d_u}$ and $\beta = \frac{f}{d_v}$. γ describes the skewness between the axes of the pixel coordinate system. If γ is zero the coordinate axis of the image coordinate system are perpendicular.

After all, our camera mapping $\mathcal{K}$ can be modelled by $\mathcal{K} = P \circ \delta \circ P_3 \circ T$. All parameters which describe the mapping $\Pi = P \circ \delta \circ P_3$ are called *intrinsic camera parameters*. We further assume the intrinsic camera parameters are known by calibration using for example the algorithm of Zhang [10].

2.2   Re-projection of Image Points

As one can see from the last section each point in the image plane determines a straight line in the reference coordinate system intersecting the projection center of the camera by re-projection. Let $A \subseteq \mathbb{R}^2$ be the image plane of the camera, then the re-projected ray of a point $i \in A$ is defined by the pre-image of i under $\mathcal{K}$. In our camera model $\mathcal{K}^{-1}(\{i\})$ is a straight line (the so called *viewing ray*).

For $(u, v) \in \mathbb{R}^2$ the set $P_3^{-1}(\{(u, v)\}) = \{s(u, v, 1)|s \in \mathbb{R}\setminus\{0\}\}$ is a straight line in $\mathbb{R}^3$ with direction $(u, v, 1)$ without the origin. In order to construct a functional section of the relation P_3^{-1} we choose a suitable representative of $P_3^{-1}(\{(u, v)\})$ by setting $P_3^{-1}((u, v)) := \frac{1}{\sqrt{u^2 + v^2 + 1}}(u, v, 1)$. Note that this representative has norm 1 and it is the direction of the viewing ray. Then we can define $\Pi^{-1} = (P \circ \delta \circ P_3)^{-1} = P_3^{-1} \circ \delta^{-1} \circ P^{-1}$.

3   Euclidean Pose Determination

3.1   Problem Formulation

For the re-projective pose estimation we are able to formulate the pose determination problem, which is optimal in the Euclidean sense. This means that the distance of the determined pose of the planar prototype points to the re-projected observed points is minimal.

Let $\mathbf{P} = \{p_1, \ldots, p_m\} \subseteq \mathbb{R}^3$ be a finite set of coplanar points with respect to the reference coordinate system. $\mathbf{P}$ is called *prototype*. Without loss of generality let $p_j = (x_j, y_j, 0)^{tr} \in \mathbb{R}^3$ for all $p_j \in \mathbf{P}$. For every $p_j \in \mathbf{P}$ we denote $i_{p_j} \in A$ for the observed projection of p_j in the image plane $A \subseteq \mathbb{R}^2$ with respect to the image coordinate system and define $n_j = \mathit{\Pi}^{-1}(i_{p_j})$. Furthermore, we assume that not all i_{p_j} are equal and $\mathbf{P}$ is not collinear.

For a direction $n \in \mathcal{S}^2 := \{x \in \mathbb{R}^3 \mid \|x\| = 1\}$ we define $L_n := \{\alpha n \mid \alpha \in \mathbb{R}\}$ as the *line with direction n containing the origin*. It is easy to show that for every point $p \in \mathbb{R}^3$

$$\mathrm{dist}(p, L_n)^2 = \|(I - N_n)p\|^2 \tag{3}$$

holds, where N_n is the observed line-of-sight projection matrix defined as $N_n = n\, n^{tr}$ and I is the 3×3 identity matrix. The Euclidean pose estimation problem is to obtain $R \in \mathrm{SO}(3) := \{U \in \mathbb{R}^{3\times3} \mid \det(U) = 1 \wedge UU^{tr} = I\}$ and $t \in \mathbb{R}^3$ minimizing the least-squares sum

$$E_{os}(R, t) = \sum_{j=1}^{m} \mathrm{dist}(Rp_j + t, L_{n_j})^2 . \tag{4}$$

E_{os} is called *object-space error function*. With r_1, r_2, r_3 denoting the column vectors of R, $r = \binom{r_1}{r_2} \in \mathbb{R}^6$ and

$$Q_j = \begin{pmatrix} x_j & 0 & 0 & y_j & 0 & 0 \\ 0 & x_j & 0 & 0 & y_j & 0 \\ 0 & 0 & x_j & 0 & 0 & y_j \end{pmatrix} \in \mathbb{R}^{3\times6} \tag{5}$$

we have $Rp_j + t = x_j \cdot r_1 + y_j \cdot r_2 + t = Q_j r + t$. Using (3) we get

$$\sum_{j=1}^{m} \mathrm{dist}(Rp_j + t, L_{n_j})^2 = \sum_{j=1}^{m} \|(I - N_{n_j})(Q_j r + t)\|^2$$

$$= \sum_{j=1}^{m} (r^{tr} Q_j^{tr} + t^{tr})(I - N_{n_j})(Q_j r + t) = r^{tr} \left(\sum_{j=1}^{m} Q_j^{tr}(I - N_{n_j})Q_j \right) r +$$

$$2r^{tr} \left(\sum_{j=1}^{m} Q_j^{tr}(I - N_{n_j}) \right) t + t^{tr} \left(\sum_{j=1}^{m} (I - N_{n_j}) \right) t \tag{6}$$

since $I - N_{n_j}$ is symmetric and idempotent. Using the abbreviations

$$M_1 = \sum_{j=1}^{m} Q_j^{tr}(I - N_{n_j})Q_j, \quad M_2 = \sum_{j=1}^{m} Q_j^{tr}(I - N_{n_j}), \quad M_3 = \sum_{j=1}^{m} (I - N_{n_j}) \tag{7}$$

the Euclidean pose determination problem is to obtain $r = \binom{r_1}{r_2}$ with $r_1, r_2 \in \mathcal{S}^2$, $r_1^{tr} r_2 = 0$ and $t \in \mathbb{R}^3$ minimizing

$$r^{tr} M_1 r + 2 r^{tr} M_2 t + t^{tr} M_3 t \ . \tag{8}$$

Since this minimization problem is quadratic in t, given a fixed vector r the optimal value for t can be computed in closed form as

$$t = -M_3^{-1} M_2^{tr} r \ . \tag{9}$$

For (9) to be well-defined, M_3 must be positive definite, which can be verified as follows: Let $\|X\|$ denote the Frobenius norm of a matrix $X \in \mathbb{R}^{3 \times 3}$. For any $x \in \mathbb{R}^3 \setminus \{0\}$

$$x^{tr} M_3 x = x^{tr} \left(\sum_{j=1}^{m} (I - N_{n_j}) \right) x = \sum_{j=1}^{m} (\|x\|^2 - x^{tr} N_{n_j}^{tr} N_{n_j} x) =$$

$$= \sum_{j=1}^{m} (\|x\|^2 - \|N_{n_j} x\|^2) \geq \sum_{j=1}^{m} (\|x\|^2 - \|N_{n_j}\|^2 \cdot \|x\|^2) = 0. \tag{10}$$

Since $\|x\| = \|N_{n_j} x\|$ implies that x is a eigenvector to the eigenvalue 1 of N_{n_j} not all terms $\|x\|^2 - \|N_{n_j} x\|^2$ can be equal to zero unless all image points i_{p_j} are equal. Since this case is excluded, $x^{tr} M_3 x$ is strictly greater than zero in every case. Therefore positive definiteness of M_3 follows.

Using equation (9) and the symmetry of M_3 we have

$$r^{tr} M_1 r + 2 r^{tr} M_2 (-M_3^{-1} M_2^{tr} r) + (-M_3^{-1} M_2^{tr} r)^{tr} M_3 (-M_3^{-1} M_2^{tr} r)$$

$$= r^{tr} M_1 r - 2 r^{tr} M_2 M_3^{-1} M_2^{tr} r + r^{tr} M_2 M_3^{-1} M_3 M_3^{-1} M_2^{tr} r$$

$$= r^{tr} M_1 r - r^{tr} M_2 M_3^{-1} M_2^{tr} r = r^{tr} (M_1 - M_2 M_3^{-1} M_2^{tr}) r \ . \tag{11}$$

Since $M_1 - M_2 M_3^{-1} M_2^{tr} \in \mathbb{R}^{6 \times 6}$ is also symmetric, there are symmetric matrices $A, C \in \mathbb{R}^{3 \times 3}$ and a matrix $B \in \mathbb{R}^{3 \times 3}$ with

$$M_1 - M_2 M_3^{-1} M_2^{tr} = \begin{pmatrix} A & B \\ B^{tr} & C \end{pmatrix} \tag{12}$$

and problem (8) is equivalent to the determination of $r_1, r_2 \in \mathcal{S}^2$, $r_1^{tr} r_2 = 0$, such that

$$r_1^{tr} A r_1 + 2 r_1^{tr} B r_2 + r_2^{tr} C r_2 \ . \tag{13}$$

becomes minimal.

3.2 Estimation of a Global Minimum for r_2

Before we start describing our new approach to get a minimum of problem (13), we will first explain how an optimal $r_2 \in \mathcal{S}^2$ with $r_1^{tr} r_2 = 0$ can be computed for every $r_1 \in \mathcal{S}^2$. Since problem (13) is symmetric in r_1 and r_2 this method can also be applied to compute an optimal $r_1 \in \mathcal{S}^2$ with $r_1^{tr} r_2 = 0$ for every $r_2 \in \mathcal{S}^2$.

Assume $r_1 \in \mathcal{S}^2$ is fixed then there are $v_1, v_2 \in \mathcal{S}^2$ with $v_1^{tr} r_1 = v_2^{tr} r_1 = v_1^{tr} v_2 = 0$. Then every $r_2 \in \mathcal{S}^2$ with $r_1^{tr} r_2 = 0$ can be written as $r_2 = \lambda v_1 + \mu v_2$ with $\lambda^2 + \mu^2 = 1$. Therefore

$$r_1^{tr} A r_1 + 2 r_1^{tr} B r_2 + r_2^{tr} C r_2 =$$
$$r_1^{tr} A r_1 + 2 (r_1^{tr} B v_1, r_1^{tr} B v_2) \begin{pmatrix} \lambda \\ \mu \end{pmatrix} + (\lambda, \mu) \begin{pmatrix} v_1^{tr} C v_1 & v_1^{tr} C v_2 \\ v_1^{tr} C v_2 & v_2^{tr} C v_2 \end{pmatrix} \begin{pmatrix} \lambda \\ \mu \end{pmatrix} =$$
$$(\lambda, \mu, 1) \begin{pmatrix} v_1^{tr} C v_1 & v_1^{tr} C v_2 & r_1^{tr} B v_1 \\ v_1^{tr} C v_2 & v_2^{tr} C v_2 & r_1^{tr} B v_2 \\ r_1^{tr} B v_1 & r_1^{tr} B v_2 & r_1^{tr} A r_1 \end{pmatrix} \begin{pmatrix} \lambda \\ \mu \\ 1 \end{pmatrix} =: (\lambda, \mu, 1) E \begin{pmatrix} \lambda \\ \mu \\ 1 \end{pmatrix}. \qquad (14)$$

Since $r_1^{tr} A r_1 + 2 r_1^{tr} B r_2 + r_2^{tr} C r_2 \geq 0$ for all $r_1, r_2 \in \mathbb{R}^3$ the matrix E is positive semi-definite. Then there is an unitary matrix $U \in \mathbb{R}^{3 \times 3}$ and a diagonal matrix $D \in \mathbb{R}^{3 \times 3}$ with non-negative diagonal elements such that $E = U^{tr} D U$. D contains the eigenvalues of E and U the corresponding eigenvectors with norm 1. Let

$$D = \begin{pmatrix} d_1 & 0 & 0 \\ 0 & d_2 & 0 \\ 0 & 0 & d_3 \end{pmatrix}, \; D^{\frac{1}{2}} = \begin{pmatrix} \sqrt{d_1} & 0 & 0 \\ 0 & \sqrt{d_2} & 0 \\ 0 & 0 & \sqrt{d_3} \end{pmatrix}, \; z_{\lambda,\mu} = \begin{pmatrix} \lambda \\ \mu \\ 1 \end{pmatrix} \qquad (15)$$

we obtain λ, μ with $\lambda^2 + \mu^2 = 1$ minimizing

$$z_{\lambda,\mu}^{tr} E z_{\lambda,\mu} = z_{\lambda,\mu}^{tr} U^{tr} D^{\frac{1}{2}} D^{\frac{1}{2}} U z_{\lambda,\mu} = \left\| D^{\frac{1}{2}} U z_{\lambda,\mu} \right\|^2. \qquad (16)$$

Let $U = (u_1, u_2, u_3)$ then $\lambda D^{\frac{1}{2}} u_1 + \mu D^{\frac{1}{2}} u_2$ describes an ellipse in the plane spanned by the vectors $\{D^{\frac{1}{2}} u_1, D^{\frac{1}{2}} u_2\}$ through the point $D^{\frac{1}{2}} u_3$. Let $n \in \mathcal{S}^2$ be a normal vector of this plane. Then $\left\{ x \in \mathbb{R}^3 \mid x^{tr} n = (D^{\frac{1}{2}} u_3)^{tr} n \right\}$ is the set of all points in this plane and $((D^{\frac{1}{2}} u_3)^{tr} n) n$ is the orthogonal projection of the origin into this plane, hence

$$\left\| D^{\frac{1}{2}} U z_{\lambda,\mu} \right\|^2 = \left\| D^{\frac{1}{2}} U z_{\lambda,\mu} - ((D^{\frac{1}{2}} u_3)^{tr} n) n \right\|^2 + \left\| ((D^{\frac{1}{2}} u_3)^{tr} n) n \right\|^2. \qquad (17)$$

The left term is minimal with respect to $\lambda^2 + \mu^2 = 1$ if the point $\lambda D^{\frac{1}{2}} u_1 + \mu D^{\frac{1}{2}} u_2 + D^{\frac{1}{2}} u_3$ is the point on the ellipse with minimal distance to $((D^{\frac{1}{2}} u_3)^{tr} n) n$. This problem can be solved easily and numerical stable by computing the largest root of a quartic polynomial ([11]).

So for every $r_1 \in \mathcal{S}^2$ a $r_2 \in \mathcal{S}^2$ can be computed to minimize problem (13). This defines a function $r_2(r_1)$. Now consider $\widetilde{E_{os}} : \mathcal{S}^2 \to \mathbb{R}, \, r_1 \mapsto r_1^{tr} A r_1 + 2 r_1^{tr} B r_2(r_1) + (r_2(r_1))^{tr} C r_2(r_1)$. To estimate local minima of $\widetilde{E_{os}}$ the vector r_1 can be parameterized as follows: $r_1 = (\sin(\theta) \cos(\varphi), \sin(\theta) \sin(\varphi), \cos(\theta))$ for

$\theta \in [0, \pi]$, $\varphi \in [0, 2\pi]$. Since if r_1 is a local minimum of $\widetilde{E_{os}}$ then $-r_1$ is also a local minimum. From this w.l.o.g. let $\cos(\theta) \in [0, 1]$ and therefore it is sufficient to analyze the optimization problem for $\theta \in [0, \frac{\pi}{2}]$. Since there are only two degrees of freedom the graph of the error function $\widetilde{E_{os}}$ can be visualized graphically. We want to mention explicitly that the original problem (4) is identical to the transformed $\widetilde{E_{os}}$. Figure 1 shows the graph representing an example.

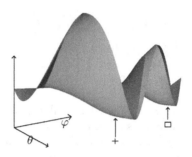

Fig. 1. Graph of a typical object space error function $\widetilde{E_{os}}$ which now depends only from $\varphi \in [0, 2\pi]$ and $\theta \in [0, \frac{\pi}{2}]$. This function has one local minimum ($\square$) and one global minimum in front ($+$).

To estimate the local minima of $\widetilde{E_{os}}$ the intervals $[0, \frac{\pi}{2}]$ and $[0, 2\pi]$ are divided into 20 subintervals and $\widetilde{E_{os}}$ is evaluated at the border of the intervals. The corresponding rotation matrix and translation vector to the minimum values are chosen as initial value for an iterative minimization algorithm. In our experiments we used the iterative algorithm proposed by [6]. As a final step, the global minimum is selected.

3.3   New Robust Pose Estimation Algorithm

In order to get a fast pose estimation algorithm a set of collinear points is generated. If $\mathbf{P} = \{p_1, \ldots, p_m\}$ is our prototype w.l.o.g. let p_1 and p_2 be the two points with maximal distance between each other, i.e.

$$\|p_1 - p_2\| = \max \{\|p_i - p_j\| \mid i, j = 1, \ldots, m\} \ .$$

Furthermore let $l(p_1, p_2)$ be the line through p_1 and p_2 and w.l.o.g. p_m be the point with maximal distance from $l(p_1, p_2)$. If this distance is 0 all points are already collinear. Otherwise let $\overline{p_j}$ be the point of intersection between $l(p_1, p_2)$ and $l(p_m, p_j)$ for $j = 1, \ldots, m-1$.

Since for a line l the image $P_3(l)$ is also a line the same construction can also be made for the observed projection points $q_j := (\delta^{-1} \circ P^{-1})(i_{p_j})$: let $l(q_1, q_2)$ be the line through q_1 and q_2 and $\overline{q_j}$ the point of intersection between $l(q_1, q_2)$ and $l(q_m, q_j)$ for $j = 1, \ldots, m-1$.

For collinear points a closed-form solution can be formulated ([12]). Without loss of generality let $\overline{p_j} = (x_j, 0, 0)^{tr} \in \mathbb{R}^3$ for all $j \in \{1, \ldots, m-1\}$. Using $\overline{n_j} = P_3^{-1}(\overline{q_j})$, (6) and (7) we get the following problem

$$\overline{E_{os}}(R, t) = \sum_{j=1}^{m-1} \text{dist}(R\overline{p_j} + t, L_{\overline{n_j}})^2 = M_1 - M_2 M_3^{-1} M_2^{tr}, \tag{18}$$

where

$$M_1 = \sum_{j=1}^{m-1} \binom{x_j I}{0} (I - N_{\overline{n_j}})(x_j I, 0) = \begin{pmatrix} \sum_{j=1}^{m-1} x_j^2 (I - N_{\overline{n_j}}) & 0 \\ 0 & 0 \end{pmatrix}, \tag{19}$$

$$M_2 = \begin{pmatrix} \sum_{j=1}^{m-1} x_j (I - N_{\overline{n_j}}) \\ 0 \end{pmatrix} \quad \text{and} \quad M_3 = \sum_{j=1}^{m-1} (I - N_{\overline{n_j}}).$$

Then using $M_4 = \sum_{j=1}^{m-1} x_j^2 (I - N_{\overline{n_j}})$ and $M_5 = \sum_{j=1}^{m-1} x_j (I - N_{\overline{n_j}})$ it holds

$$M_1 - M_2 M_3^{-1} M_2^{tr} = \begin{pmatrix} M_4 - M_5 M_3^{-1} M_5 & 0 \\ 0 & 0 \end{pmatrix}. \tag{20}$$

With the notation of the last section we get $B = C = 0$. The optimization problem can be rewritten as

$$\min_{r_1 \in \mathcal{S}^2} r_1^{tr} (M_4 - M_5 M_3^{-1} M_5) r_1. \tag{21}$$

Since $M_4 - M_5 M_3^{-1} M_5$ is a symmetric matrix, this is a eigenvalue problem, where a normalized eigenvector to the smallest eigenvalue of $M_4 - M_5 M_3^{-1} M_5$ is a solution. Fast and numerical stable algorithms are wellknown ([13]).

This leads to our new pose estimation algorithm:

1. Estimate the first column r_1 of the rotation matrix R by (21) using the set of generated collinear points.
2. Compute an optimal r_2 with respect to r_1 applying the algorithm of section 3.2 using all points of the planar prototype.
3. Compute $r_3 = r_1 \times r_2$ and the translation vector t (equation (9)).
4. Use $R = (r_1, r_2, r_3), t$ as an initial value for an iterative pose estimation algorithm. In our experiments we used the iterative algorithm proposed by [6].

4   Experimental Results

In order to assess the performance of the pose estimation algorithm, both synthetic and real-world tests were carried out.

For all experiments with synthetic data, we used the following setup:

1. For each test, we generated a random prototype consisting of 10 coplanar points $p_j = (x_j, y_j, 0)^{tr}$ for $j \in \{1, \ldots, 10\}$ in the range $x_j \in [-1.0, 1.0]$ and $y_j \in [-1.0, 1.0]$.
2. For each test, we generated a random rotation matrix R using Euler angles and a random translation vector $t = (t_1, t_2, t_3)^{tr}$ in the range $t_1, t_2 \in [-5.0, 5.0]$ and $t_3 \in [1.0, 10.0]$.
3. To compute each image point $i_{p_j} = \mathcal{K}(Rp_j + t)$ the camera mapping of section 2 is used. Furthermore Gaussian noise was added and the result was rounded to a nearby integer value.

The intrinsic camera parameters for the camera mapping $\mathcal{K}$ are set as follows: $\alpha = 240$, $\beta = 240$, $\gamma = 0.0$, $u_0 = 320$, $v_0 = 240$, $D = 2$, $k_1 = -0.2$, $k_2 = 0.18$.

4.1   New Pose Estimation Algorithm

To test the method presented in the last section we considered 13 different noise levels reaching from zero to 6 pixels. For each noise level, we generated 1000 random models and poses. In Figure 2, the results of our algorithm are compared to two other state-of-the-art methods. The solid line shows the results of our algorithm. In the noise-free case (Gaussian noise = 0 pixels) we obtain a rate of 99 percent to find the correct pose (the global minimum of E_{os}). With increasing noise, the rate decreases down to 97 percent for 6 pixels Gaussian noise. Results for an existing iterative pose estimation algorithm ([6]) and an algorithm locating two local minima ([7]) are given. The error function E_{os} is minimized by both methods. Because there are most of the time two local minima of E_{os}, the rate of finding the correct pose with an iterative method is just about 67 percent. Only the algorithm of Schweighofer [7] maintains two solutions. But this method is less robust since the rate decreases from 92 down to 89 percent for 6 pixels Gaussian noise.

In Figure 2, we plot also the number of iterations required for the 13 different noise levels. Each point represents results averaged over 1000 random models

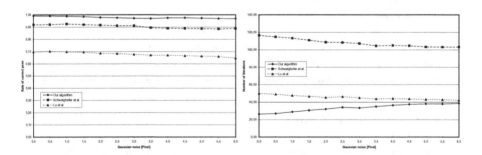

Fig. 2. Rate of choosing the correct pose (left) and number of iterations needed (right)

and poses. The iterative algorithm proposed by [6] is initialized with a weak-perspective approximation. In our case this algorithm is initialized with the results of Step 1 to Step 3 of our method. Since in the algorithm of [7] the iterative method must be applied twice the number of iterations required is significantly higher. As a result, for low noise a better initial estimate can be computed by our algorithm and thus the number of iterations needed during minimization are reduced.

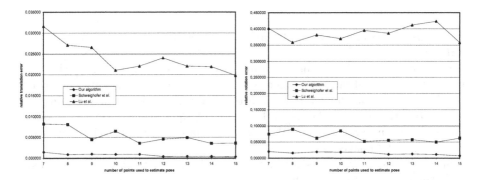

Fig. 3. Rotation and translation error versus number of points used for pose estimation with 2 pixels Gaussian noise

Figure 3 shows the relative translation and rotation error for our algorithm and the two others where the number of points used for pose estimation vary from 7 to 15. We add 2 pixels Gaussian noise to all images. If the correct pose is given by R and t and the determined pose is given by R_x and t_x then the relative translation error is computed as $2\frac{\|t-t_x\|}{\|t\|+\|t_x\|}$. If q is the unit quaternion of R and q_x is the unit quaternion of R_x, respectively, then the relative rotation error is computed as the absolute error in the unit quaternion $\|q - q_x\|$. Note, that our algorithm outperforms both others.

4.2 Real Data

To test the new method with real data we compared our algorithm with a standard technique to obtain a starting value for the non-linear pose calculation problem where the observed homography is exploited. Since the points of the prototype **P** are coplanar the following approach can be applied.

Let $H \in \mathbb{R}^{3\times3}$ be the homography describing the movement of the prototype **P** to the observed image points (i.e. H minimizes the error function $\sum_{j=1}^{m} \|P_3(H\tilde{p}_j) - i_p\|^2$ with $\tilde{p}_j = (x_j, y_j, 1)^{tr}$). Obviously H can be determined only up to scale factor. The determination of H requires a non-linear optimization to achieve an appropriate solution (see for example [10]). For cameras satisfying the pinhole equation we have $\mu H = \tilde{P}(r_1, r_2, t)$ for a $\mu \in \mathbb{R}$ (see e. g. [14]), where $R = (r_1, r_2, r_3)$ and t determine the transformation of the observed prototype **P**. For a calibrated camera we know $\tilde{P} = \begin{pmatrix} \alpha & \beta & u_0 \\ 0 & \gamma & v_0 \\ 0 & 0 & 1 \end{pmatrix}$ (see (2)) yielding

$(r_1, r_2, t) = \mu \tilde{P}^{-1} H =: A$ for a $\mu \in \mathbb{R}$. Since the first two columns of (r_1, r_2, t) must be orthonormal and $t_z \geq 0$ should hold, we set $c = \sqrt{A_{1,1}^2 + A_{2,1}^2 + A_{3,1}^2}$ and

$$\tilde{A} = \begin{cases} \frac{1}{c} A & , \text{ if } A_{3,3} > 0 \text{ holds} \\ -\frac{1}{c} A & , \text{ otherwise} \end{cases} . \tag{22}$$

The first two columns of R and t can be obtained by the equality $(r_1, r_2, t) = \tilde{A}$ and $r_3 = r_1 \times r_2$.

For our experiments we chose a standard CCD camera with a 1/3" chip and 8mm lenses. To calibrate our camera we use the calibration algorithm described by Zhang ([10]) including radial distortion parameters k_1, k_2 determining the radial distortion $\delta(u, v)$. We use a 9×5 grid of equidistant (5 cm) tiny points, which define our prototype **P**. We use a standard point extraction algorithm ([15]) to get the image points.

Since the correct pose was not known the following test is made: We compare the estimated transformation from Step 3 of our algorithm and the transformation obtained by the standard approach with the ground truth: The ground truth is defined by the transformation minimizing $(R, t) \mapsto \sum_{p \in \mathbf{P}} \| i_p - \Pi(Rp + t) \|^2$ which we determine using the non-linear optimization method of Levenberg and Marquardt. In our coordinate system the x- and y-axis are the axis of the camera plane. The z-axis (depth) is perpendicular to them. We used 12 different positions of the prototype **P**. Let R_s, t_s and R_n, t_n be the solution of the initial value of the standard pose estimation and the new method, respectively. The reconstruction error is defined as $\frac{1}{m} \sum_{j=1}^{m} \| (Rp_j + t) - (R_x p_j + t_x) \|^2$ for $x \in \{s, n\}$.

Figure 4 shows the reconstruction error for the standard technique and the proposed method. We obtained the parameters in the matrix $\tilde{P}$ and the radial distortion parameters by the calibration algorithm due to Zhang [10]. The estimated radial distortion parameters are: $k_1 = -0.19612$, $k_2 = 0.17438$.

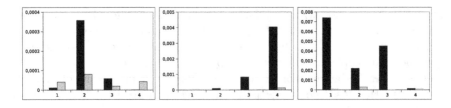

Fig. 4. Reconstruction errors for the setup for 12 different positions, where the prototype plane is nearly parallel to the camera plane (left). The middle and right diagram are associated with a little and big skewness between prototype plane and camera plane. Left columns: standard pose estimation, right columns: proposed method.

One can see that for almost all cases the proposed method leads to a rotation matrix and a translation vector such that the transformed prototype **P** is closer to the ground truth than the standard approach does. The values show that the reconstruction error is increasing if the angle between the z-axis and the camera plane is decreasing.

5 Conclusions

Our goal was to develop an efficient and robust pose estimation algorithm for a finite planar point set with known relative positions. The new method is based on the construction of collinear points, where a closed form solution can be formulated. This solution is a good initial value for an iterative method. We show results that our algorithm is robust and avoid the problems associated with local minima for iterative algorithms. This new algorithm should be relevant for many applications in computer vision.

References

1. Grimson, W.: Object Recognition by Computer. MIT Press, Cambridge (1990)
2. Malik, S., Roth, G., McDonald, C.: Robust 2D tracking for real-time augmented reality. In: Proc. Conf. Vision Interface, pp. 399–406 (2002)
3. Horaud, R., Conio, B., Leboulleux, O., Lacolle, B.: An analytic solution for the perspective 4-point problem. Computer Vision, Graphics and Image Processing 47, 34–44 (1989)
4. Haralick, R., Lee, C., Ottenberg, K., Nolle, M.: Analysis and solutions of the three point perspective pose estimation problem. In: Proc. IEEE Conf. Computer Vision and Pattern Recognition, pp. 592–598 (1991)
5. Lowe, D.: Three-dimensional object recognition from single two-dimensional image. Artifical Intelligence 31, 355–395 (1987)
6. Lu, C.P., Hager, G.D., Mjolsness, E.: Fast and globally convergent pose estimation from video images. IEEE Trans. Pattern Analysis and Machine Intelligence 22, 610–622 (2000)
7. Schweighofer, G., Pinz, A.: Robust pose estimation from a planar target. IEEE Trans. Pattern Analysis and Machine Intelligence 28, 2024–2030 (2006)
8. Oberkampf, D., DeMenthon, D., Davis, L.: Iterative pose estimation using coplanar feature points. Computer Vision and Image Understanding 63, 495–511 (1996)
9. Ansar, A., Daniilidis, K.: Linear pose estimation from points and lines. IEEE Transactions on Pattern Analysis and Machine Intelligence 25, 578–589 (2003)
10. Zhang, Z.: A Flexible new technique for camera calibration. Technical report, Microsoft Research, Technical Report MSR-TR-98-71 (1998)
11. Eberly, D.H.: 3D Game Engine Design. Morgan Kaufmann, San Francisco (2001)
12. Pisinger, G., Hanning, T.: Closed form monocular re-projective pose estimation. In: International Conference of Image Processing (ICIP), pp. V–197–200 (2007)
13. Golub, G.H., Loan, C.F.: Matrix Computations. John Hopkins UP, Baltimore (1989)
14. Faugeras, O.D.: Three-dimensional computer vision. a geometric viewpoint. MIT Press, Cambridge (1993)
15. Harris, C., Stephens, M.: A combined corner and edge detector. In: Proc. of the 4th Alvey Vision Conference, pp. 147–151 (1988)

Tracking and Retexturing Cloth for Real-Time Virtual Clothing Applications

Anna Hilsmann and Peter Eisert

Fraunhofer Heinrich Hertz Institute,
Einsteinufer 37, 10587 Berlin, Germany
{anna.hilsmann,peter.eisert}@hhi.fraunhofer.de
http://iphome.hhi.de/

Abstract. In this paper, we describe a dynamic texture overlay method from monocular images for real-time visualization of garments in a virtual mirror environment. Similar to looking into a mirror when trying on clothes, we create the same impression but for virtually textured garments. The mirror is replaced by a large display that shows the mirrored image of a camera capturing e.g. the upper body part of a person. By estimating the elastic deformations of the cloth from a single camera in the 2D image plane and recovering the illumination of the textured surface of a shirt in real time, an arbitrary virtual texture can be realistically augmented onto the moving garment such that the person seems to wear the virtual clothing. The result is a combination of the real video and the new augmented model yielding a realistic impression of the virtual piece of cloth.

Keywords: Cloth Tracking and Augmentation, Augmented Reality, Virtual Clothing.

1 Introduction

Merging computer generated objects with real video sequences is an important technique for many applications such as special effects in movies and augmented reality applications. In many cases this is time-consuming and processing is done off-line. Real-time applications would make processing easier and open new areas of application. We developed a method for real-time dynamic texture overlay for garments. A single camera captures a moving person in front of a display that shows the mirrored input image. Camera and display are mounted such that the user has the impression of looking into a mirror showing his upper part of the body. Color and texture of the moving garment are exchanged by a virtual one, such that the user seems to wear the virtual cloth in the mirror (see Figure 1). For that purpose we developed a dynamic retexturing method for garments that estimates elastic movements of a surface from monocular images and renders a virtual texture onto the moving garment with correct deformation and realistic illumination.

Deformable tracking and retexturing of surfaces from monocular sequences without 3D reconstruction of the surface encounters the problem of self-occlusions during tracking and realistic lighting during retexturing. We propose a real-time method based on optical-flow that can handle those problems. We address the important problem of

A. Gagalowicz and W. Philips (Eds.): MIRAGE 2009, LNCS 5496, pp. 94–105, 2009.

Fig. 1. The Virtual Mirror system replaces color and texture of a t-shirt in real time

self-occlusions during tracking of deformations in 2D using a two-dimensional motion model to regularize the optical flow field and define smoothing constraints locally according to the self-occlusion of a region. Occlusion estimates are established from shrinking regions in the mesh. In order to achieve realistic lighting, we exploit the fact that the original image contains all shadows and illumination to be rendered onto the virtual texture. Hence, shadows and wrinkles are realistically recovered from the original video frame using a simple impainting algorithm. This enhances the realistic appearance of the augmented garment. Additionally, we segment the shirt and can change the color of the non-textured part of the apparel.

One important characteristic of our system is that we achieve very realistic results in real-time with simple hardware, like a regular PC, a single camera and a display. The deformation estimation is very robust and correct illumination and shading enhance the realistic experience such that the user has the impression of actually wearing the augmented shirt.

The remainder of this paper is structured as follows. Section 2 briefly sums up the related work in garment tracking and retexturing and states the contribution of this paper. Section 3 describes our system as a whole. Section 4 explains the automatic detection and segmentation of the shirt. Section 5 describes the tracking approach for deformable surfaces in monocular sequences used for garment tracking. Section 6 describes the recovery of shading and illumination and the rendering approach. Section 7 presents experimental results.

2   Related Work and Contribution

Garment tracking and retexturing has been addressed lately by a number of researchers. Most researchers focused on multi-view systems [11], [4], [15], [19] for cloth tracking but little research has been done in real-time garment tracking from monocular sequences.

Pritchard and Heidrich [11] introduced a feature-based approach to cloth motion capturing with a calibrated stereo camera pair to acquire 3D. They recovered geometry from stereo correspondences using SIFT features. Scholz and Magnor [15] presented a method that uses optical flow to calculate three dimensional scene flow in a multi-view system. They use a cloth model that is matched to the surface, minimizing the deformation energy of the patch. Drift is countered by constraining the edge of the simulation to the silhouette of the real cloth. Guskov *et al.* [4] introduced a multi-view real-time system based on color-coded markers with a limited size of codewords. Their system needs small motions and the markers have to be quite large. Scholz *et al.* [16] improved upon Guskov *et al.* by using a color-code with more codewords using a pseudo random colored dots pattern. The algorithm identifies the exact position of colored ellipses on the cloth by examining their local neighborhood and makes use of a-priori knowledge on the surface connectivity. The three-dimensional geometry is reconstructed using a multi-view setup. Their approach achieves remarkable results and can cope with fast motion. White *et al.* [21] presented an extension to the approach based on a stereo-setup to reconstruct a random pattern of colored triangles printed on a cloth. Hasler *et al.* [5] used an analysis-by-synthesis approach and they, too, relay on markers in the form of the patterned cloth. Recently, Bradley *et al.* [1] introduced a marker-free approach to capturing garments from sixteen views. However, their method needs some user-assisted key-framing.

All these multi-view methods yield impressive results in 3D reconstruction. However, they need sophisticated acquisition setups and have a high computational cost. We strive for a different goal, i.e. to create an augmented reality in real-time with as little hardware as possible. For that purpose we are interested in deformable tracking from single-view video. Little research has been done in monocular garment tracking and augmentation. One approach in tracking deformable surfaces in monocular sequences is to formulate deformations in 3D [13], [18]. However, recovering 3D position and deformation from monocular surfaces is an ill-posed problem. Another approach is to make use of deformation models in 2D and track the elastic deformations in the image plane. In this case special care has to be taken considering self-occlusion e.g. if a surface is bent or deformed in a way that parts of it occlude other parts or due to the image projection. Furthermore, to retexture the surface shadows and illumination have to be recovered from single images.

Scholz and Magnor [14] modified their approach for multi-view setups using the color-code for garment tracking in monocular sequences. To recover shading maps they interpolate the color coded pattern using thin-plate splines. However, their system needs user interaction and the computation time is quite high.

White and Forsyth [20] presented a method for retexturing non-rigid objects from a single viewpoint without reconstructing 3D geometry using color markers. They limited their method to recover irradiance to screen printing techniques with a finite number of colors. Pilet *et al.* [9] proposed a feature-based method for deformable object detection and tracking in monocular image sequences that uses a wide baseline matching algorithm to establish correspondences. They continued their work by taking shadows and illumination into account by assuming Lambertian illumination in the reference frame [10]. Gay-Bellile *et al.* [3] proposed a direct non-rigid registration method and address the problem of self-occlusions by detecting them as shrinking areas in the 2D warp. The warp is forced to shrink by penalizing a variation in the sign of the partial derivatives of the warp along some directions. A binary decision excludes self-occluded pixels from consideration in the error function.

Some researchers have also addressed virtual clothing scenarios [12], [17], [2]. Saito and Hoshino [12] propose a technique to merge computer generated apparels with human video by 3D human pose estimation in off-line video sequences. Taguchi *et al.* [17] use two-dimensional plane models and the 2D shape of the garment to recover the 3D shape. Ehara and Saito [2] proposed a texture overlay method for virtual clothing systems. The deformation of the surface of the shirt was estimated from its silhouette in the input image based on a pre-collected database of a number of shapes of the same shirt. Their system needs homogeneous background and markers on the shirt. Unfortunately, they do not take illumination into account.

In [6] we presented an optical-flow based approach for single-view video using a mesh-based motion model together with smoothing constraints that are formulated locally on the mesh. Self-occlusions are taken into account by weighting the smoothness constraints locally according to the occlusion of a region. Thereby we force the mesh to shrink instead of fold in presence of self-occlusion.

The main contribution of this paper lies in a real-time tracking and retexturing method for elastically deforming surfaces in monocular image sequences. The method uses direct image information instead of features to track the deformation in the image plane. The markerless tracking approach can be applied to any surface that is rich enough in detail to exploit optical flow constraints if the initial state of the surface is known. Therefore, we combine this method with a simple detection method for initialization that is based only on a-priori knowledge of the color of the shirt and the knowledge that there is a textured region on the shirt. Additionally, we recover illumination and shadows from the original image. The new virtual texture is deformed in real-time in the 2D image plane and relighted with the recovered shading map. The result is a real-time system with little hardware that exchanges the color and texture of the shirt very realistically while the person in front of the system can move freely.

3   System Overview

The system consists of a regular PC, a single camera and a display that shows a combination of real video and computer generated content. The XGA-firewire camera is mounted on top of the display and captures the upper part of the user's body. The placement of the display and the viewing direction of the camera are chosen such that an average sized person sees about the same as he/she would expect when looking in a real mirror located at the same position as the display. Lights are mounted on either side of the display to be more independent from surrounding lighting conditions. As user interface we use a touchscreen on which the user can modify the augmentation result and choose between different colors and textures. In the test scenarios the user wears a green shirt with a rectangular drawing from Picasso as textured region (see Figure 2) to be more independent from the background. Our system does not require that specific pattern. The only assumption we make is that the shirt is green and there is a highly textured rectangular region on the shirt. Other suitable assumptions are possible. Once the line pattern is detected all its movements and deformations are tracked and replaced by any virtual texture while the person can move freely in front of the display. Additionally, we recover illumination and shadows from the original image using a simple impainting method.

The algorithm consists of several parts that all run on a single PC in real-time. Basically, we distinguish two modes. In the Detection Mode the shirt is detected and

Fig. 2. Work Flow: Input image, mesh used for tracking, mirrored augmented output images (left to right). The texture on the shirt is exchanged with a new one and the shirt is re-colored.

the mesh that is used for tracking is initialized. Once the shirt is detected the system switches into the Tracking Mode. This mode consists of several parts, that will be described in the following sections: Shirt segmentation and image preprocessing (Section 4), deformable surface tracking (Section 5) and illumination recovery and rendering (Section 6).

4   Shirt Segmentation and Image Preprocessing

During the Detection Mode the algorithm searches for the shirt in the image using the a-priori knowledge of the color of the shirt and the assumption that it contains a rectangular highly textured region. Additionally, in this mode the mesh that is used as motion model to estimate the deformation of the texture in the 2D image plane (see Section 5) is initialized. During the Tracking Mode we segment the shirt region in order to give it a new color. We assume that if the shirt is in the image, it is the largest area of that color and all parts of the shirt are connected. With these assumptions we can use a very simple but efficient approach to detect and segment the shirt in the image that is robust against illumination changes and changing background.

First, we transform all pixels in the image into the normalized RGB-colorspace which is more invariant to changes of surface orientation relatively to the light source. We find the largest blob by evaluating the normalized pixel values and fill all holes in that region. Using this very general method for segmentation ensures that all parts of the shirt are segmented, including dark shadows or highlighted regions. During the Tracking Mode we use this segmentation mask to recolor the untextured part of the shirt.

In the Detection Mode we proceed further by searching for a highly textured region in the detected green region to initialize the mesh used for tracking. We find and sample the contours of the texture defining the positions of the border mesh vertices and interpolate all inner vertex position points (see Figure 2). By this method we assure that if the texture is already deformed in the model frame, we get an idea about the deformation at the beginning knowing the rectangular shape of the undeformed texture. The mesh and the texture in this frame are used as model in the optical-flow-based tracking method (see Section 5).

5 Deformable Surface Tracking

Once the shirt has been detected and the mesh has been initialized we track the textured region of the shirt in an optical-flow based approach. We exploit the optical flow constraint equation [8] and regularize the optical flow field with a predefined mesh-based motion model. The best transformation can then be determined by minimizing a quadratic error:

$$E = \sum_{i=1}^{n} \left(\nabla I\left(x_i, y_i\right) \cdot \mathbf{d}\left(x_i, y_i\right) + \frac{\partial I}{\partial t}\left(x_i, y_i\right) \right)^2 \tag{1}$$

where $\nabla I\left(x_i, y_i\right)$ denotes the spatial derivatives of the image I at pixel position $[x_i, y_i]^T$ and $\frac{\partial I}{\partial t}\left(x_i, y_i\right)$ denotes the temporal gradient between two images. $\mathbf{d}\left(x_i, y_i\right)$ denotes the displacement vector at position $[x_i, y_i]^T$ and is defined by the motion model described in Section 5.1. n is the number of pixels selected for contribution to the error function, i.e. pixels where the gradient is non-zero.

The optical flow equation has some requirements that have to be fulfilled. First, it is valid only for small displacements between two successive frames because it is derived assuming the image intensity to be linear. To account for larger displacements we use a hierarchical framework.

Second, the optical flow equation assumes uniform Lambertian illumination. Illumination changes are taken into account by using highpass-filtered images. By that, high frequencies like sharp edges used by the optical flow algorithm are preserved whereas low frequencies like soft intensity changes due to illumination changes are filtered out and the difference between the two frame are mostly due to motion.

The aperture problem will be addressed by the motion model and the smoothing constraints in the following section.

Drift is countered by using the frame, in which the mesh was initialized, as model represented by the initial mesh. For each frame we estimate the elastic motion from the model frame, that is warped with the previous estimate of the previous frame, to the current frame. Using this model-to-frame tracking approach instead of a frame-to-frame tracking approach avoids error accumulation in the estimation and allows us to recover from small inaccuracies.

5.1 Motion Parameterization

The optical flow equation provides one equation for two unknowns, i.e. the displacement in $x-$ and $y-$ direction at each pixel position. To overcome this problem we need additional constraints on the pixel displacement. Therefore, we introduce a 2D triangulated regular mesh with K vertices $\mathbf{v_k}, (k = 1...K)$, as motion model to regularize the optical flow field. The position of each vertex $\mathbf{v_k}$ is given by its image coordinates $\mathbf{p_k} = [x_k, y_k]^T$. Each pixel $\mathbf{p_i} = [x_i, y_i]^T$ in the image can be represented by the barycentric coordinates of the enclosing three vertices.

$$[x_i, y_i]^T = \sum_{\substack{j=1 \\ \mathbf{v_j} \in \mathbf{v_k}}}^{3} B_{jt}\left(x_i, y_i\right) \cdot \mathbf{v_j}, \sum_{j=1}^{3} B_j\left(x_i, y_i\right) = 1, \ 0 \leq B_j \leq 1 \tag{2}$$

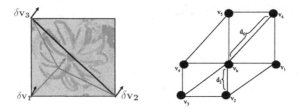

Fig. 3. Pixel displacement parameterization used as regularization of the optical flow field (left) and neighborhood of a vertex in the model (right)

where $B_j (x_i, y_i)$, $(j = 1, 2, 3)$ are the three barycentric coordinates of pixel $[x_i, y_i]^T$ and $\mathbf{v_j}$ are the three enclosing vertices. With a mesh deformation, $[x_i, y_i]^T$ is mapped onto $[x'_i, y'_i]^T$. Thus, we are looking for a deformation of the mesh, i.e. a displacement of each vertex $\mathbf{v_k}$ to $\mathbf{v'_k}$ such that the barycentric coordinates of $[x'_i, y'_i]^T$ are those of $[x_i, y_i]^T$. The deformation model then yields [7]:

$$\mathbf{d} (x_i, y_i) = \sum_{\substack{j=1 \\ t_j \in t_k}}^{3} B_j (x_i, y_i) \cdot \delta\mathbf{v_j} \tag{3}$$

where $\delta\mathbf{v_j}$ are the three vertex displacements of the enclosing triangle. Inserting the motion models into equation (1) leads to an overdetermined linear equation system that can be efficiently solved in a linear least-squares sense.

Additionally, we incorporate smoothing constraints for the vertex displacement field. This yields the following error functional to be minimized:

$$E = \sum_{i=1}^{n} \left(\nabla I (x_i, y_i) \cdot \mathbf{d} (x_i, y_i) + \frac{\partial I}{\partial t} (x_i, y_i) \right)^2 + \lambda \sum_{k=1}^{K} w_k E_s (\delta\mathbf{v_k}) \tag{4}$$

where K is the number of vertices and λ is the regularization parameter. $E_s(\delta\mathbf{v_k})$ is a local smoothing function for the displacement $\delta\mathbf{v_k}$ of a vertex $\mathbf{v_k}$ between two successive frames weighted by w_k. We choose $E_s(\delta\mathbf{v_k})$ to be

$$E_s(\delta\mathbf{v_k}) = \left(\delta\mathbf{v_k} - \frac{1}{\sum_{n \in N_k} \frac{1}{d_n}} \sum_{n \in N_k} \frac{1}{d_n} \cdot \delta\mathbf{v_n} \right)^2 \tag{5}$$

where N_k denotes the neighborhood of vertex $\mathbf{v_k}$ and $\delta\mathbf{v_n}$ denote the neighboring vertex displacements. d_n denotes the distance between vertex $\mathbf{v_k}$ and the neighbor vertex $\mathbf{v_n}$. Hereby, nearer neighbors have a higher influence on the displacement of vertex $\mathbf{v_k}$ than neighbors with a larger distance. For example, in Figure 3 vertex $\mathbf{v_2}$ has a larger impact on the displacement of vertex $\mathbf{v_k}$ than vertex $\mathbf{v_6}$ due to its smaller distance to vertex $\mathbf{v_k}$. $E_s(\delta\mathbf{v_k})$ is a measure of vertex displacement deviation to the displacements of its neighbors. It regularizes the optical flow field. Especially, it is necessary if

Fig. 4. Occlusion handling. Original image, occlusion map, detail of the deformed mesh without and with occlusion handling (left to right). Without occlusion handling the 2D mesh folds at the occlusion boundary which leads to inaccuracies during tracking because the triangles at the occlusion boundary contain wrong texture regions. Dark colors in the occlusion map mark occluded regions.

a vertex displacement is completely unconstrained by the optical flow equation due to lack of image gradient in the surrounding triangles. The less a vertex displacement is constrained by the optical flow equation the more it is constrained by the smoothness constraint. Furthermore, defining it locally on the mesh allows us to weight it according to the occlusion of a region to overcome the problem of self-occlusion that usually appears during deformable tracking in 2D (see Section 5.2).

5.2   Handling Self-occlusions

2-dimensional deformation estimation in presence of self-occlusion is a very challenging problem. In the image plane self-occlusion can also appear in parts that are not occluded in 3D due to projection. Naturally, a 2D mesh folds under these conditions which causes inaccuracies during tracking because the mesh triangles contain wrong texture regions (see Figure 4). This is an undefined status in 2D because a texture point cannot uniquely be assigned to one triangle in the mesh. Especially, when the surface is unfolded again the mesh is not able to unfold having reached this undefined status in 2D. We addressed this problem in our previous work [6] but repeat it briefly here because it is essential for the robustness of the presented real-time scenario. One solution to the problem of self-occlusion is to force the mesh to shrink instead of fold in occluded regions [3] because the mesh tracks the visible surface in the 2D image projection plane. We account for the problem of self-occlusions by weighting the smoothness constraints locally according to the occlusion of a region. In Section 5.1 we defined the smoothing constraints locally on the mesh. This allows us to weight them according to the occlusion of a region. This penalizes fold overs and forces the mesh to shrink instead of fold as this results in a smoother deformation field [6].

We estimate an occlusion map from shrinking areas in the deformed mesh (see Figure 4). For each vertex $\mathbf{v_k}$ in the mesh we calculate the average distance to its vertical and horizontal neighbors and scale it by the initial vertex distance in the reference mesh:

$$D_k = \frac{1}{2 |N_{vk}| D_v} \sum_{n \in N_{vk}} \|\mathbf{v_k} - \mathbf{v_n}\|_2 + \frac{1}{2 |N_{hk}| D_h} \sum_{n \in N_{hk}} \|\mathbf{v_k} - \mathbf{v_n}\|_2 \quad (6)$$

where N_{vk} and N_{hk} are the vertical and horizontal neighborhoods of vertex $\mathbf{v_k}$ and D_v and D_h denote the initial vertical and horizontal distances between two neighboring vertices in the regular reference mesh. By interpolating the average distances that present local mesh shrinking estimates over the entire surface we can establish an estimate of occluded regions in an occlusion map. The occlusion maps are used to adapt the weight w_k in equation (4) for vertex $\mathbf{v_k}$ to the degree of its occlusion. Vertices in occluded regions are assigned a higher weight to the smoothness constraint than vertices in non-occluded regions, i.e. the smaller the distance of a vertex to its neighbors the more its displacement is constrained by the surrounding displacements. Hereby, we counter foldings because vertices in mesh shrinking regions are forced to behave like their neighbors. Additionally, when unfolding the surface the increased smoothing weight in occluded regions causes the shrunk (i.e. occluded) region of the mesh to be stretched by the vertices at the occlusion boundary whose displacements are constrained by the optical flow equation. We choose w_k to be

$$w_k \propto \frac{1}{D'_k} \tag{7}$$

where D'_k equals D_k after the vector of all D_k is normalized so that the maximum of all D_k is one and zero-values have been set to a value ϵ close to zero. Hereby, we do not adapt the weight to the smoothness constraint if the mesh expands or shrinks uniformly, e.g. due to a movement toward or away from the camera.

6   Illumination Recovery and Rendering

As we use a monocular video sequence without 3D reconstruction of the surface an estimation of shading and illumination is needed to enhance the augmented reality experience. For the real-time scenario we use a green t-shirt with a texture that consists of black curves. This allows us to establish a shading map in a very simple but efficient way by interpolating the texture pixels from neighboring pixels while preserving main wrinkles and fold overs. Here, we use the fact that the real-world illumination and shadows that should be cast onto the virtual texture are already visible in the original image. That means, the input image exhibits the exact shadows to be rendered onto the virtual texture in the image intensities. The idea is to determine a shading map from the intensity of the original image after the texture lines have been removed. As we know the position of the texture from the tracking process we can easily identify the pixels belonging to the line pattern. We identify the texture pixels as pixels that differ from the mean color of the pixel neighborhood. The intensities of these pixels are then iteratively interpolated from neighboring pixels resulting in a smooth shading map that preserves shadows at main wrinkles and fold overs. Hereby, we preserve smooth intensity changes that result from illumination and shading and filter out the sharp edges of the texture. Figure 5 shows an example where the intensities are represented as height fields. The left height field represents the intensities of the original image. The right height field illustrates the intensities after the texture pixels have been interpolated from neighboring pixels iteratively. It shows a smooth intensity field that still exhibits smooth intensity changes at wrinkles and fold overs. Finally, the texture is rendered into the real scene using an OpenGL shader where the intensity of the filtered image is used to modulate the RGB/A texture.

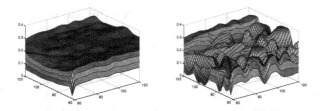

Fig. 5. Recovering shading and illumination. Representation of intensities as height fields before (left) and after (right) interpolation. The result yields a smooth intensity field and preserves smooth intensity changes at main folds and wrinkles.

7   Results and Application

Our system has already been tested in our lab and in public at an exhibition. In the environment at the exhibition hall the system was tested under natural conditions like lighting changes, changing background and users unfamiliar with the system. We use

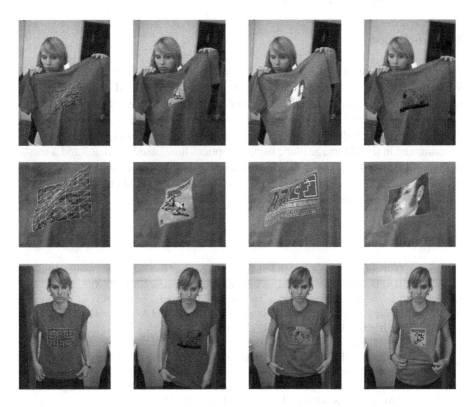

Fig. 6. Real-time cloth tracking: Mesh on the moving garment (left) and virtually augmented textures and colors. The addition of real lighting increases the perception that the cloth is truly exhibiting the virtual texture.

a XGA-Firewire camera with a resolution of 1024×768 and process 25 frames per second. The system is robust against lighting changes and changing backgrounds. Users can move freely in front of the system and perform elastic deformations with the cloth, i.e. stretching and bending it. Rotation is possible as long as the texture on the shirt is visible. Figure 2 shows the work flow of the deformation estimation. The first image shows the input image of the camera. The second image depicts the position of the mesh and the last two images show tracking and retexturing results. The method is able to recover even strong deformations and incorporating the shading maps enhances the realistic impression of the augmented texture.

Figure 6 shows tracking and retexturing results of different deformations in the image plane. These examples demonstrate that although the elastic deformation is estimated in the 2D image plane the result is a three-dimensional impression. The first two rows of Figure 6 demonstrate realistic augmentation results under self-occlusion. The addition of realistic lighting increases the perception of spatial relations between real and virtual objects. This is even more visible in the second row of Figure 6 that shows a few close ups of the virtually augmented textures demonstrating the correct deformation and illumination of these textures. Note that in Figure 2 and Figure 6 also the segmentation result is visible from the re-colored shirt.

8   Conclusion

We presented a robust real-time deformable surface tracking and retexturing approach for monocular image sequences that is incorporated into a Virtual Mirror setup. The system exchanges the color and the texture of a shirt while the person wearing the shirt can move freely in front of the mirror and even perform elastic deformations of the cloth like stretching and bending or move toward or away from the camera. The deformations are estimated in the 2D image plane using an optical-flow based approach and a mesh-based motion model. By taking illumination and shadows into account we achieve very realistic augmentation results. Besides these results, one important characteristic of our system is that it uses very simple hardware.

References

1. Bradley, D., Popa, T., Sheffer, A., Heidrich, W., Boubekeur, T.: Markerless Garment Capture. In: SIGGRAPH, ACM Transactions on Graphics, Los Angeles, USA (2008)
2. Ehara, J., Saito, H.: Texture Overlay for Virtual Clothing Based on PCA of Silhouettes. In: 5th IEEE and ACM International Symposium on Mixed and Augmented Reality, ISMAR 2006, Santa Barbara, USA, pp. 139–142 (2006)
3. Gay-Bellile, V., Bartoli, A., Sayd, P.: Deformable Surface Augmentation in Spite of Self-Occlusions. In: 6th IEEE and ACM International Symposium on Mixed and Augmented Reality, ISMAR 2007, Nara, Japan (2007)
4. Guskov, I., Klibanov, S., Bryant, B.: Trackable Surfaces. In: Proc. ACM/EG Symposium on Computer Animation, San Diego, USA, pp. 251–257 (2003)
5. Hasler, N., Asbach, M., Rosenhahn, B., Ohm, J.-R., Seidel, H.-P.: Physically Based Tracking of Cloth. In: Proc. of the International Workshop on Vision, Modeling, and Visualization, VMV 2006, Aachen, Germany, pp. 49–56 (2006)

6. Hilsmann, A., Eisert, P.: Tracking Deformable Surfaces with Optical Flow in the Presence of Self Occlusions in Monocular Image Sequences. In: CVPR Workshop on Non-Rigid Shape Analysis and Deformable Image Alignment, Anchorage, USA (2008)

7. Hilsmann, A., Eisert, P.: Deformable Object Tracking Using Optical Flow Constraints. In: Proc. 4th Int. Conf. on Visual Media Production CVMP 2007, London, UK (2007)

8. Horn, B.K.P., Schunck, B.G.: Determining Optical Flow, Technical report,Massachusetts Institute of Technology, Cambridge, MA, USA (1980)

9. Pilet, J., Lepetit, V., Fua, P.: Real-Time Non-Rigid Surface Detection. In: Proc. Conference on Computer Vision and Pattern Recognition, CVPR 2005, San Diego, USA (2005)

10. Pilet, P., Lepetit, V., Fua, P.: Fast Non-Rigid Surface Detection, Registration and Realistic Augmentation. Int. Journal of Computer Vision 96(2), 109–122 (2008)

11. Pritchard, D., Heidrich, W.: Cloth Motion Capture. In: Proc. Eurographics (Computer Graphics Forum), Granada, Spain, pp. 263–271 (2003)

12. Saito, H., Hoshino, J.: A Match Moving Technique for Merging CG and Human Video Sequences. In: Proc. of the Acoustics, Speech, and Signal Processing, ICASSP 2001, Salt Lake City, USA, pp. 1589–1592 (2001)

13. Salzmann, M., Pilet, J., Ilic, S., Fua, P.: Surface Deformation Models for Non-Rigid 3–D Shape Recovery. IEEE Trans. on Pattern Analysis and Machine Intelligence 29(8), 1481–1487 (2007)

14. Scholz, V., Magnor, M.: Texture Replacement of Garments in Monocular Video Sequences. In: Proc. of Eurographics Symposium on Rendering, Nicosia, Cyprus, pp. 305–312 (2006)

15. Scholz, V., Magnor, M.: Cloth Motion from Optical Flow. In: Proc. Vision, Modeling and Visualization, VMV 2004, Stanford, USA (2004)

16. Scholz, V., Stich, T., Keckeisen, M., Wacker, M., Magnor, M.: Garment Motion Capture Using Color-Coded Patterns. In: Proc. Eurographics (Computer Graphics Forum), Dublin, Ireland, pp. 439–448 (2005)

17. Taguchi, A., Aoki, T., Yasuda, H.: A Study on Real-time Virtual Clothing System Based on Two-Dimensional Plane Model. In: Proc. 6th Asia-Pacific Symposium on Information and Telecommunication Technologies, APSITT 2005 (2005)

18. Torresani, L., Yang, D., Alexander, E., Bregler, C.: Tracking and Modeling Non-Rigid Objects with Rank Constraints. In: Proc. Computer Vision and Pattern Recognition CVPR 2001, Hawaii, USA (2001)

19. White, R., Crane, K., Forsyth, D.: Capturing and Animating Occluded Cloth. In: SIGGRAPH, ACM Transactions on Graphics, San Diego, USA (2007)

20. White, R., Forsyth, D.: Retexturing Single Views Using Texture and Shading. In: Leonardis, A., Bischof, H., Pinz, A. (eds.) ECCV 2006. LNCS, vol. 3954, pp. 70–81. Springer, Heidelberg (2006)

21. White, R., Forsyth, D., Vasanth, J.: Capturing Real Folds in Cloth, Technical report, EECS Department, University of California, Berkley (2006)

A Novel Approach to Spatio-Temporal Video Analysis and Retrieval

Sameer Singh[1], Wei Ren[2], and Maneesha Singh[1]

[1] Research School of Informatics, Loughborough University, Loughborough
LE11 3TU, United Kingdom
[2] Faulty of Computer and Information Engineering, ShenZhen Graduate School of Peking
University, ShenZhen 518055, China
s.singh@lboro.ac.uk, w.ren@123.com, m.singh@lboro.ac.uk

Abstract. In this paper, we propose a novel Spatio-Temporal Analysis and Retrieval model to extract attributes for video category classification. First, the spatial relationships and temporal nature of the video object in a frame is coded as the sequence of binary string –VRstring. Then, the similarity between shots is matched as sequential features in hyperspaces. The results show that VRstring allows us to define higher level semantic features capturing the main narrative structures of the video. We also compare our algorithm with state of the art longest common substring finding video retrieval model by Adjeroh et.al.[1] on the Minerva international video benchmark.

1 Introduction

Advances in video compression technology have brought new perspectives as well as new aspirations. With decreasing cost of capture and storage devices, many new digital video applications and services are increasingly emerging such as digital TV, digital video conferencing, mobile TV, interactive multimedia, video surveillance, intelligent traffic control etc. The size of multimedia data collection has drastically expanded, not only for the professional and commercial content providers' repositories but also for the personal archives. Accessing and manipulating the information with such a tremendously large amount of data has become a challenging and timely issue. There are various increasing real requirements for the effective techniques for indexing large video achieves. The key to these techniques involve the similarity measure among the videos. The similarity comparison based on low-level visual information has been intensively addressed, while the similarity based on spatio-temporal information remains a difficult problem that has not yet been fully explored.

Video retrieval is the task of finding a set of most similar videos from a database to a query. We can classify existing video similarity comparison approaches into two types:

(a) *Feature based similarity*: In such systems, each video frame (or the key-frames only) is analysed based on those low-level features in terms of colour, texture, shape and motion features. Retrieval techniques work on indexing video by treating video

A. Gagalowicz and W. Philips (Eds.): MIRAGE 2009, LNCS 5496, pp. 106–115, 2009.

sequences as collections of still images, extracting relevant key-frames, and comparing their low-level features. Image matching methods are used to compare the scenes of the videos. The match between two videos becomes as the comparison of two images. In the past decade, a large number of researches including most video retrieval systems well established focus on such approaches. The examples include QBIC [1], Virage [2], VideoQ [3], WebSEEk [4] and Video Scout [5]. There is a big issue in here. A key characteristic of video data is its associated spatial and temporal information that delivers semantically coherent narrative. The temporal ordering of frames in the shot cannot be left out. The better solutions are needed that match all frames across videos. However, a relatively few algorithms for general-purpose video retrieval that consider temporal constraints and use sequence matching have been investigated. Kim and Park [6] evaluate the similarity between video sequences by employing the directed divergence on Y, U and V colour histogram. Adjeroh *et al.* [7] quantize and map low-level features of each of frames into symbols. The sequence of frames is transformed as a string - *v*strings. Then they used edit distance to measure the dissimilarity between videos by sequence-to-sequence matching.

(b) *Region based similarity, i.e. video object similarity.* VOs (Video objects) are formally introduced in international video coding standard MPEG4 and Video description standard MPEG7. VOs activities in a video sequence can provide an accurate cue to scene description and content-based semantic interpretation. It is realised as powerful tooling for video analysis and retrieval. However, current researches only investigate VO motion trajectory such as [8], [9],[10], and Netra-V [11]. The evolution of object motions represent VO trajectory, while evolution of spatial constraints among VOs can be further interpreted as high-level semantic events and delivers story. However, it is not addressed yet.

In this paper, a novel model STVR is proposed to retrieve video by fusing spatio-temporal information. Spatio-temporal information fusion can be expressed as the temporal evolution of an object which changes its position and/or extent over time, and the evolution of spatial relations among objects over time. To facilitate subsequent user-centric browsing, searching and retrieval, more important is, our focus has been shifted from designing sophisticated low-level feature based system to attempting toward filling the 'semantic gap' from the visual descriptions of videos to the high-level semantic concepts. We also compare the retrieval results with the state-of-the-art Adjeroh's video retrieval model [1].

This paper is organized as follows: The next section describes a model of spatio-temporal reasoning –*VR*string. Section 3 details our STVR model. Section 4 details Adjeroh's[1] model and section 5 demonstrates the experimental results and finally in section 6 we conclude the findings.

2 Feature Description of STVR Model - *VR*string

Spatio-temporal modelling in video retrieval is a crucial step for deciphering semantic information. This information can be embedded in the video representation and used for similarity computation between query and database videos. However, how to effectively model and represent spatio-temporal information is not straightforward. In this paper a model is presented to capture the spatial and temporal information, which integrates and formats them as binary string *VRstring* for video retrieval. Our model first identifies the physical structure of the video and then partition the video into physically meaningful units (shots). This is followed by modelling the spatial relationships among objects in each frame. A final step analyses the temporal evolution of spatial relationships among objects over temporal intervals in each shot as well as in the whole video sequence. The model is detailed as follows.

2.1 Temporal Model Using Machine Leaning

We believe that such temporal information is of vital importance for video retrieval. In order to obtain this information we develop an automated machine-learning based predictive system [12].

We extract a set of colour, texture, shape, motion and statistical features by comparing difference between frame pairs to train a system that can predict transitions on unseen test data. Machine learning methods such as neural network and k-NN (Nearest Neighbour) are used to automatically detect video transition between video scenes such as cut, fade-in, fade-out, dissolve; at same time, camera motion: pan-left, pan-right, tilt-up, tilt-down, camera-still are captured and analysed. By estimating camera movement and acquiring local object motion, we are able to track the trajectories of multiple objects and automatically label the corresponding regions.

2.2 Spatial Model Using Spatial Reasoning

Our spatial model addresses the handling of video object-to-object relationships that are crucial in video content description. In this case, which objects (VOs) present in the video are first identified (Detailed in [13]). In most scenarios, including our study, developing a completely automated classifier for recognising complex objects in general-purpose videos is impossible, since it is short of prior knowledge to build models for machine learning. Thus, we use a semi-automatic object labelling scheme. First, an automated video segmentation identification scheme is applied to extract key-frames. Each key-frame then is first segmented and each object is manually labelled. An automated object tracking scheme thereafter transfers the labels across frames between two key-frames by using spatial attributions such as colour, texture, shape, motion, and temporal hierarchical attributes among video objects.

A total of eight topological relations are designed to use between two object pairs (p, q) with eight bits *{Up, Down, Left, Right, Touch, Front, Contain, Overlap}*, which correspond to eight binary values. The value is '1' if the relation holds and '0' otherwise.

For each frame, a concatenated string can be calculated that encodes the spatial, temporal and object information in that frame. A given frame in video is represented by a *VR*string feature vector g . We can transform the sequence of frames in a video clip into *VRstring*s using these string vectors.

Given that we have a total of N image objects within the video, the number of possible spatial relationships between all object pairs is $4N(N-1)$. The length of g is set to $4N(N-1)+11+K$, where the number 11 are obtained from the temporal model predication for test data. And the number K of values are used to weight objects turning up as object part bits. Therefore, $g = (b_1, b_2, ..., b_t)$, where each binary value can be 0 or 1. In a given frame, not all objects appear simultaneously, and thus the multidimensional representation of g for all frames is fairly sparse.

3 Video Retrieval Based on STVR

A spatio-temporal model, more importantly, should suggest a practical solution for effective indexing and comparison. The following we detail how to use our STVR model for video indexing.

3.1 STVR Video Retrieval Algorithm

Task: To find the closest video to query video V_q from a database video set.

1) Given a query video V_q, and database videos $V_{d_1}, ..., V_{d_N}$.

2) Segment the videos into scenes (shots), so that:

V_q : Scenes are: $S_1, S_2, ... S_m$

V_{d_1} : Scenes are: $T_{d_11}, T_{d_12}, ... T_{d_1 n_1}$

and so on till V_{d_N} Scenes are: $T_{d_N 1}, T_{d_N 2}, ..., T_{d_N n_N}$

3) If we wish to find similarity between videos V_q and V_{d_j}, we need to find the similarity between their shot-pairs and match shots between two videos. Calculate each shot pair similarity $\vartheta_{S_i T_k}$ by MCF, $\forall \ 1 \leq i \leq m, \ 1 \leq k \leq n_j$ and $1 \leq j \leq N$. It will be detailed in section 3.2.

4) Calculate the similarity between video pairs (V_q, V_{d_1}), (V_q, V_{d_2}), ..., (V_q, V_{d_N}). It is taken two steps as follows.

Step 1. If the similarity between shot pair $(S_{q_i}, T_{d_j k})$, is given by $\vartheta_{ik}(S_q)$, then the spatio-temporal similarity between video pair (V_q, V_{d_j}) can be further computed by MCF based on considering their shot sets as bipartite network. Let us denote this distance as. $\vartheta_{V_q V_j}(V_q)$

Step 2. The overall similarity $\Psi(V_q, V_{d_j})$ between videos V_q and V_{d_j}, which should includes spatio-temporal part and object part similarities, is given by:

$$\Psi(V_q, V_{d_j}) = \omega_1(X_v) + \omega_2(1 - Z_v) \tag{1}$$

where ω_1 and ω_2 are weights. X_v is defined as the overall spatio-temporal similarity between videos over all shot pairs.

$$X_v = \vartheta_{V_q V_j}(V_q)$$

Z_V is the average object dissimilarity between videos. Suppose that video and, V_q has $(n_a + n_b)$ objects and video V_{d_j} has $(n_a + n_c)$ objects. n_a is the number of common objects in the two videos. Object dissimilarity Z_v between two videos can be calculated as follows.

$$Z_v = \frac{n_b + n_c}{2 * n_a + n_b + n_c} \tag{2}$$

3.2 Shot Similarity Algorithm

Each shot in a clip is considered as a collection of discrete feature points. Hence, the similarity between shots is further characterised as a distance between discrete feature point sets in a hyperspace. The shot similarity measure problem can be formalised as bipartite network problem and be solved by finding Minimum Cost Flow (MCF) between the networks. The MCF problem is to find a cheapest possible way to send a certain amount of shipment through a flow network. It is most commonly used to dispose the various type management problems. It is introduced to video retrieval here. STVR similarity takes these two matrices (or two point sets) as input and generates the similarity ϑ_{ST} as output. The ϑ_{ST} is calculated by first discussing the nature of MCF.

Consider shot $\vec{S}$ and $\vec{T}$ as two observations, containing a total of N_1 and N_2 finite point sets (or observations), respectively.

$$\vec{S} = \left\{ (\vec{x}_1, \rho_1), (\vec{x}_2, \rho_2), \dots (\vec{x}_i, \rho_i), \dots, (\vec{x}_m, \rho_m) \right\}$$
$$\vec{T} = \left\{ (\vec{y}_1, \lambda_1), (\vec{y}_2, \lambda_2), \dots, (\vec{y}_j, \lambda_j), \dots, (\vec{y}_n, \lambda_n) \right\}$$

The observation $\vec{S}$ is assumed to be a set of sources, associated with an amount of supply or weight ρ_i of source $\vec{x}_i$, $\forall\ i = 1, \dots, m$. The $\vec{T}$ observation is assumed to be a set of destinations with the total capacity or a weight λ_j of destination $\vec{y}_j$, $\forall\ j = 1, 2, \dots, n$. While c_{ij} is the cost of shipping a unit of supply from $x_i \in \vec{X}$ to $y_j \in \vec{Y}$.

We want to find a set of flows f_{ij} that minimises the overall transportation cost:

$$\vartheta = \sum_{i}^{m} \sum_{j}^{n} c_{ij} f_{ij} \tag{3}$$

This is subject to the following constraints:

(1) $f_{ij} \geq 0$;

(2) $\sum_{j}^{n} f_{ij} \leq \rho_i$;

(3) $\sum_{i}^{m} f_{ij} \le \lambda_j$; and

(4) $\sum_{i}^{m} \sum_{j}^{n} f_{ij} = \min \left(\sum_{i}^{m} \rho_i, \sum_{j}^{n} \lambda_j \right)$.

To balance among different shipments, the overall transportation cost is normalized by the total flows:

$$\overline{\vartheta} = \frac{\sum_{i}^{m} \sum_{j}^{n} c_{ij} f_{ij}}{\min \left(\sum_{i}^{m} \rho_i, \sum_{j}^{n} \lambda_j \right)} \tag{4}$$

4 Adjeroh's Video Retrieval Model

Adjeroh et al. [1] video retrieval model uses 23 features which include 18 colour features extracted from Lab colour (Plataniotis and Venetsanopoulos, 2000) histogram with 18 bins ($2 \times 3 \times 3$) and 5 texture features extracted from co-occurrence matrix (Haralick et al., 1973). The co-occurrence matrix P is calculated for four orientations ($0°$, $45°$, $90°$ and $135°$). This is an $N_g \times N_g$ matrix (with N_g representing the number of distinct grey levels in the quantised image) and P_{ij} is defined as the joint probability of occurrence of a pair of neighbouring pixels, one with grey-level i and the other with grey-level j, occur in the image under the four specific angular relationships. From each matrix, five statistical features are determined (angular second moment, contrast, correlation, variance and entropy). The average values of these features over the four orientations are the final features.

The first step is to transform features into of symbols using quantisation. This involves the following:

Let Σ be the number of bins or number of alphabet symbols, f_v be a feature value, and max f_v and min f_v be the respective maximum and minimum values for a given index feature. The quantisation step size is given by

$$\Delta = \frac{\max f_v - \min f_v}{\Sigma}$$

The quantisation level to which a given f_v belongs is then obtained using:

$$q(f_v) = i \text{ if } (i-1) \cdot \Delta \le f_v < i \cdot \Delta ; \quad i = 1, 2, ..., \Sigma.$$

If a feature value belongs to the ith quantisation level, we assign the ith symbol to it. A given sample can now be represented as a string of these quantised symbols.

4.1 Video Retrieval Based on String Matching

As the result of quantisation, a video frame is represented by set of M symbols for M features. A query video and database videos are transferred to string representation.

Videos need to be matched as a two step process. We first match shots between two videos. Consider the problem of matching two videos: V_1, V_2

$$V_1 \quad \text{shots are: } S_1, S_2, ... S_{M_1} \text{ \&}$$

$$V_2 \quad \text{shots are: } T_1, T_2, T_3, ..., T_{M_2}$$

Video Level Similarity

The similarity between two videos, ignoring temporal information at video level, is given by

$$X_v = \frac{1}{Q} \sum_{s=1}^{Q} X_s \text{ , where } Q \text{ is the number of shot pairs. } Q = M_1 \text{ if } M_1 < M_2 \text{, or}$$

$Q = M_2$ if $M_2 < M_1$. X_s is the average shot level similarity across the shot pair s, where $1 \leq s \leq Q$.

Shot Level Similarity

How to find similarity u between two shots S and T ?

$$S = (x_1, x_2, ... x_i, x_{N_1}) ; T = (y_1, y_2, ..., y_j, ... y_{N_2})$$

where x and y are the key-frames in each scene, $N_1 \leq N_2$.

We can represent two shots of length N_1 and N_2 as matrices of size $N_1 \times M$ and $N_2 \times M$ for M features. Each column of these matrices is now compared for similarity using edit distance [1]. The average edit distance across all columns is calculated and used as X_s.

Given two strings A: $a_1 a_2 ... a_n$ and B: $b_1 b_2 ... b_m$, which represent a feature column of shot S and shot T, respectively, over an alphabet Σ and a set of allowed edit operations, the edit distance between A and B is calculated by the minimum number of edit operations needed to transform A into B. Three basic types of edit operations are used: *insertion, delete, and substitution.*

Category 1 News   Category 2 Wildlife

Category 3 City Tour   Category 4 Seaside

Fig. 1. Representative sample frames from videos of Minerva database

5 Experimental Results

The international video benchmark MINERVA is used for our analysis. MINERVA includes 250 videos with ten categories (See Fig.1): C1: News; C2: Wildlife; C3: City Tour; C4: Seaside; C5: Sport; C6: Garden; C7: Stage Performance; C8: Train Station; C9: Car-Boot Sale; and C10: Traffic.

24 low-level features are extracted for automatic video segmentation. A trained neural network is used to automatically classify video transition between different scenes and, at same time, to identify camera movements. Adaptive thresholds were used to automatically select key frames.

Our experiments employ a five fold cross validation strategy based on random sampling in each category to determine the true performance of the retrieval system. Validation data is used for optimising retrieval system parameters. Before sampling, data in each category must be randomised. We split our data for test, validation and training as partitions of size 20%, 20% and 60%, respectively. By using the validation set, we find the optimised weight scheme, which indicates video similarity metric should integrate 2/3 spatio-temporal information with the remaining 1/3 information from VOs presenting.

A total of 250 videos are used as query by using five fold cross-validation, and the percentages are based on how many videos matched from 25 per category are allocated to each of the 10 semantic classes. From the Table 1, we can find that the diagonal elements of STVR model are very high. The fact is that STVR does rather well on News and Sports videos but performs poorly on City Tour videos. From a retrieval perspective, we assume that a user querying with News Video for example should always get a News video as the best match. Finally, in Table 2 we find that the Adjeroh's model also makes substantial mistakes. In particular, the performance on Seaside at 4% is the worst one. All other categories have variable performances ranging between 24% to 76%. Considering these two models together, it appears that Car Boot videos are the most easy to retrieve whereas other categories are easier with SVTR model and highly variable with the baseline model. The other reason that the

Table 1. Confusion matrix for **STVR** best matching video (%)

Category	News	Wild-life	City Tour	Sea-side	Sports	Gard-en	Stage Perf.	Train Stn	Car boot	Traffic
News	100%	0	0	0	0	0	0	0	0	0
Wildlife	0	88%	0	0	4%	8%	0	0	0	0
City Tour	4%	0	80%	0	4%	4%	0	0	0	8%
Seaside	0	0	0	100%	0	0	0	0	0	0
Sports	0	4%	0	0	88%	0	0	0	0	8%
Garden	0	4%	0	0	0	92%	0	0	0	4%
Performance Stage	0	0	0	0	0	0	100%	0	0	0
Train Station	0	0	0	0	0	0	0	96%	0	4%
Car boot	0	0	0	0	0	0	0	0	100%	0
Traffic	0	0	0	0	0	0	0	0	0	100%

Adjeroh's model represents bad results is because the longest common substring is not good enough to measure temporal similarity. Therefore, their poor performances are actually not that bad because texture and colour features vary tremendously across videos and should only be useful for a certain type of video, e.g. Sports where most videos show a green coloured pitch. Purely visual information is obviously not sufficient as a basis for high-level semantic video scene categorisation.

Table 2. Confusion matrix for Adjeroh *et al.* [1] (baseline model) best matching video (%)

Category	News	Wild-life	City Tour	Sea-side	Sports	Gar-den	Stage Perf.	Train St.	Car boot	Traffic
News	44%	0	0	0	20%	4%	0	0	32%	0
Wildlife	8%	24%	16%	4%	12%	24%	0	0	0	12%
City Tour	16%	4%	60%	0	8%	4%	0	0	0	8%
Seaside	8%	16%	32%	4%	4%	4%	0	12%	0	20%
Sports	28%	0	0	0	64%	0	0	4%	4.%	0
Garden	8%	0	0	0	32%	56%	0	0	0	4%
Performance Stage	0	0	8%	0	4%	4%	56%	8%	20%	0
Train Station	16%	0	0	0	8%	0	0	36%	20%	20%
Car boot	0	0	0	0	12%	0	0	0	76%	12%
Traffic	8%	0	12%	4.0%	8%	0	0	0	0	68%

6   Conclusions

In the paper, we propose a novel spatio-temporal model STVR to view on temporal continuity of video events and actions into high-level semantic categories based on evolutions of VOs' spatial and temporal constraints. The fusion of the object relationships and their temporal natures in a video are formulised into the sequence of string *VR*string. Then, video matching problem is coded as bipartite network problem. The similarity between videos is computed as finding minimum cost flows between networks. The experimental result shows that our model is much superior to the state-of-the-art approaches based on longest common string.

References

1. Niblack, W., Zhu, X., Hafner, J.L., Bruel, T., Ponceleon, D.B., Petkovic, D., Flickner, M., Upfal, E., Nin, S.I., Sull, S., Dom, B.E.: Updates to the QBIC system. In: Proceedings of IS&T SPIE, Storage and Retrieval for Image and Video Databases VI, San Jose, vol. 3312, pp. 150–161 (1998)
2. Hampapur, A., Gupta, B., Horowitz, C.-F., Shu, C., Fuller, J.R., Bach, M., Gorkani, R.: Virage video engine. In: Proceedings of SPIE Storage and Retrieval for Image and Video Databases V, vol. 3022, pp. 188–198 (1997)
3. Chang, S.-F., Chen, W., Meng, H.J., Sundaram, H., Zhong, D.: A fully automated content-based video search engine supporting spatio-temporal queries. IEEE Transactions on Circuits and Systems for Video Technology 8(5), 602–615 (1998)

4. Smith, J.R., Chang, S.-F.: An image and video search engine for the world-wide web. In: Proceedings of Symposium on Electronic Imaging: Science and Technology - Storage & Retrieval for Image and Video Databases V, vol. 3022, pp. 84–95 (1997)
5. Jasinschi, R.S., et al.: Integrated multimedia processing for topic segmentation and classification. In: Proceedings of IEEE International Conference Image Processing (ICIP 2001). IEEE CS Press, Los Alamitos (2001)
6. Kim, S.H., Park, R.-H.: An efficient algorithm for video sequence matching using the modified Hausdorff distance and the directed divergence. IEEE Transactions on Circuits and Systems for Video Technology 12(7), 592–596 (2002)
7. Adjeroh, D.A., Lee, M.C., King, I.: A distance measure for video sequences. Computer Vision and Image Understanding 75(1/2), 25–45 (1999)
8. Hsieh, J.-W., Yu, S.-L., Chen, Y.-S.: Motion-based video retrieval by trajectory matching. IEEE Transactions on Circuits and Systems for Video Technology 16(3), 396–409 (2006)
9. Dao, M.-S., DeNatale, F.G.B., Massa, A.: Video retrieval using video object-trajectory and edge potential function. In: Proceedings of 2004 International Symposium on Intelligent Multimedia, Video and Speech Processing, October 20-22, pp. 454–457 (2004)
10. Lie, W.-N., Hsiao, W.-C.: Content-based video retrieval based on object motion trajectory. In: Proceeding of IEEE Workshop on Multimedia Signal Processing, December 9-11, pp. 237–240 (2002)
11. Deng, Y., Manjunath, B.S.: NeTra-V: Towards an object-based video representation. IEEE Transactions on Circuits and Systems for Video Technology 8, 616–627 (1998)
12. Ren, W., Singh, S.: Automatic video segmentation using machine learning. In: Yang, Z.R., Yin, H., Everson, R.M. (eds.) IDEAL 2004. LNCS, vol. 3177, pp. 285–292. Springer, Heidelberg (2004)
13. Ren, W., Singh, S.: An Automatic Video Annotation System. In: Proceedings of 3rd International Conference on Advances in Pattern Recognition (ICAPR), Bath, UK (August 2005)
14. Bimbo, D., Vicario, E., Zingoni, D.: Symbolic description and visual querying of image sequences using spatiotemporal logic. IEEE Transactions in Knowledge Data Engineering 7, 609–622 (1995)
15. Lee, S., Hsu, F.: Spatial reasoning and similarity retrieval of images using 2D C-string knowledge representation. Pattern Recognition 25, 305–318 (1992)

A *Bag of Words* Approach for 3D Object Categorization

Roberto Toldo, Umberto Castellani, and Andrea Fusiello

Dipartimento di Informatica, Università di Verona,
Strada Le Grazie 15, 37134 Verona, Italy
roberto.toldo@univr.it
umberto.castellani@univr.it
andrea.fusiello@univr.it

Abstract. In this paper we propose a novel framework for 3D object categorization. The object is modeled it in terms of its sub-parts as an histogram of 3D visual word occurrences. We introduce an effective method for hierarchical 3D object segmentation driven by the minima rule that combines spectral clustering – for the selection of seed-regions – with region growing based on fast marching. The front propagation is driven by local geometry features, namely the Shape Index. Finally, after the coding of each object according to the Bag-of-Words paradigm, a Support Vector Machine is learnt to classify different objects categories. Several examples on two different datasets are shown which evidence the effectiveness of the proposed framework.

1 Introduction

The availability of large collections of 3D models has increased the interest in content-based 3D search and retrieval [1–3]. Typical object retrieval systems require the user to define a query-model which output is a set of its most similar objects in the database. In general, such approach requires the comparison of the query-model with *all* the objects in the dataset according with a given matching criterion, after the coding of the object with respect to some indexing technique. Shape signatures [4] are commonly utilized as a fast indexing mechanism for shape retrieval.

In this paper we present a 3D object categorization method based on a *learning-by-example* approach [5]. Geometric features representing the query-model are fed into a Support Vector Machine (SVM) which, after a learning stage, is able to assign a *category* (or a *class*) to the query-model without an explicit comparison with all the models of the dataset. Our approach is inspired to the *Bag-of-Words* framework for textual document classification and retrieval. In this approach, a text is represented as an unordered collection of words, disregarding grammar and even word order.

The extension of such approach to non-textual data requires the building of a *visual vocabulary*, i.e., the set of all the visual analog of words. For example, in [6] images are encoded by collecting interest points which represent local salient

A. Gagalowicz and W. Philips (Eds.): MIRAGE 2009, LNCS 5496, pp. 116–127, 2009.
© Springer-Verlag Berlin Heidelberg 2009

regions. This approach has been extended in [7] by introducing the concept of *pyramid* kernel matching. Instead of building a fixed vocabulary, the visual words are organized in a hierarchical fashion in order to reduce the conditioning of the free parameter definition (i.e., the number of bins of the histogram). Recently, in [8] the Bag-of-Words paradigm has been introduced for human actions categorization from real movies. In this case, the visual words are the vector quantization of spatiotemporal local features. The extension to 3D objects have been proposed in few work [9, 10], to the best of our knowledge. In [9] range images are synthetically generated from the full 3D model, then salient points are extracted as for the 2D (intensity) images. In [10] Spin Images are chosen as local shape descriptors after a random samples of the mesh vertices.

In our approach a 3D visual vocabulary is defined by extracting and grouping the geometric features of the object sub-parts from the dataset, after a hierarchical 3D object segmentation. Thank to this *part-based* representation of the object we achieve pose invariance, i.e., insensitivity to transformation which change the skeletal articulations of the 3D object [11]. Moreover, our approach is able to discriminate objects with similar skeletons, a feature that is shared by very few other works [12]. Its main steps are:

Object segmentation (Sec. 2). Spectral clustering is used for the selection of seed-regions. Being inspired by the *minima-rule* [13], the adjacency matrix is defined purposely in order to allow convex regions to belong to the same segment. Furthermore, a multiple-region growing approach is introduced to expand the selected seed-regions. In particular, a weighted fast marching is proposed by guiding the front propagation according to local geometry properties. In practice, the main idea consist on reducing the speed of the front for concave areas which are more likely to belong to the region boundaries. Then, the hierarchical segmentation is recovered by combining recursively the seeds selection and the region-growing steps.

Object sub-parts description (Sec. 3). Local region signature are introduced to define a compact representation of each sub-part. Working at the part level, as opposed to the whole object level, enables a more flexible class representation and allows scenarios in which the query model is significantly transformed (e.g., deformed) to be classified. We focus on region signatures easy to compute and partially available from the previous step (see [4] for an exhaustive overview of shape descriptors).

3D visual vocabulary construction (Sec. 4). The set of region descriptors are properly clustered in order to obtain a fixed number of 3D visual *words* (i.e., the set of clusters centroids). In practice, the clustering defines a vector quantization of the whole region descriptor space. Note that the vocabulary should be large enough to distinguish relevant changes in image parts, but not so large as to distinguish irrelevant variations such as noise.

Object categorization by SVM (Sec. 5). Each 3D object is encoded by assigning to each object sub-part the corresponding visual word. Indeed, a Bag-of-Words representation is defined by counting the number of object sub-parts assigned to each word. In practice, a histogram of visual words

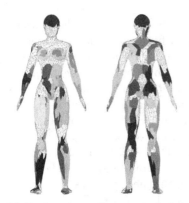
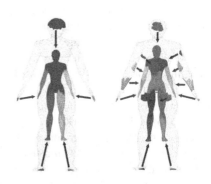

(a) Seed regions after spectral clustering.

(b) Segmentation at depth 2 (left) and depth 3 (right) of the hierarchy, with corresponding seed regions.

Fig. 1. Segmentation

occurrences is build for each 3D object which represent its *global* signature [6]. Then, a SVM is trained by adopting a learning by example approach. In particular, a suitable kernel function is defined in order to implicitly implement the sub-part matching.

2   Objects Segmentation

Due to its wide ranging applications, 3D object segmentation has received a great attention lately. The recent survey by [14] and the comparative study by [15] have thoroughly covered the several different approaches developed in literature.

In the following we present a novel mesh segmentation technique that provides a consistent segmentation of similar meshes complying with the cognitive *minima rule* [13]. In addition, the final segmentation is extracted in a hierarchical structure in order to improve the flexibility in modeling the object sub-parts.

The segmentation proceeds top-down: starting with a root node corresponding to the whole mesh, the segmentation is recursively created by partitioning the current leaf nodes into two or more child nodes. The minima rule states that human perception usually divides a surface into parts along the concave discontinuity of the tangent plane [13]. Therefore this suggest to cluster in the same set convex regions and to detect boundary parts as concave ones. A concise way to express the type of shape in terms of principal curvatures is given by the *Shape Index* (SI) [16].

$$SI = -\frac{2}{\pi} \arctan \left(\frac{k_1 + k_2}{k_1 - k_2} \right) \quad k_1 > k_2 \tag{1}$$

where k_1, k_2 are the principal curvatures of a generic vertex $x \in V$. The SI varies in $[-1, 1]$: a negative value corresponds to concavities, whereas a positive value represents a convex surface.

The key idea behind our algorithm is the synergy between two main phases: (i) the detection of similar connected convex regions as *seed*-region, and (ii) the expansion of these seed-regions using a multiple region growing approach. According to the minima-rule the SI is employed for both the phases.

2.1   Seed-Regions Detection by Spectral Clustering

The extraction of the seed-regions is accomplished with Normalized Graph Cuts [17]. It has been firstly applied to image segmentation although it is stated as a general clustering method on weighted graphs. In our case, the weight matrix is built using the SI at each vertex:

$$W(x_i, x_j) = e^{-|SI(x_i) - SI(x_j)|} \qquad (2)$$

where the vertices with negative SI – i.e., those corresponding to concave regions – have been previously discarded. In this way we cluster together vertices representing the same convex shape.

Final clusters are not guaranteed to be connected. This happens because we don't take into account any (geodesic) distance information at this stage. Hence, we impose connection as a post-processing step: the final seed regions are found as connected components in the mesh graph, with vertices belonging to the same cluster. An example of seed regions found by the algorithm is shown in Fig. 1(a).

2.2   Multiple Region Growing by Weighted Fast Marching

Once the overall seed regions are found we must establish a criteria to select the starting seed regions of each node of the hierarchical segmentation tree. For each tree node we consider only the seed regions that are contained in the parent segmentation. We firstly find the two farthest seed regions. We then add more regions until the distance from the regions already added is less than half the two farthest seed regions. As explained next, the distance between two regions can be efficiently calculated with the Fast Marching algorithm [18, 19]. In particular, when the seed regions of the current tree node are found, we expand them using a *weighted* geodesic distance. In formulae, given two vertices $x_0, x_1 \in V$, we define the *weighted geodesic distance* $d(x_0, x_1)$ as

$$d(x_0, x_1) = min_\gamma \left\{ \int_0^1 \|\gamma'\| w(\gamma(t)) dt \right\} \qquad (3)$$

where $w(\cdot) =$ is a weight function (if $w(\cdot) = 1$ this is the classic geodesic distance) and γ is a piecewise regular curve with $\gamma(0) = x_0$ and $\gamma(1) = x_1$. Our weight function is based on the Shape Index SI:

$$w(x) = e^{\alpha SI(x)} \qquad (4)$$

Fig. 2. Examples of segmentation extracted on several meshes of the Aim@Shape Dataset

where α is an arbitrary constant. An high α value heavily slow down the front propagation where the concavity are more prominent. In our categorization paradigm we used a fixed $\alpha = 5$ to obtain consistent segmentations.

An example segmentation along with starting seed regions is shown in Fig. 1(b). Several other examples of segmentation on different objects are shown in Fig. 2.

The overall hierarchical algorithm is summarized below:

Algorithm 1. Hierarchical clustering

---

1. Find all seed-regions S.
2. Initialize C as the entire mesh and place in the priority queue Q.
3. Get the current top cluster $C \in Q$ and remove it from Q.
4. Find starting regions $S_C \in S \bigcap C$.
   If the starting regions are more than one go to next step else go to step 6.
5. Find final cluster starting from S_C trough weighted geodesic distance and add them to Q.
   These are child cluster of C in the hierarchical tree.
6. If Q is empty stop, else go to step 3.

---

3   Segment Descriptors

We chose four type of descriptors to represent each extracted region. The first two are based on SI and *Curvedness* [16]. Both encode local surface geometry properties for each vertex. In particular, the SI allows the classification of the surface among peek, valley, saddle, and so on. The Curvedness CU instead, is a concise way to measure the size of a local patch:

$$CU = \frac{2}{\pi} \ln \sqrt{\frac{k_1^2 + k_2^2}{2}} \qquad (5)$$

The two descriptors SIH and CUH are defined as the normalized histograms of the observed SI and CU values in the region vertices, respectively.

The other two descriptors are normalized region histograms of vertex-distances derived directly from our segmentation algorithm. The idea is to describe the shape of a region in relationship with its starting seed. In practice, we compute the geodesic distance and the weighted geodesic distance of each vertex of a segment to its seed region. The point-to-seed-region distance is defined as the geodesic distance between the point itself and its closest point belonging to the seed region. The two descriptors GD and GDW are the normalized histograms of such distances (respectively) over the vertices of the segment.

Note that GD can be interpreted as an approximation of the eccentricity [20], and GDW, implicitly encodes also the local surface geometry information since the weight function depends on the SI, according to Eq. (4).

Note further, that the number of bins chosen for each histogram is a critical choice. A small number reduce the capability of the region descriptor in discriminating among different segments. On the other hand, a high number increases the noise conditioning. Hence we introduce, for each descriptor, histograms with different number of bins in order to obtain a *coarse-to-fine* regions representation.

4   3D Visual Vocabulary Construction

The different sets of region descriptors must be clustered in order to obtain several visual words. Since we start with a hierarchical segmentation and different types of descriptors, we adopted a multi-clustering approach rather than merging descriptors in a bigger set. Before the clusterization, the sets of descriptors are thus split in different subsets as illustrated in Fig. 3. The final clusters are obtained with a k-means algorithm. Again, instead of setting a fixed free parameter k, namely the number of cluster, we carry out different clusterizations while varying its value.

Once the different clusters are found we retain only their centroids, which are our *visual words*. In Fig. 4 an example of descriptors subset clusterization with relative distance from centroid is shown. Note that object sub-parts from different categories may fall in the same cluster since they share similar shape.

More in details, at the end of this phase we obtain the set of visual vocabularies $V_{l,s}^{d,b,c}$, where:

- l identifies the region level of the hierarchical 3D segmentation ($l \in \{2, 3\}$),
- s identifies the index of the multiple 3D segmentation (variable segmentation parameter $s \in \{4, 8, 12\}$),

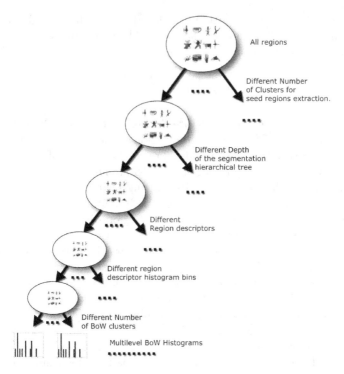

All regions

Different Number
of Clusters for
seed regions extraction.

Different Depth
of the segmentation
hierarchical tree

Different
Region descriptors

Different region
descriptor histogram bins

Different Number
of BoW clusters

Multilevel BoW Histograms

Fig. 3. The Vocabulary construction is performed in a multilevel way. At the beginning we have all region extracted for different numbers of seed regions (variable segmentation parameter). The regions are divided by the different segmentations and by the different depth of the segmentation tree. For every region, different descriptors are attached. The different region descriptors are divided by the type of descriptor and its number of bins. The final clusterizations are obtained with varying number of clusters. At the end of the process we obtain different Bag-of-Words histograms for each mesh.

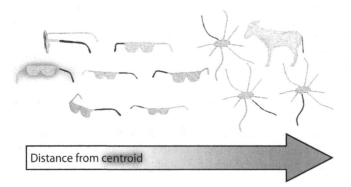

Distance from centroid

Fig. 4. Example of a Bag-of-Words cluster for SI descriptors. The centroid is highlighted with red and others region in the same cluster are sorted by distance from centroid. Note that sub-parts of meshes from different categories may fall in the same cluster since they share similar shape.

- d identifies the region descriptor types ($d \in \{SIH, CUH, GD, GDW\}$),
- b identifies the refined level of the region descriptor (number of histogram bins $b \in \{10, 25, 50, 100, 200\}$),
- c identifies the refined level of the vocabulary construction (number of clusters $c \in \{20, 40, 60, 80\}$).

In order to construct a Bag-of-Words histogram of a new 3D object, we compare its regions descriptors with the visual words of the associated visual vocabularies.

5   Object Categorization by SVM

One of the most powerful classifier for object categorization is the Support Vector Machine (SVM) (see [21] for a tutorial). The SVM works in a vector space, hence the Bag-of-Words approach fits very well, since it provides a vector representation for objects. In our case, since we work with multiple vocabularies, we define the following positive-semi-definite kernel function:

$$K(A, B) = \sum_{l,s,d,b,c} k(\phi_{l,s}^{d,b,c}(A), \phi_{l,s}^{d,b,c}(B)), \tag{6}$$

where (A, B) is a pair of 3D models, and $\phi_{l,s}^{d,b,c}(\cdot)$ is a function which returns the Bag-of-Words histogram with respect to the visual vocabulary $V_{l,s}^{d,b,c}$. The function $k(\cdot, \cdot)$ is in turn a kernel which measures the similarity between histograms h^A, h^B:

$$k(h^A, h^B) = \sum_{i=1}^{c} min(h_i^A, h_i^B), \tag{7}$$

where h_i^A denotes the count of the ith bin of the histogram h^A with c bins. Such kernel is called *histogram intersection* function and it is shown to be a valid kernel [7]. Histograms are assumed to be normalized such that $\sum_{i=1}^{n} h_i = 1$. Note that, as observed in [7] the proposed kernel implicitly encodes the sub-parts matching since corresponding segments are likely to belong to the same histogram bin. Indeed, the histogram intersection function counts the number of sub-parts matching being intermediated by the visual vocabulary. Finally, since the SVM is a binary classifier, in order to obtain an extension to a multi-class framework, a one-against-all approach [5] is followed.

6   Results

We tested our categorization paradigm with two different datasets. For each dataset we performed a Leave-One-Out cross validation [5].

6.1 TOSCA Non Rigid Shape Dataset

The TOSCA dataset [22–24], publicly available[1], contains various non-rigid shapes in a variety of poses divided by category. The dataset is composed by: 9 cats, 8 men, 9 dogs, 21 gorillas, 17 horses and 9 women. The meshes used are shown in Fig. 5. Please note that each category is composed by the same model with different pose. Furthermore, some classes are very similar, e.g. men and women, and contains a number of elements very variable.

In this case, our categorization algorithm works perfectly in each category with a rate of success of 100%. This experiment shows that our system copes finely with the categorization of objects that present high inter-class similarity. Nevertheless, the methods is robust with objects that appear with different poses, by varying strongly their skeletal articulations (e.g., the gorillas).

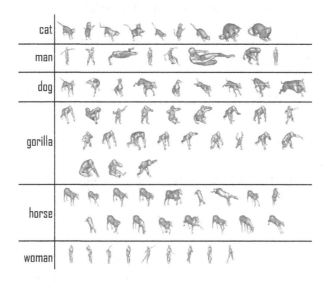

Fig. 5. TOSCA Non rigid Shape Dataset models, divided by category. Overall success rate of categorization is **100%**.

6.2 Aim@Shape Watertight

The Aim@Shape Watertight dataset has been used for various retrieval contests [25]. This dataset contains 20 categories each composed of 20 meshes. The entire dataset is shown in Fig. 6 together with the categorization results. In this case the algorithm fails with some meshes, but the overall rate of success is still fairly good. The dataset is tough since there are many categories and objects inside the same category can be very different. We can notice that the system is less accurate when the shapes are CAD-like (e.g. *mechanics, bearings* and *tables*).

---

[1] http://tosca.cs.technion.ac.il/data_3d.html

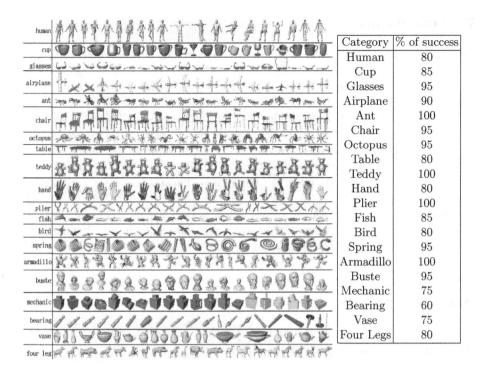

Category	% of success
Human	80
Cup	85
Glasses	95
Airplane	90
Ant	100
Chair	95
Octopus	95
Table	80
Teddy	100
Hand	80
Plier	100
Fish	85
Bird	80
Spring	95
Armadillo	100
Buste	95
Mechanic	75
Bearing	60
Vase	75
Four Legs	80

Fig. 6. Aim@Shape Watertight Dataset objects, divided by category and success rate of categorization. Overall the rate is **87.25%**.

This suggests that the descriptors based on curvature may not discriminate enough these kind of regions. Future improvements of the system can be obtained by adding more descriptors.

6.3 Timing

The categorization pipeline is computationally efficient in each sub-part. We used an entry level laptop at $1.66Ghz$ to perform tests. The code is written in Matlab with some parts in C. An entire mesh segmentation of 3500 vertices, with a maximum hierarchical depth of four is computed in less than a minute. Precisely, $\sim 8s$ are necessary to extract all the seed regions, while $\sim 50s$ are needed to compute the entire hierarchical segmentation. Region descriptors are computed efficiently. On the average it only takes $\sim 0.2s$ to extract all the four descriptors of a single region. Also the k-means clusterizations are not time consuming. For example 10 clusters for 300 points each composed of 200 feature are extracted in less than one second. Finally, the time needed to train a SVM with 400 elements is $\sim 80s$, while the time needed to validate a single element is about $\sim 2s$.

7 Conclusions

In this paper a new approach for 3D object categorization is introduced basing on the Bag-of-Words paradigm. The main steps of the involved categorization pipeline have been carefully designed by focusing on both the effectiveness and efficiency.

The Bag-of-Words approach allows naturally the object sub-parts encoding by combining effectively sub-part descriptors into several visual vocabularies. Moreover, we have proposed a Learning-by-Example approach by introducing a local kernel which implicitly performs the object sub-parts matching. In particular, the object categories are inferred without an exhaustive pairwise comparison between all the models.

The experimental results are encouraging. In particular, our framework is able to categorize objects which heavily deform their shape and change significantly their pose. Nevertheless, the method is able to distinguish also categories with the same skeletal structure (e.g., a man from a woman).

Acknowledgments

This paper was partially supported by PRIN 2006 project 3-SHIRT. Thanks to the anonymous reviewers for useful remarks and suggestions.

References

1. Iyer, N., Jayanti, S., Lou, K., Kalynaraman, Y., Ramani, K.: Three dimensional shape searching: State-of-the-art review and future trend. Computer Aided Design 5(37), 509–530 (2005)
2. Funkhouser, T., Kazhdan, M., Patrick, M., Shilane, P.: Shape-based retrieval and analysis of 3D models. Communications of the ACM 48(6), 58–64 (2005)
3. Tangelder, J.W., Veltkamp, R.C.: A survey of content based 3d shape retrieval methods. In: International Conference on Shape Modelling and Applications (2004)
4. Shilane, P., Funkhouser, T.: Selecting distinctive 3D shape descriptors for similarity retrieval. In: International Conference on Shape Modelling and Applications. IEEE Computer Society Press, Los Alamitos (2006)
5. Duda, R., Hart, P., Stork, D.: Pattern Classification, 2nd edn. John Wiley and Sons, Chichester (2001)
6. Cruska, G., Dance, C.R., Fan, L., Willamowski, J., Bray, C.: Visual categorization with bags of keypoints. In: ECCV Workshop on Statistical Learning in Computer Vision (2004)
7. Grauman, K., Darrell, T.: The pyramid match kernel: Efficient learning with sets of features. Journal of Machine Learning Research 8(2), 725–760 (2007)
8. Laptev, I., Marsza, M., Schmid, C., Rozenfeld, B.: Learning realistic human actions from movies. In: IEEE Conference on Computer Vision and Pattern Recognition (2008)
9. Ohbuchi, R., Osada, k., Furuya, T., Banno, T.: Salient local visual features for shape-based 3d model retrieval. In: International Conference on Shape Modelling and Applications (2008)

10. Li, Y., Zha, H., Qin, H.: Sapetopics: A compact representation and new algorithm for 3d partial shape retrieval. In: International Conference on Computer Vision and Pattern Recognition (2006)
11. Gal, R., Shamir, A., Cohen-Or, D.: Pose-oblivious shape signature. IEEE Transaction on Visualization and Computer Graphics 13(2), 261–271 (2007)
12. Tam, G.K.L., Lau, W.H.R.: Deformable model retrieval based on topological and geometric signatures. IEEE Transaction on Visualization and Computer Graphics 13(3), 470–482 (2007)
13. Hoffman, D.D., Richards, W.A.: Parts of recognition. Cognition, 65–96 (1987)
14. Shamir, A.: A survey on mesh segmentation techniques. Computer Graphics Forum 27, 1539–1556 (2008)
15. Attene, M., Katz, S., Mortara, M., Patane, G., Spagnuolo, M., Tal, A.: Mesh segmentation - a comparative study. In: Proceedings of the IEEE International Conference on Shape Modeling and Applications. IEEE Computer Society Press, Los Alamitos (2006)
16. Petitjean, S.: A survey of methods for recovering quadrics in triangle meshes. ACM Computing Surveys 34(2), 211–265 (2002)
17. Shi, J., Malik, J.: Normalized cuts and image segmentation. IEEE Transactions on Pattern Analysis and Machine Intelligence 22(8), 888–905 (2000)
18. Sethian, J.: A fast marching level set method for monotonically advancing fronts. In: Proceedings of the National Academy of Sciences, vol. 93 (1996)
19. Kimmel, R., Sethian, J.: Computing geodesic paths on manifolds. In: Proceedings of the National Academy of Sciences, vol. 95 (1998)
20. Ion, A., Artner, N.M., Peyr, G., Marmol, S.B.L., Kropatsch, W.G., Cohen, L.: 3d shape matching by geodesic eccentricity. In: Proceedings of S3D Workshop (2008)
21. Burges, C.: A tutorial on support vector machine for pattern recognition. Data Mining and Knowledge Discovery 2, 121–167 (1998)
22. Bronstein, A.M., Bronstein, M.M., Kimmel, R.: Numerical geometry of non-rigid shapes. Springer, Heidelberg (2007)
23. Bronstein, A.M., Bronstein, M.M., Kimmel, R.: Calculus of non-rigid surfaces for geometry and texture manipulation. Transactions on Visualization and Computer Graphics 13(5), 902–913 (2007)
24. Bronstein, A.M., Bronstein, M.M., Kimmel, R.: Efficient computation of isometry-invariant distances between surfaces. SIAM Journal of Scientific Computing 28(5), 1812–1836 (2006)
25. Veltkamp, R.C., ter Haar, F.B.: Shrec 2007 3d retrieval contest. Technical Report UU-CS-2007-015, Department of Information and Computing Sciences (2007)

An Improved Structured Light Inspection of Specular Surfaces Based on Quaternary Coding

Chengkun Xue and Yankui Sun

Department of Computer Science and Technology, Tsinghua University,
Beijing 100084, P.R. China
syk@mails.tsinghua.edu.cn

Abstract. Structured light techniques with binary coding are practical to inspect the specular surfaces. The structured light approaches use a scanned array of point sources and images of the resulting reflected highlights to compute local surface orientation. Binary coding scheme is the classic scheme for efficiently coding the light sources. This paper proposes a novel quaternary coding scheme which is much more efficient than the classic binary coding scheme. In this scheme, polychromatic light sources are utilized and coded in quaternary scheme. Our experimental system is described in detail. The problem caused by the polychromatic light sources is discussed too. To solve the problem, we drew lesson from the erosion operator from the Mathematical Morphology and designed an effective algorithm. The experiment results show the new quaternary coding scheme not only keeps a very high accuracy, but also greatly improves the efficiency of the inspection of specular surface.

1 Introduction

Many practical tasks in inspection require interpretation of images of specular surfaces where the perceived brightness becomes a very strong function of viewing direction due to highlights or reflections from the source. Many works have been done in the area of inspecting or estimating non-lambertian surfaces from specularities [1~7]. For a purely specular surface, light is reflected such that the angle of incidence equals the angle of reflection. Therefore, illumination of a specular surface using a point source of light does not produce smooth shading on the surface. Camera images of such surfaces are difficult to interpret because they are characterized by bright points or highlights, and inspection and reconstruction of surface shape are challenging tasks.

The structured light method includes both projected coded light and sinusoidal fringe [8~9] techniques have been proposed to address these problems. [10~11] use a scanned array of point sources and images of the resulting reflected highlights to compute local surface height and orientation. Some applications prove structured light techniques are practical for many industrial tasks including inspection of machined parts and inspection of solder joints.

Because multiple scans can't be avoided in the structure light techniques, the efficiency of these kinds of systems is critical as it is usually under the constraint of the

A. Gagalowicz and W. Philips (Eds.): MIRAGE 2009, LNCS 5496, pp. 128–139, 2009.
© Springer-Verlag Berlin Heidelberg 2009

speed of the product-line. As we all know, the exposure time is usually a constant factor after the camera and environment lighting are determined. Having fixed number of cameras in a structure light system, the total shooting time can be reduced only if the scans can be reduced. To reduce the scans while at the same time keep the inspection accurate is an interesting topic. In this paper, we introduced polychromatic light sources and proposed a quaternary coding scheme. With this coding scheme, we can collect more surface orientation information with the same number of input images or collect the same surface orientation information with less input images. As polychromatic sources are introduced, new problem occurs. To solve the problem, we drew lesson from the erosion operator from the Mathematical Morphology and design an effective algorithm.

The rest of this paper is organized as follows. Sec.2 introduces the principle of the quaternary coding scheme; Sec.3 gives the implementation of the quaternary coding scheme and the problem & solution. Experimental results and evaluation are given in Sec. 4 and conclusions are given in Sec. 5.

2 Quaternary Coding

The classic binary coding scheme [11] was firstly proposed by Arthur C. In his scheme, monochromatic point sources are distributed on a semi-sphere and the binary-coded point sources are scanned, and highlights on the object surface resulting from each point source are used to derive local surface orientation. Comparing with the sequential scanning, the most significant advantage of the binary coding is that it is far more efficient. Quaternary coding scheme, which will be introduced in this article, is a brand new coding scheme which again greatly improves the coding efficiency even comparing with the binary scheme.

2.1 Algorithm Principle

In fact, one point source can give off variety of colored light (usually in different conditions, such as variety of voltage). Assuming one point light source is able to give off r colors (under this assuming, the original binary encoding approach is the special case when r = 1). Take r = 3 as an example, suppose the light sources can emit 3 different colors: Green, Blue and Red. We map the color Green to 1, Blue to 2 and Red to 3, while 0 represents non-luminous. Fig.1 shows the principle of the quaternary coding scheme. Two images are required to be grabbed in order to get the orientation information of 15 points on the surface.

In this example, 15 point sources are shown. The source numbers are converted into their corresponding quaternary codes. The numbers from 1 through 15 can each be uniquely expressed in quaternary by using 2 digits, namely Digit (1) and Digit (2), For the first scan, namely Scan (1), all point sources that have value of 1, 2 and 3 in Digit (1) are controlled to turn Green, Blue and Red respectively, and the remaining sources which have value of 0 for this digit are turned off. An image of the surface is grabbed into the frame buffer of the computer and converted to a 4-level gray image (A pixel in this image can only have a value from 0, 1, 2 and 3). The converting

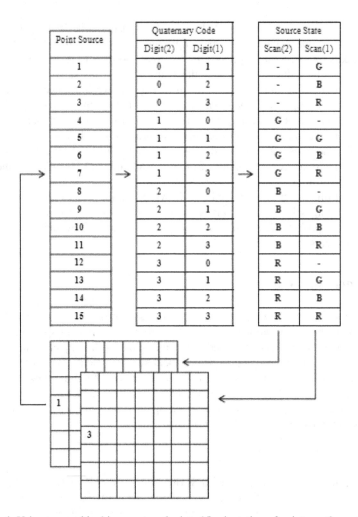

Fig. 1. Using two grabbed images to calculate 15 orientation of points on the surface

principles are: each pixel $p(i, j)$ in the image has a RGB representation (r, g, b), firstly the dominant color component whose intensity is higher than other two color comments is selected as the color of this pixel.

Then a threshold is used. If the color of a pixel is green and its green component is greater than the threshold, then this pixel is set to '1' in the 4-level gray image corresponding to a green highlight. With the similar principle, a pixel can be set to '2' in the 4-level gray image corresponding to a blue highlight and a pixel can be set to '3' corresponding to a red highlight. If a pixel is not set to any value among '1', '2' and '3', the '0' is set which means no highlight. The 4-level gray image corresponding to Scan (1) is obtained after we traverse all the pixels. In a similar manner, the 4-level gray image (2) is obtained for Scan (2). By reading the contents of the same pixel at location (i, j) in the two 4-level images, we obtain a 2-digit pattern, namely, (1, 3) in the example.

Here we assume only a single point source can generate a highlight at any particular point on the surface due to high surface specularity. Therefore, the digit pattern (1, 3) in the 4-level gray images could result only if the surface point corresponding to image point (i, j) reflected light from point 7 into the camera. The surface orientation at point (i, j) is then computed by using the knowing direction of source 7, the viewing direction of the camera, and the specular reflectance model.

It is can be proved that taking n photos, we can get at most $(r+1)^n - 1$ orientations on the surface.

2.2 Coding Efficiency

It is obviously we can get more surface normal information using quaternary code comparing to binary code with the same count of input images. As a deduction of the theory in [11], using r

$$r \bullet (2^n - 1) \tag{1}$$

Using the same r color and n images but quaternary code, the maximum count of surface normals is:

$$(r+1)^n - 1 \tag{2}$$

Define λ to be the ratio of the above expression:

$$\lambda = \frac{(2)}{(1)} = \frac{(r+1)^n - 1}{r \cdot (2^n - 1)} \tag{3}$$

It is easy to prove for any $r >= 2$ and $n >= 2$, $\lambda > 1$. Table 1 shows some typical values of λ. The λ exponentially increases with the n proved that the quarternay coding scheme is much more efficient than binary coding scheme.

Table 1. The λ exponentially increase with the n

r \ n	2	3	4	5	6	7	8	9
2	1.33	1.86	2.67	3.90	5.78	8.61	12.9	19.3
3	1.67	3.00	5.67	11.0	21.7	43.0	85.7	171

Here we limit r equal or less than 3 because in a RGB color space, we have only 3 independent color components R, G and B. If we use more than 3 colors in the system, at least one color can be represented by the linear combination of the other three colors. Then in some case we can't find the correspondence between the highlight in the specular object and the source point because there may be more than one candidate source points.

2.3 Industrial Practical

It is need to point out that the efficiency is essentially important for a structure lights solution in industry. If each frame grab takes about 30 ms (30fps, a typical time for most CCDs), then sequential scanning of 252 point sources would take a total of 7.56 s. Using the binary coding scheme, the total time is reduced to 0.24s; Using quaternary coding scheme, the total time is reduced to 0.12s. For many high-speed applications such as solder inspection, the speed of the production-line can't be change, the quicker the inspection system works, the wider ranges of production-lines the system can be applied.

In the other hand, the lighting engineering develops fast. The RGB LED [12~13] and multi-color LED [14] is industrial practical. RGB LEDs usually contain red, green and blue emitters, generally using a four-wire connection with one common lead (anode or cathode). In the other hand, the color CCD technique improved rapidly these years [15], some commercial products like SONY XCL-U1000C and Dalsa 4M60 can be employed to carry out the quaternary coding scheme. In a word, it is realistic for us to realize the quaternary code system.

3 Implementation

3.1 System Configuration

Fig.2 shows the system configuration. The specular object is centered at the origin; the point sources are uniformly distributed around the object and a single camera is used to view the reflected highlights. The point sources are activated, and highlights on the object surface are used to compute local surface orientations.

For a uniform distribution of the light sources in the virtual semi-sphere, we place them in such a manner: from the top to the bottom of the semi-sphere, we place 7 layers of the source lights, each layer can be imagined as a plane which is parallel to the XY-plane. Each layer intersects with the virtual semi-sphere and the points of intersection form a circle, namely, $C_i, i \in [1,2,3,4,5,6,7]$. In the Euclidean coordinate, the radius of C_i is:

$$R_i = 2\pi R \cos \theta_i \tag{4}$$

Where the R is the radius of the virtual semi-sphere, and θ_i is the angle between vector v_i and the XY-plane. v_i is the vector from the origin to a point source in C_i, because C_i is parallel to the XY-plane, the θ_i for all the point sources in C_i is the same. Meanwhile, for any $i \neq j \Rightarrow \theta_i \neq \theta_j$. From equation 4 we can see, for any $\theta_i > \theta_j$, we get $R_i < R_j$. So we place less source points in the top circles and more in

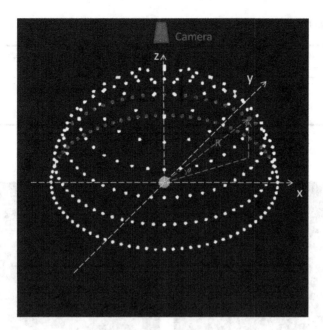

Fig. 2. System configuration

the bottom circles. In our experiment, from the top to the bottom, each circle has a total number of point source of 4, 8, 16, 32, 48, 64, 80 respectively. So the total point sources are 252.

We assume that the point sources are distant from the specular surface, that the surface is at a fixed reference height and that the extent of the specular surface from the origin is much smaller than the distance to the source. Under this assumption, the angle of incidence of illumination is determined only by the position of the source and does not depend on the relative position of illumination on the surface.

During the scanning process, each light ray emitted by a point source is reflected off the specular surface such that the angle of incidence equals the angle of reflection. The fixed camera on the top of the specular object images the reflected light ray only if it is positioned and oriented such that it is admitted by the camera's projective axis, the orthographic projection is used here to simplify the process, for practical use, the perspective projection can also be used with the additional camera calibration. Once the reflected light ray is observed as a highlight in the camera image, and the direction of the incident ray is known, the orientation of the surface element where the light was reflected can be found.

The Radiance software [16] is used to render the specular object and image it. In our experiment, a top-half metal sphere is created and the martial of the sphere surface is listed in Table 2. The low diffuse, high specular and low coarseness parameters make the object a high specular object. Also we listed two trials with different coarseness in order to show how the coarseness of the specular object affects of the final image.

Table 2. The martial parameters for the specular object in Radiance

trials \ params	Reflectance (R,G,B)	Diffuse	Specular	Coarseness
1	(0.6 0.62 0.64)	0.05	0.95	0.01
2	(0.6 0.62 0.64)	0.05	0.95	0.02

The resulting reflectance maps for some scans are shown in Fig.3.

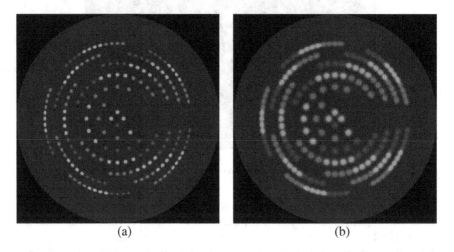

(a)   (b)

Fig. 3. (a) The grabbed image with the object coarseness = 0.01 (b) The grabbed image with the object coarseness = 0.02

The two sample grabbed images show the distribution of the elliptical contours depend on the surface specularity and the spacing of the discrete point sources. An effective way is to properly arrange point sources to span the reflectance map, effectively sampling the space of orientations of the target specular surface. Although we have done this by places all our point sources in 7 space layers, and accurately calculate the numbers of point sources for each layer, we can't control the coarseness of the inspected object. Fig.3 shows two inspected objects with different coarseness; it can be seen the more specular the object is the smaller elliptical contours we get. A small elliptical contour is less likely to mix with other elliptical contour, while at the same time a too small elliptical contour may make difficulties for the camera to capture it.

3.2 Obtain the Surface Orientation

Consider a series of scanning images in one experiment shown in Fig. 4, we expect to recover 252 surface orientations from them based on quaternary coding scheme.

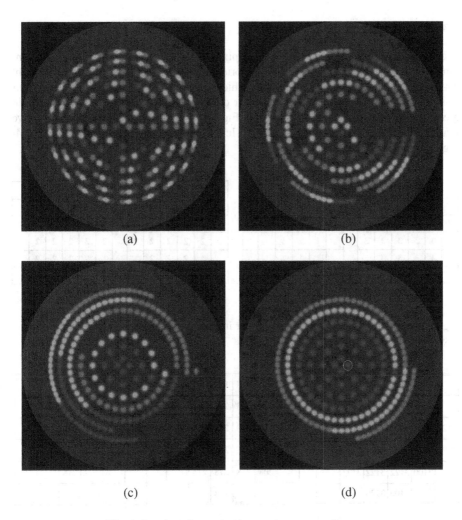

(a)   (b)

(c)   (d)

Fig. 4. A series of scanning images in one experiment

Take the highlight elliptical contours surrounded with the white circle near center in Fig.4(d) for example, let's see how to get the surface orientation of this small area. Note this small elliptical contours is blue in Fig.4(d), blue in Fig.4(c), green in Fig.4(b) and red in Fig.4(a). This means the point source that produces this highlight is blue in scan (1), blue in scan (2), green in scan (3) and red in scan (4). This series of states imply this point source's quaternary code is '3122' (remember we mapped the color Green to 1, Blue to 2 and Red to 3). So the correspondent decimal number is $3 \times 64 + 1 \times 16 + 2 \times 4 + 2 \times 1 = 218$, as we know the position of the point source numbered with 218, we can calculate the surface orientation in this small area.

3.3 Problem and Solution

If every highlight area only occupies one pixel in the same coordinate in every scanning image, we know this pixel's surface orientation now. However, because the light source illuminates an elliptical contour which contains several pixels in all the four images we get four set of pixels. The four corresponding small areas after threshold are extracted and magnified as shown in Fig.5. To obtain the surface orientation we read the values of each pixel in the four 4-level images, for each pixel we obtain a 4-digit pattern as shown in Fig.6(a) .

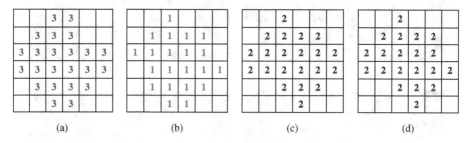

Fig. 5. The same small areas extract from four images shown in Fig. 4, each cell represents a pixel in the 4-level image

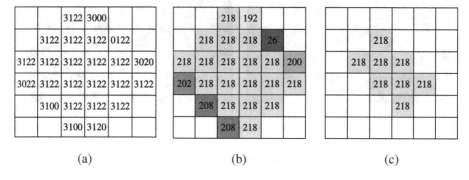

Fig. 6. (a) Overlap the four areas of Fig. 5, the codes in each pixel are in quaternary representation. (b) decimal code, some pixels near the border are wrongly coded (c) the result of the erosion operator.

In ideal condition, all the four images in Fig.5 should have the same pixel set which are illuminated by light 218 because the light source never change its position during the scans. But there are three reasons for the different illuminated pixel set: 1) The light source change its colors during scans, the inspected object may have different reflectance coefficient for different colors; 2) the color CCD usually has different response curves for different colors; 3) System noise. As a result, there will always be variance in the border of any illuminated area as shown in Fig.6(b). This kind of variance will decrease the system accuracy. To solve this problem, we drew lesson from the erosion operator from the Mathematical Morphology [17]. The classic erosion operator is used in binary image. Let E be an Euclidean space or an integer grid, the erosion of the binary image A in E by the structuring element B is defined by:

$$A \ominus B = \{z \in E \mid B_z \subseteq A\} \tag{5}$$

where B_z is the translation of B by the vector z, i.e. $B_z = \{b + z \mid b \in B\}, \forall z \in E$.

In our application, the image is actually a 4-level image, so we extend the algorithm in such a manner: each pixel in the image is detected, if the pixel is the same with all its 4 neighbors, we preserve this pixel; otherwise we set this pixel to be 0. By doing this, we actually set the structure element B to be a 5 pixel cross. The result of Fig.6(b) after the erosion operator is shown in Fig.6(c). Our experiment results (Table 3) show that with the erosion operator, the average angle reduced much.

4 Result Evaluations

4.1 Evaluation Method

To evaluation the result, we measured the error between real surface normal and the recovered surface normal on each highlight point. Let

$$E = <v_r, v_n> \tag{6}$$

where v_r and v_n denote the recovered surface normal and the real surface normal of the same highlight point on the surface respectively; $<,>$ means dot product of two 3D vectors. So the average normal error is,

$$E_{avg} = \frac{\sum_i <v_{ir}, v_{in}>}{N}, \forall i \in [0, N) \tag{7}$$

where v_{ir} is the recovery normal vector for i^{th} highlight, v_{in} is the real normal vector for i^{th} highlight. The average angle error is,

$$E_{avgangle} = \frac{\sum_i \arccos(<v_{ir}, v_{in}>)}{N}, \forall i \in [0, N) \tag{8}$$

And the Maximum angular error is,

$$E_{max\,angle} = \max_i(\arccos(<v_{ir}, v_{in}>)), \forall i \in [0, N) \tag{9}$$

4.2 Experiment Results

Firstly, the new quaternary scheme recovered 252 surface normals with only 4 input images. In [11], with the classic binary coding scheme, 127 surface normals are recovered with 7 input images. The new quaternary scheme is proved to be much more efficient than the classic binary coding scheme.

Secondly, we use the evaluation approach described in Sec.4.1 to evaluate the accuracy of our experiments. Table 3 shows that the recovered surface orientations are in a high precision. The results also show that with the erosion operator, the average angle error reduced from 4.0697 degree to 0.8459 degree for trial 1 and from 5.8874 degree to 0.8465 degree for trial 2. The worst angle errors are effectively controlled into a reasonable range.

These results prove that the new coding scheme not only keeps a very high accuracy, but also greatly improves the efficiency of the inspection of specular surface.

Table 3. The accuracy of our experiments, the parameters for each trials are shown in Table 2

trials err	E_{avg}	$E_{avgangle}$ (degree)	$E_{max\,angle}$ (degree)
1 (no erosion)	0.997471	4.0697	22.7946
2 (no erosion)	0.994713	5.8874	30.4242
1 (with erosion)	0.999891	0.8459	1.1459
2 (with erosion)	0. 999891	0.8459	1.1459

5   Conclusions

Structured light techniques with the classic binary coding are practical and efficient to inspect the specular surfaces. With introducing the polychromatic light sources, we proposed a novel quaternary coding scheme which is much more efficient than the classic binary coding scheme. We also designed an effective algorithm to solve the problem caused by the polychromatic light sources. The experiment results show the new coding scheme not only keeps a very high accuracy, but also greatly improves the efficiency of the inspection of specular surface.

Acknowledgment

This work was supported in part by the National Natural Science Foundation of China under Grant No. 60873249 and 03470487, 863 Project of China under Grand 2006AA02Z472 and 2008AA01Z419.

References

1. Blake, A., Brelstaff, G.: Geometry from specularities. In: Second International Conference on Computer Vision, pp. 394–403 (1988)
2. Zisserman, A., Giblin, P., Blake, A.: The information available to a moving observer from specularities. Image and Vision Computing 7(1), 38–42 (1989)

3. Schultz, H.: Retrieving shape information from multiple images of a specular surface. IEEE Transactions on Pattern Analysis and Machine Intelligence 16(2), 195–201 (1994)
4. Bonfort, T., Sturm, P.: Voxel carving for specular surfaces. In: Int. Conf. Computer Vision, Nice, France, pp. 591–596 (2003)
5. Yang, R., Pollefeys, M., Welch, G.: Dealing with textureless regions and specular highlights—a progressive space carving scheme using a novel photo-consistency measure. In: Int. Conf. Computer Vision, Nice, France, pp. 576–584 (2003)
6. Solem, J.E., Aanæs, H., Heyden, A.: Pde based shape from specularities. In: Griffin, L.D., Lillholm, M. (eds.) Scale-Space 2003. LNCS, vol. 2695. Springer, Heidelberg (2003)
7. Solem, J.E., Aanæs, H., Heyden, A.: A variational analysis of shape from specularities using sparse data 3D Data Processing, Visualization and Transmission. In: Proceedings of 2nd International Symposium on 3DPVT 2004, September 6-9, pp. 26–33 (2004)
8. Jalkio, J.A., Kim, R.C., Case, S.K.: Three dimensional inspection using multistripe structured light. Opt. Eng. 24(6), 966–974 (1985)
9. Wahl, F.: A coded light approach for depth map acquisition,Muskererkennung 86, Informatik Fachberichte 125. Springer, Heidelberg (1986)
10. Sanderson, A.C., Weiss, L.E., Nayar, S.K.: Structured Highlight Inspection of Specular Surfaces. IEEE Transactions On Pattern Analysis And Machine Intelligence 10(1), 44–55 (1988)
11. Nayar, S.K., Sanderson, A.C., Weiss, L.E., et al.: Specular surface inspection using structured highlight and Gaussian images. IEEE Transactions On Pattern Analysis And Machine Intelligence 10(1), 208–218 (1990)
12. Wisniewski, A.: Digital control methods of LED and LED lamps. Przeglad elektrotechniczny 84(8), 178–181 (2008)
13. Gilewski, M., Karpiuk, A.: An electronic control of light RGB LEDs. Przeglad elektrotechniczny 84(8), 194–198 (2008)
14. Muller, C.D.: Reckefuss, Multi-color polymeric OLEDs by solution processing. In: Organic light-emitting materials and devices VII, vol. 5214, pp. 21–30 (2004)
15. Allan, G.R.: High-speed, high-resolution, color CCD image sensor. In: Proceedings of the society of photo-optical instrumentation engineers (SPIE), vol. 3965, pp. 70–79 (2000)
16. Debevec, P.: Image-Based Lighting. IEEE Computer Graphics and Applications 22(2), 26–34 (2002)
17. Soille, P.: Morphological Image Analysis: Principles and Applications, 2nd edn. (2003) ISBN:3540429883

Robust Detection and Tracking of Multiple Moving Objects with 3D Features by an Uncalibrated Monocular Camera

Ho Shan Poon, Fei Mai, Yeung Sam Hung, and Graziano Chesi

Department of Electrical and Electronic Engineering
The University of Hong Kong
Pofulam Road, Hong Kong
{hspoon,feimai,yshung,chesi}@eee.hku.hk

Abstract. This paper presents an algorithm for detecting multiple moving objects in an uncalibrated image sequence by integrating their 2D and 3D information. The result describes the moving objects in terms of their number, relative position and motion. First, the objects are represented by image feature points, and the major group of point correspondences over two consecutive images is established by Random Sample Consensus (RANSAC). Then, their corresponding 3D points are reconstructed and clustering is performed on them to validate those belonging to the same object. This process is repeated until all objects are detected. This method is reliable on tracking multiple moving objects, even with partial occlusions and similar motions. Experiments on real image sequences are presented to validate the proposed algorithm. Applications of interest are video surveillance, augmented reality, robot navigation and scene recognition.

1 Introduction

Tracking of multiple moving objects means detection of their trajectories in image sequences. Its main steps are the following two: first, detect moving features and second, cluster them to validate the number of moving objects available.

Different kinds of feature tracking algorithms have been proposed such as tracking of affine invariant pieces of level lines (AIPLL) [1], scale invariant feature transform (SIFT) [2] and Kanade-Lucas-Tomasi feature (KLT) [3]. These characteristic image features are local and allow us to cope with partial occlusion, mild intensity change of light source, and moderate projective deformation. However, since the tracking principle is based only on pixel information to find out apparent image features between an image pair, it is possible that such methods track apparent image features which are out of the original objects. Therefore, feature tracking requires not only accuracy in pixel information but also geometric property.

Other algorithms have been proposed, such as that by Helmholtz, Desolneux, Moisan and Morel [4] where an unsupervised detection principle without parameter

A. Gagalowicz and W. Philips (Eds.): MIRAGE 2009, LNCS 5496, pp. 140–149, 2009.

tuning is described. Thomas Veit, Frédéric Cao and Patrick Bouthemy [5] have applied
it into tracking. They placed image position coordinate (**x,y**) and polar form velocity
(**r,θ**) into "a contrario" [4] framework to perform clustering. They even clustered overlapped objects moving in opposite direction successfully. However, with overlapped
objects moving in the same direction, the method [5] has difficulty in distinguishing
them because of lacking in depth information. Therefore, clustering of moving feature
points in 3D Euclidian space is needed to cope with this situation.

In this paper, we propose an algorithm to track multiple moving objects, which is
able to cope also with partial occlusions and similar motions. This paper is organized
as follows: Section 2 briefly explains the overall strategy of the proposed algorithm.
Section 3 describes the extraction of features. Section 4 describes the robust matching
by RANSAC. Section 5 describes the 3D reconstruction. Section 6 describes the "a
contrario" clustering. The experimental results are reported in Section 7.

2   Overall Strategy

Given a sequence of images, we aim to extract feature points of moving objects from
successive images and cluster them into groups. The strategy we propose is an iterative process including four major steps, as shown in Figure 1.

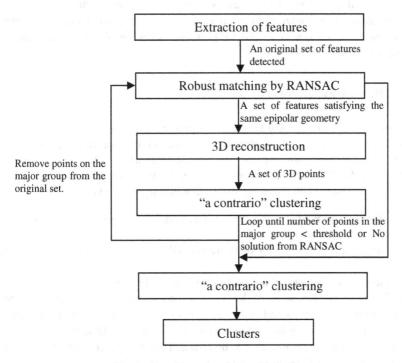

Fig. 1. Algorithm of multiple objects detection

The role of these four steps is as follows:

1) Extraction of features: Objects are represented by feature points, and SIFT [2] is applied to extract an original set of feature points because it can withstand partial occlusions, mild intensity change, and similarity deformation satisfactorily.
2) Robust matching by RANSAC: Robust point correspondences over two consecutive images are established by Random Sample Consensus (RANSAC) [6], and a major set of matched feature points satisfying the same epipolar geometry is detected.
3) 3D reconstruction: With matched feature points, projective reconstruction is performed [7] to find out their relative coordinates in 3D projective space, and then the projective reconstruction is upgraded to Euclidean space. Thus a set of 3D points is obtained.
4) "a contrario" clustering: The reconstructed 3D points usually contain outliers. To cluster valid groups of points together and remove outliers, "a contrario" framework [4] is adopted and a major group of points representing a moving object is finally detected.

Once a moving object is detected, the corresponding set of points is removed from the original set, and steps 2) – 4) are repeated to detect another moving object. By iterations, the matched features in the original set will gradually reduce. The algorithm stops when there are no valid solutions from RANSAC or the number of points in the major group after clustering is smaller than a threshold. After iterations, there are several major groups detected, and we perform "a contrario" clustering again to validate the number of valid objects.

3 Extraction of Features

Features extracted should be local enough to cope with partial occlusion, mild intensity change of light source, and moderate projective deformation through image sequences. SIFT [2] feature point is chosen because it is invariant to the above criterion in certain extent. SIFT feature points are extracted by the convolution of the difference of Gaussian (DoG), with the image $\mathbf{I(x,y)}$ at multiple scales (σ):

$$\mathbf{D(x,y,\sigma) = (G(x,y,k\sigma) - G(x,y,\sigma)) * I(x,y)}$$ (1)

$\mathbf{k}$ is a constant multiplicative factor of Gaussian, $\mathbf{G(x,y,\sigma)}$, and * is the convolution operator.

The extrema around 26 neighbors in 3×3 regions at the current and two adjacent scales are chosen. However, some of them should be rejected because they are sensitive to noise with low contrast and strong edge response. Therefore, the interpolated location of the maximum is performed to improve the stability of feature points.

The interpolation is performed by the quadratic Taylor expansion of (1) with the candidate feature point as the origin:

$$D(x) = D + \frac{\partial D^T}{\partial x} x + \frac{1}{2} x^T \frac{\partial^2 D}{\partial x^2} x \qquad (2)$$

In (2) $x = (x, y)$. The location of feature point x' is determined by the zeros of the derivative of (2):

$$x' = -\frac{\partial^2 D^{-1}}{\partial x^2} \frac{\partial D}{\partial x} \qquad (3)$$

Substitute (3) into (2) to obtain the function value of feature point:

$$D(x') = D + \frac{1}{2} \frac{\partial D^T}{\partial x} x' \qquad (4)$$

If $|D(x')|$ is less than a threshold, feature point x' is rejected.

By (1) – (4), feature points sensitive to noise with low contrast are rejected. In order to remove those with strong edge response, a 2×2 Hessian matrix, H, is computed at the position of feature points:

$$H = \begin{bmatrix} D_{xx} & D_{xy} \\ D_{xy} & D_{yy} \end{bmatrix} \qquad (5)$$

D_{xx}, D_{xy} and D_{yy} are the respective second derivatives of (1) and feature points with strong edge response are identified with the method proposed by Harris and Stephens [8]:

$$C = \frac{\text{trace}(H)^2}{\det(H)} \qquad (6)$$

If C is less than a threshold, the corresponding feature point is rejected.

By (1) – (6), SIFT feature points with high contrast and weak edge response are extracted. Then, the description vector representing the image property of the corresponding feature point is obtained by computing the gradient of the 16×16 sample array around it. Then a 4×4 descriptor array of histograms with 8 orientations bins is computed from the sample array. Therefore, every description vector consists of 4×4×8 = 128 elements.

A set of SIFT feature points $s_i(x_i, y_i, \sigma_i, v_i)$ containing image position (x_i, y_i), scale (σ_i) and description vector (v_i) is computed.

4   Robust Matching by RANSAC

Once the image features in the two images are extracted, it is required to find out the correspondence between them, i.e. to know where one image point in one image is in the other image. This can be done via RANSAC by identification of a majority of points satisfying the same epipolar geometry with a fundamental matrix F.

First, $\| \mathbf{v}_i - \mathbf{v}_j \|$, the Euclidean norm of the difference of the corresponding description vector of SIFT feature points $\mathbf{s}_i$ and $\mathbf{s}_j$ is chosen as the putative match. Then, randomly select 7 pairs of point correspondences to form a set of linear equations:

$$\mathbf{Af} = \begin{bmatrix} \mathbf{x'}_1\mathbf{x}_1 & \mathbf{x'}_1\mathbf{y}_1 & \mathbf{x'}_1 & \mathbf{y'}_1\mathbf{x}_1 & \mathbf{y'}_1\mathbf{y}_1 & \mathbf{y'}_1 & \mathbf{x}_1 & \mathbf{y}_1 & 1 \\ \vdots & \vdots & \vdots & \vdots & \vdots & \vdots & \vdots & \vdots & \vdots \\ \mathbf{x'}_7\mathbf{x}_7 & \mathbf{x'}_7\mathbf{y}_7 & \mathbf{x'}_7 & \mathbf{y'}_7\mathbf{x}_7 & \mathbf{y'}_7\mathbf{y}_7 & \mathbf{y'}_7 & \mathbf{x}_7 & \mathbf{y}_7 & 1 \end{bmatrix} \mathbf{f} = 0 \qquad (7)$$

(x_i,y_i) and (x'_i,y'_i) represent the i-th pair of feature point correspondence and the 9×1 vector $\mathbf{f}$ represents the entries of $\mathbf{F}$. By singular value decomposition (SVD) of matrix $\mathbf{A} = \mathbf{UDV}^T$, the last column of $\mathbf{V}$ is chosen as the least-squares solution for $\mathbf{f}$.

Then, Sampson approximate error $\mathbf{d}_{i\perp}$ representing the reprojection error of point correspondence (x_i,x'_i) is computed:

$$\mathbf{d}_{i\perp} = \frac{(\mathbf{x'}_i^{\mathrm{T}} \mathbf{Fx}_i)^2}{(\mathbf{Fx}_i)_1^2 + (\mathbf{Fx}_i)_2^2 + (\mathbf{Fx'}_i)_1^2 + (\mathbf{Fx'}_i)_2^2} \qquad (8)$$

$(Fx_i)_j^2$ is the square of the j-th entry of vector Fx_i. If $\mathbf{d}_{i\perp}$ is less than a threshold, that point is considered as an inlier of $\mathbf{F}$ and an outlier otherwise.

A set of inliers is used to re-estimate $\mathbf{F}$ by minimizing the sum of $\mathbf{d}_{i\perp}$ for all inliers with Levenberg–Marquardt algorithm [9]. With the re-estimated $\mathbf{F}$, a search strip about the epipolar line is defined to determine further interested point correspondences. By iterations, $\mathbf{F}$ is repeatedly estimated and a set of inliers is also repeatedly updated. The number of iteration, $\mathbf{k}$, is defined as follow:

$$\mathbf{k} = \frac{\log(1-\mathbf{p})}{\log(1-\mathbf{w}^n)} \qquad (9)$$

$\mathbf{n}$ is the least number of points required to estimate $\mathbf{F}$ so that $\mathbf{n} = 7$. $\mathbf{p}$ is the probability of producing a valid result and $\mathbf{w}$ is the ratio of the number of inliers to the number of whole data points. After iterations, a set of points satisfying the same epipolar geometry with $\mathbf{F}$ is extracted.

5 3D Reconstruction

RANSAC does not guarantee that the set of points extracted represents the same object because RANSAC includes as many point correspondences as possible to satisfy the same $\mathbf{F}$ representing relative motion between them. Therefore, if points of multiple objects have similar relative motions, there will be a $\mathbf{F}$ satisfying their epipolar geometry. Therefore, in order to distinguish them and remove outliers, their relative position in 3D Euclidian space should be obtained by 3D reconstruction. In this paper, we apply the algorithm proposed in [7] to establish the projective reconstruction over two views with minimization of the 2D reprojection error. This algorithm reformulates the projective reconstruction problem into a sequence of weighted least-squares problems, where a control parameter is gradually increased to force the final solution to approach a minimum point of the 2D reprojection error. By solving this minimization problem, the

3D positions of points and the projection matrices can be recovered simultaneously. Furthermore, the algorithm does not require any information from camera calibration. In order to measure the relative positions of points in 3D Euclidian space, it is necessary to upgrade the reconstructed scene from the projective frame to the Euclidean frame. This is performed by means of a linear subspace algorithm [10].

6 "a contrario" Clustering

After 3D reconstruction, a set of 3D points $\{x_1, x_2,...,x_M\}$ is obtained. "a contrario" framework [4] is adopted to find out valid groups of points from a hierarchical binary tree. Its principle is to compare the density of the distribution of the set of points with a given independent and identical distributed (i.i.d) background model. If it has higher density than that model, this set will be considered as a valid group.

This framework can answer two questions about the distribution of 3D points.

1) Is a candidate group valid in the binary tree?
2) If two sibling groups and their parent group are all valid in the binary tree, which one will be more valid?

For 1), the validity of a candidate group G in the binary tree is defined as follow:

$$NFA(G) = M^2 \cdot |\Re| \min_{\substack{x \in G, R \in \Re \\ G \subset x+R}} B(M - 1, n - 1, \pi(x + R))$$

$$B(M-1, n-1, \pi(x+R)) = \sum_{k=n-1}^{M-1} \binom{M}{k} \pi(R)^k (1-\pi(x+R))^{M-k}$$

(10)

M is the total number of 3D points. n is the number of points in the candidate group. R is the subset of region set $\Re$ where the largest region in $\Re$ contains all 3D points. $\pi(x+R)$ is the probability distribution function centered on 3D position x in region R. $|\Re|$ is the cardinality of region set $\Re$. If $NFA(G) < 1$, the candidate group G is considered as a valid group. Lower $NFA(G)$ represents higher validity of candidate group G.

For 2) the validity of two sibling groups G_1 and G_2 in the binary tree is defined as follow:

$$NFA_g(G_1, G_2) = M^4 \cdot |\Re|^2 \min_{\substack{x_1, x_2 \in G_1 \times G_2 \\ R_1, R_2 \in \Re \\ G_1 \subset x_1 + R_1 \\ G_2 \subset x_2 + R_2}} T(M-2, k_1-1, k_2-1, \pi_1, \pi_2)$$

$$T(M-2, k_1-1, k_2-1, \pi_1, \pi_2) = \sum_{i=k_1-1}^{M-2} \sum_{j=k_2-1}^{M-2-i} \binom{M}{i,j} \pi_1^i \pi_2^j (1-\pi_1-\pi_2)^{M-i}$$

(11)

k_1 and k_2 are the number of points in candidate groups G_1 and G_2 respectively. π_1 and π_2 are their respective probability distribution function centered on (x_1, x_2) in regions (R_1, R_2) respectively. Then, we compare the validity of two sibling groups (G_1, G_2), and their parent group G, where $G_1 + G_2 \subset G$. If $NFA(G) < NFA_g(G_1, G_2)$, G is more valid and (G_1, G_2) are more valid otherwise.

By answering these two questions, the number of present groups can be validated.

7 Experimental Result

In order to validate the proposed algorithm for multiple moving objects detection and tracking, we show the experimental results on both synthetic and real image sequences. In the first synthetic image sequence generated by Pov Ray, a plane and two

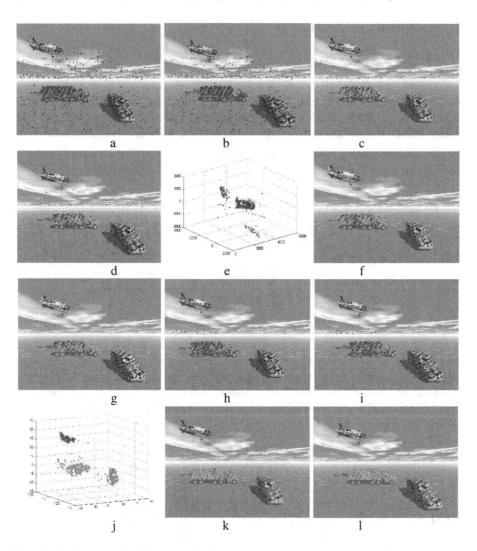

Fig. 2(a-b): Detected SIFT feature **(c-d):** Extraction of the point correspondences by RANSAC **(e):** Reconstructed 3D points and clustering in "a contrario" framework **(f-g):** Extraction of the major set of point correspondences **(h-i):** A set of point correspondences extracted after 5 iterations **(j-l):** Three objects detected after 3D reconstruction and clustering

trucks with respective orientations are moving in different directions, while in the second real image sequence, two moving cars are detected, and in the third real image sequence, two cars are moving together like one rigid object. The experimental results show that all moving objects are detected and tracked successfully.

7.1 Detection of Objects Moving in Different Directions

In Figure 2, a truck is moving towards right; another truck is moving forwards with right turn and a plane is moving backwards with left turn. Extracted features belong to two trucks, the plane and background respectively, as shown in Figures 2 (a) and 2 (b). After robust matching by RANSAC, a set of point correspondences satisfying the same F is extracted, as shown in Figures 2 (c) and (d) which show that point correspondences come from two trucks, the plane and the background respectively. There fore, by reconstructing the 3D feature points and clustering them in 3D space by

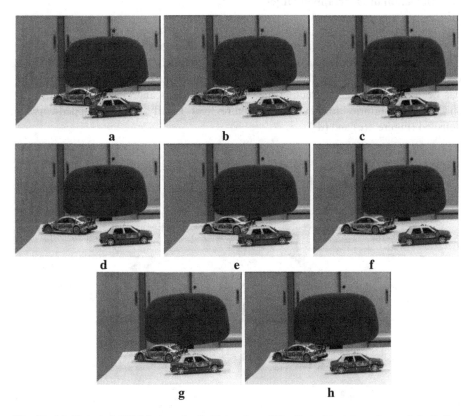

Fig. 3(a-b): Detected SIFT feature **(c-d)**: Extraction of the first object (a racing car) **(e-f)**: Repeated extraction of the second object (a taxi) **(g-h)**: Two objects with different motions detected after 5 iterations

"a contrario" framework [5] (Figure 2(e)), the major set of point correspondences is extracted (Figure 2(f) and (g)). After five iterations, a set of point correspondences is extracted (Figure 2(h) and (i)). With reconstruction of 3D feature points and clustering in "a contrario" framework [5] (Figure 2(j)), all moving objects (two trucks and the plane) are detected (Figure 2(k) and (l)).

In Figure 3, a racing car (left) is moving towards left and a taxi (right) is moving towards right in two consecutive images. Extracted features belong to the racing car, taxi and background respectively, as shown in Figures 3 (a) and 3 (b). After the first iteration, the major group representing the racing car is detected, as shown in Figures 2 (c) and 2 (d). Then, remove the feature points representing the racing car and repeat the process, the taxi is detected afterwards (Figures 2 (e) and 2 (f)). The process is iterated until there are no meaningful groups detected. Finally, two moving objects, the racing car and taxi, were detected (Figures 2 (g) and 2 (h)).

7.2 Detection of Overlapped Objects

The major advantage of the algorithm proposed in this paper is capable of detecting overlapped objects moving in the same direction. In Figure 4 (a) and (b), the racing car and taxi moving in the same direction are overlapped. Therefore, method [5] has difficulty in distinguishing them due to lacking in depth information. This problem has been solved by reconstructing the 3D feature points and clustering them in 3D space by "a contrario" framework [5] (Figure 3(e)). Finally, two overlapped cars are detected (Figure 3(c) and (d)).

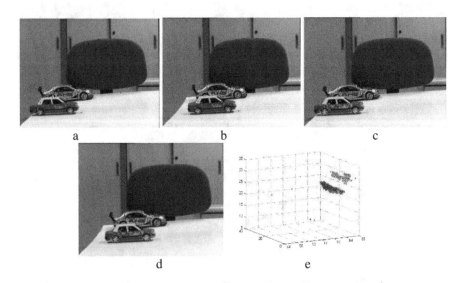

Fig. 4(a-b): Both racing car and taxi detected by RANSAC at the same time due to similar motion **(c-d)**: Two overlapped objects distinguished by "a contrario" clustering in 3D space. **(e)**: Reconstructed 3D points

8 Conclusion

This paper presents an algorithm to detect and track multiple moving objects in image sequences. It involves feature extraction, feature matching, 3D reconstruction and clustering. The result is the descriptions of dynamic contents in terms of moving objects, their number, relative position and motion in 3D space without being distracted by overlapping of objects and background features. This multiple objects detection is able to apply to surveillance, augmented reality, robot navigation and scene recognition.

A defect of this algorithm is the randomness of the fundamental matrix estimation by RANSAC which causes the extraction of the point correspondences not deterministic. A deterministic estimation of fundamental matrix is the further improvement of this algorithm.

Acknowledgement

The work described in this paper was supported by the Research Grants Council of Hong Kong Special Administrative Region, China (Project Nos. HKU711208E and HKU712808E).

References

[1] Lisani, J., Moisan, L., Monasse, P., Morel, J.: On the theory of planar shape. SIAM Multiscale Modeling and Simulation 1(1), 1–24 (2003)

[2] Lowe, D.: Distinctive image features from scale-invariant keypoints. International Journal of Computer Vision 60(2), 91–110 (2004)

[3] Shi, J., Tomasi, C.: Good features to track. In: IEEE Int. Conf. on Computer Vision and Pattern Recognition, Seattle, June 1994, pp. 593–600 (1994)

[4] Desolneux, A., Moisan, L., Morel, J.: A grouping principle and four applications. IEEE Trans. on Pattern Analysis and Machine Intelligence 25(4), 508–513 (2003)

[5] Veit, T., Cao, F., Bouthemy, P.: Space-time A Contrario Clustering for Detecting Coherent Motions. In: IEEE International Conference on Robotics and Automation (2007)

[6] Fischler, M.A., Bolles, R.C.: Random Sample Consensus: A Paradigm for Model Fitting with Applications to Image Analysis and Automated Cartography. Communications of the ACM 24(6), 381–395 (1981)

[7] Hung, Y.S., Tang, W.K.: Projective Reconstruction from Multiple Views with Minimization of 2D Reprojection Error. International Journal of Computer Vision 66(3), 305–317 (2006)

[8] Harris, C., Stephens, M.: A combined corner and edge detector. In: Fourth Alvey Vision Conference, Manchester, UK, pp. 147–151 (1988)

[9] Hartley, R., Zisserman, A.: Multiple View Geometry. In: Computer Vision, 2nd edn., pp. 297–298. Cambridge University Press, Cambridge (2003)

[10] Tang, W.K.: A factorization-based approach to 3D reconstruction from multiple uncalibraed images. Ph.D. dissertation, Department of Electrical and Electronic Engineering, The University of Hong Kong (2004)

Automatic Golf Ball Trajectory
Reconstruction and Visualization

Tadej Zupančič and Aleš Jaklič

University of Ljubljana, Faculty of Computer and Information Science,
Computer Vision Laboratory,
Tržaška 25, 1000 Ljubljana, Slovenia
{tadej.zupancic,ales.jaklic}@fri.uni-lj.si

Abstract. The article presents the steps required to reconstruct a 3D trajectory of a golf ball flight, bounces and roll in short game. Two video cameras were used to capture the last parts of the trajectories including the bounces and roll. Each video sequence is processed and the ball is detected and tracked until is stops. Detected positions from both video sequences are then matched and 3D trajectory is obtained and presented as an X3D model.

Keywords: tracking, golf, stereo, trajectory, 3D, video.

1 Introduction

Video analysis is nowadays an important tool not only for professional athletes but also for amateurs in various sports. It gives them visual information of what they are doing and can help them improve by seeing their (not perfect) moves and the consequences, that they cause. There are a lot of golf accessories on the market, probably more than in any other sport. On the other hand, golf ball tracking articles are not so common. To detect and display the trajectory of the golf ball, usually some expensive equipment like radars is used. While there is no doubt, that the trajectory obtained that way is accurate, its price tag is out of the range of the average user.

In the area of golf video tracking there has been a research of golf club tracking during the swing [1] and tracking of the position of the golf club and the ball using markers [2]. Ball tracking in tennis [3,4], on the other hand, has been studied extensively and the successful results can be seen on the television during the tennis broadcasts. There are also quite a lot of articles about soccer ball detection [5,6], but these are not as widely used in practice as the ones from tennis.

2 Motivation

Golf is a sport full of variety. Because of different course condition, flag position, weather and tactics, every shot is different. The player is faced with the choice of what kind of shot to play before each shot. The number of options increases

A. Gagalowicz and W. Philips (Eds.): MIRAGE 2009, LNCS 5496, pp. 150–160, 2009.

when the player gets closer to the hole. When the ball lies e.g. 50 meters from the hole, the player's wish is to make a good shot, that will get the ball as close to the hole as possible to increase the possibility of a good score.

At that distance, he has many options of the shots he can make. He can use a very lofted club and fly it on a high trajectory. In that case, the high angle of impact causes the ball to stop near the point, where it touches the ground. If the flag is on the back of the green, he can make a lower flying shot, land the ball on the front of the green and let it roll to the hole(Fig. 1).

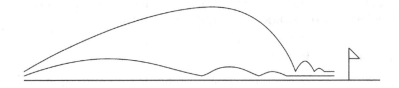

Fig. 1. Two possible approach shot trajectories

If a player wants to get close to the hole, he has to have a good understanding of how the ball bounces and rolls, to know what kind of shot to make and exactly where to land the ball to increase his chance of making a low score. This knowledge is usually absorbed through extensive practice. This process could be accelerated by showing him a 2D trajectory drawn over the actual video or by generating a 3D trajectory model, which would hopefully help him improve in much shorter time.

3   Algorithm Input

Static cameras were positioned in a way to grab only the last part of the trajectories – the part that is important for the player – angle of impact, bounces and roll. One camera was rotated approximately 90 degrees compared to the other camera and was positioned at a larger distance from the player.

Video acquisition was done using one consumer DV camcorder and one consumer HDV camcorder. DV recording was interlaced, therefore a software filter was used to deinterlace the stream. Streams were later manually synchronized on the computer, but an automatic synchronization would be possible.

4   Obtaining the Position of the Ball

4.1   Background Registration and Subtraction

Using the static cameras gives us the possibility to generate the background from multiple frames. When we subtract such background from the current frame, we obtain only the difference – the pixels that have changed. If the background was also static, the difference would contain only the moving ball. Since the nature is not static, the difference also contains e.g. some moving leafs on the trees and parts of the image, where the brightness has changed.

In this case we can not use the static background, but we have to use multiple consequent frames to generate a dynamic one and adjust it continuously to the present conditions. The pixel in image is tagged as background, if its value has not changed for a predefined number of frames. Each frame of the stream is processed and the background value of each pixel is saved, if such value already exists.

The difference image is then obtained by subtracting the background from the current frame. If a pixel does not have a background value, we use the value of the pixel at the same position in the previous frame. More detailed description of this background registration technique can be found in [7].

4.2 Problematic Pixels

Some pixels' values are constantly changing and that can cause problems if this issue is not addressed. The change can be caused by physical movement because of the wind or some other factor, but it can happen also by static objects, that diffuse light and cause the value of pixels to be different in each frame. For such pixel, a reliable background value can not be set, because any value we choose, causes a non zero pixel value in subsequent difference images. Therefore we count the number of frames in which the value of such pixel in the difference image is above some predefined threshold. If that number is too high, we tag that pixel as problematic and exclude it from further processing.

A pixel in the difference image that represents a ball can not have a high value for a long time. If the ball moves, it causes high values of the pixels in the difference image at the position of the ball. At subsequent frames, the ball is at the new positions and after a few frames the pixels from the previous frame can not have high values anymore. When the ball comes to a stop, its values in the difference image stay high until the background is updated to include the stopped ball.

4.3 Ball Hypothesis Generation

Possible ball positions are obtained by analyzing the points in the difference image. Ball is usually quite different from the background, so we check if the pixel value in the difference image is above a threshold. By calculating the weighted sum of values in the area around that pixel and a preset low threshold, we remove objects smaller than a golf ball. After that the RGB values of the image around the pixel are checked to add a color constraint. Golf balls are usually white, so we compare red, green and blue components of the pixel and test if they are approximately the same.

If one pixel is set as a ball candidate in a percentage of the frames, that is too high, that pixel is discarded. High values in the difference image are usually caused by changing light coming from reflective surface, that is bright only for some frames.

If all the above conditions are satisfied, then the pixel is tagged as a hypothesis.

4.4 Adding Hypothesis

The image of the ball consists of many pixels, that could be tagged as hypotheses. Since we want to have one hypothesis for each possible position, we group hypotheses, that belong to the same position. Each hypothesis has a group number and the pixel position. We add hypothesis A to the list according to the following pseudo-code:

```
for each hypothesis C in the list
    if A.position is close to C.position
        A.group = C.group;
        list.add(A);
        break;
    end;
end;
if A.group not defined yet
    create new group number G;
    A.group = G;
    list.add(A);
end;
```

4.5 Group Size Restriction

Previous step provides us grouped hypotheses. Each group may contain several hypotheses, which cover a certain area. Using the positions of the hypotheses we calculate the smallest non rotated rectangle(group area) that contains all the hypotheses in a group.

To remove objects that are too large to be a golf ball, we check the size of the group area. In case it is too large, the whole group of hypotheses is discarded.

4.6 Restricting Hypothesis Search Space Using Kalman Filter

Searching for the ball can be made more effective by reducing the search area. The ball moves according to the laws of physics and there is no reason to search in the area, where the ball can not be. The ball can enter the frame only at one of the edges(we don't use the videos where the player hitting the ball is visible) and then travel on a quite predictable trajectory, that can be predicted using the ball speed and position in the previous frames. Using that information, the size and position of the search window is determined. The equations of the moving ball in the frame k are as follows:

$$x(k) = x(k-1) + v_\mathrm{x}(k-1) \tag{1}$$
$$y(k) = y(k-1) + v_\mathrm{y}(k-1) \tag{2}$$
$$v_\mathrm{x}(k) = drag * v_\mathrm{x}(k-1) \tag{3}$$
$$v_\mathrm{y}(k) = drag * v_\mathrm{y}(k-1) + a_\mathrm{y}(k-1) \tag{4}$$
$$a_\mathrm{y}(k) = a_\mathrm{y}(k-1) \tag{5}$$

Fig. 2. Trajectory consisting of selected hypothesis points connected by lines drawn over the image. Sand wedge from the distance approx. 30 meters was used.

Although this model is just an approximation of the real physical model, the error is small enough to get useful results. One difference between the real and this model are the equations for the velocity. In reality the velocity should be subtracted by drag, while in this model, we multiply it by drag. This simplification does not induce a large error but makes the programming easier. In this equations time interval is not used, since it is assumed to be equal to 1.

The drag and acceleration constants were defined experimentally, but their accuracy is not that important, since these equations are used only when the ball in the previous frame has not been found. We use this model to construct a Kalman filter [8,9], which is then used to predict the position of the ball in the next frame. The predicted position is used to place the search window on the image and the size of the search window can be set larger to compensate the positioning error. This model does not expect the ball to bounce, but only to move forward on the trajectory, so we have to make the size of the search window large enough to include the balls after the unpredicted bounce.

4.7 Selecting the Hypothesis

When there is no already detected moving ball present in the frame, the hypothesis list is scanned, searching for the positions at the borders of the frame. After the matching hypothesis is found, the search window is set to a large value to be able to find the ball in the next frame. Having the positions of the ball in two consecutive frames allows us to initialize the Kalman filter using the position of the ball in the previous frame and the speed in pixels the ball has traveled from one frame to another.

The next position can now be predicted using the constructed Kalman filter. We search for the ball in the search window, which size is determined by the speed of the ball in the previous frame. After the hypothesis representing the ball is found, the filter is corrected using the measurement obtained from the image (the position and the speed of the ball). Since the found position of the ball is certain, we set the Kalman filter's measurement noise covariance to be 0.

Figure 2 presents trajectory overlaid over the last frame of the sequence, after the ball stops on the green.

5   Using Two Cameras

Using two or more cameras enables us to generate a 3D trajectory model. Cameras were positioned at the right side of the green with the angle around 90 degrees between them(Fig. 3). Video was acquired from both cameras and later manually synchronized in time on a computer up to ±0.5 frame interval of 20ms (25 frames per second). Obtained ball positions of both streams were then used to generate 3D trajectories. Video processing was done using Direct-Show and Visual c++ and stereo algorithms were implemented in Matlab. 3D trajectories can then be displayed in Matlab or exported to the visually more appealing X3D model, that also includes a green with the flag.

5.1   Synchronization

Cameras were not synchronized at the acquisition time. Stream from each camera was taken separately. We needed to find some points in the trajectories of the balls, that could be taken for synchronization reference points. Since the angle

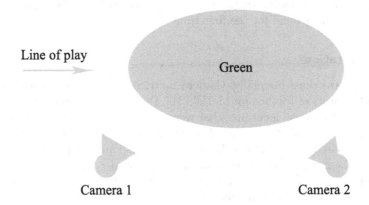

Fig. 3. Position of the cameras

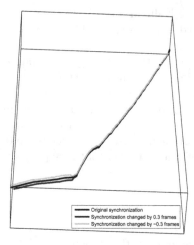

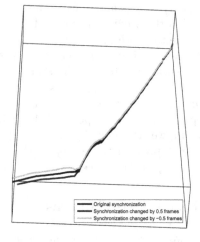

Fig. 4. Trajectories (t=0 on bottom left), where the synchronization has been changed by ±0.3 frames

Fig. 5. Trajectories (t=0 on bottom left), where the synchronization has been changed by ±0.5 frames

of the camera and field of view was different in each camera, we could not set the frame, where the ball entered the frame or the frame where the ball stopped as a reference point. For that reason the frame, where the ball first touched the ground (or just after that) was selected. Since the cameras record at only 25 frames per second, the synchronization was not perfect, but sufficient for our case. To test the effect of synchronization on the results, we interpolated the trajectory. Between each two points of the 2D trajectory of each camera, 9 new points were inserted. We changed the synchronization between the two videos by shifting one sequence of 2D points. Figures 4 and 5 show the effect of shifting one sequence by 3 and 5 points, which corresponds to 0.3 and 0.5 frames. The difference is mostly visible in the part, where the ball has the highest speed - before it hits the ground for the first time.

5.2   Calibration

Calibration was done manually. Camera matrices K_1, K_2 were obtained using Camera Calibration Toolbox for Matlab [10]. During video acquisition, we moved the flag stick to several locations, that were seen by both cameras. The flag stick had stripes of red and white color, that gave us the possibility to use more points than just top and bottom of the flag stick as correspondence points between two cameras. These were then input into the normalized 8 point algorithm [11,12] and the fundamental matrix F was obtained. Using the matrices F, K_1 and K_2 we computed the essential matrix E:

$$E = K_1^{\mathrm{T}} \cdot F \cdot K_2 \tag{6}$$

Fig. 6. Image of a trajectory as seen from the first camera

5.3   Structure Reconstruction

To compute structure, we first have to obtain rotation matrix R and translation vector t from E. This is done by computing singular value decomposition (SVD) of E:

$$E = U \cdot \Sigma \cdot V^{T} \tag{7}$$

Pairs R and $\hat{t}$ can then be obtained:[1]

$$W = \begin{bmatrix} 0 & -1 & 0 \\ 1 & 0 & 0 \\ 0 & 0 & 1 \end{bmatrix} \quad Z = \begin{bmatrix} 0 & 1 & 0 \\ -1 & 0 & 0 \\ 0 & 0 & 0 \end{bmatrix} \tag{8}$$

$$R_1 = R_3 = U \cdot W \cdot V^{T} \tag{9}$$
$$R_2 = R_4 = U \cdot W^{T} \cdot V^{T} \tag{10}$$
$$\hat{t}_1 = \hat{t}_2 = U \cdot Z \cdot U^{T} \tag{11}$$
$$\hat{t}_3 = \hat{t}_4 = -U \cdot Z \cdot U^{T} \tag{12}$$

We have to ensure, that the determinants of R matrices are positive. If it is negative, those matrices are negated. Only one pair out of these four gives us

---

[1] $\hat{\ }$ is an operator that generates a matrix $\hat{u}$ from a vector u such that $u \times v = \hat{u} \cdot v$ holds [13].

Fig. 7. Image of a trajectory as seen from the second camera

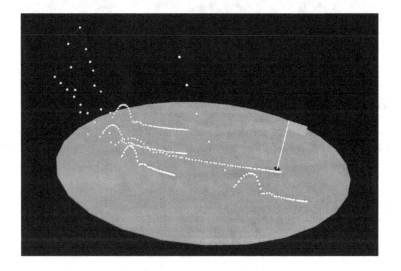

Fig. 8. Image of a view of multiple 3D trajectories in X3D viewer

the result, that places the observed points in front of both cameras. To find out which pair is the right one, we compute depth of points and check it is positive for both cameras.

Now that we have a matrix R, vector t and n pairs of correspondence points $< p_1^i, p_2^i >$, we can compute the depth of the points[14] by solving the equation:

$$M \cdot \lambda = 0 \qquad (13)$$

where the vector $\boldsymbol{\lambda}$ is defined as $\boldsymbol{\lambda} = [\lambda_1^1, \lambda_1^2, \ldots, \lambda_1^n, \gamma]$ and matrix $\boldsymbol{M}$ as:

$$
\boldsymbol{M} \doteq
\begin{bmatrix}
\widehat{\boldsymbol{p}_2^1} \cdot \boldsymbol{R} \cdot \boldsymbol{p}_1^1 & 0 & 0 & 0 & 0 & \widehat{\boldsymbol{p}_2^1} \cdot \boldsymbol{t} \\
0 & \widehat{\boldsymbol{p}_2^2} \cdot \boldsymbol{R} \cdot \boldsymbol{p}_1^2 & 0 & 0 & 0 & \widehat{\boldsymbol{p}_2^2} \cdot \boldsymbol{t} \\
0 & 0 & \ddots & 0 & 0 & \vdots \\
0 & 0 & 0 & \widehat{\boldsymbol{p}_2^{n-1}} \cdot \boldsymbol{R} \cdot \boldsymbol{p}_1^{n-1} & 0 & \widehat{\boldsymbol{p}_2^{n-1}} \cdot \boldsymbol{t} \\
0 & 0 & 0 & 0 & \widehat{\boldsymbol{p}_2^n} \cdot \boldsymbol{R} \cdot \boldsymbol{p}_1^n & \widehat{\boldsymbol{p}_2^n} \cdot \boldsymbol{t}
\end{bmatrix}
\tag{14}
$$

The equation is solved by computing the eigenvector of $\boldsymbol{M}^{\mathrm{T}} \cdot \boldsymbol{M}$ that corresponds to its smallest eigenvalue, which gives us the least-squares estimate of $\boldsymbol{\lambda}$. 3-D point coordinate $\boldsymbol{X}^i$ of a point $\boldsymbol{p}_1^i$ accurate up to a scale is computed as:

$$
\boldsymbol{X}^i = \lambda^i \boldsymbol{p}_1^i
\tag{15}
$$

The images from both cameras for one trajectory can be seen on Figs. 6,7. Image of the reconstructed X3D model with multiple trajectories is shown on Fig. 8.

6   Results

Using the described algorithm we successfully obtained ball positions in each video. There were some problems with detection in the areas, where the background is very similar to the ball. In that case Kalman filter gave us the estimate, which was then used as a ball position. In order to make detection work, some parameters needed to be set manually. Since the size of the ball is dependent on the camera position and its view angle, those parameters include the size of the ball, as well as the ball/background contrast and the ball color.

To test the synchronization effect on the 3D trajectory we interpolated the points between the captured ones in both cameras to make them more dense and shifted the synchronization by a few points. There was no major visual effect noticed. There was a minor change in the trajectory at the points where the ball was moving fast – before it hit the ground, but that was not relevant for our case as the angle of impact, bounces and roll remained visually the same.

7   Future Work

Setting the parameters manually takes some time, especially if we have videos shot from different positions or at different light conditions. We are researching a possibility of a semi-automatic parameter discovery, that would reduce the effort needed.

Visualization could be used for replays of short game shots on the tournaments. We could show different trajectories on the same image to show the viewer or to study the different tactics, that were used by different players.

Using some known length, e.g. the length of a flag stick, we could use the results to measure distances. In that case many usable statistics could be obtained. We could make games like closer to the hole, measure the average and deviation of the accuracy of different golfers, the distance between the point of impact and the point, where the ball stopped etc.

8   Conclusion

The article has described a useful application of the computer vision in golf. Players trying to improve their short game accuracy and consistency can see the trajectory of their shot and learn from it.

Displaying the ball trajectory in 3D gives them the possibility to view the shot from different angles. In that case they can see how their shot curved, looking from the view position they want. Visual trajectory representation is important for easier understanding of the bounces and rolling of the ball and surely helps getting the right feeling for selection of the right type of shot, when faced with the same situation on the golf course.

References

1. Urtasun, R., Fleet, D.J., Fua, P.: Monocular 3D Tracking of the Golf Swing. Computer Vision and Pattern Recognition 2, 932–938 (2005)
2. Woodward, A., Delmas, P.: Computer Vision for Low Cost 3-D Golf Ball and Club Tracking. In: Proc. Image and Vision Computing, New Zealand (2005)
3. Owens, N., Harris, C., Stennett, C.: Hawk-eye Tennis System. In: International Conference on Visual Information Engineering, pp. 182–185 (2003)
4. Pingali, G., Opalach, A., Jean, Y.: Ball Tracking and Virtual Replays for Innovative Tennis Broadcasts. In: 15th International Conference on Pattern Recognition, vol. 4, pp. 152–156 (2000)
5. Yu, X., Leong, H.W., Xu, C., Tian, Q.: Trajectory-Based Ball Detection and Tracking in Broadcast Soccer Video. IEEE Transactions on Multimedia 8(6), 1164–1178 (2006)
6. Ren, J., Orwell, J., Jones, G.A., Xu, M.: Tracking the soccer ball using multiple fixed cameras. Computer Vision and Image Understanding (in press)
7. Chien, S., Ma, S., Chen, L.: Efficient Moving Object Segmentation Algorithm Using Background Registration Technique. IEEE Transactions on Circuits and Systems for Video Technology 12(7), 577–586 (2002)
8. Petrie, T.: Tracking Bouncing Balls Using Kalman Filters and Condensation, http://www.marcad.com/cs584/Tracking.html
9. Maybeck, P.S.: Stochastic models, estimation and control, vol. 1. Academic Press, London (1979)
10. Bouguet, J.-Y.: Camera Calibration Toolbox for Matlab, http://www.vision.caltech.edu/bouguetj/calib_doc/
11. Kovesi, P.D.: MATLAB and Octave Functions for Computer Vision and Image Processing. School of Computer Science & Software Engineering, The University of Western Australia, http://www.csse.uwa.edu.au/~pk/research/matlabfns/
12. Hartley, R., Zisserman, A.: Multiple View Geometry in Computer Vision. Cambridge University Press, Cambridge (2003)
13. Wikipedia: Hat operator, http://en.wikipedia.org/wiki/Hat_operator
14. Ma, Y., Soatto, S., Košecká, J., Sastry, S.S.: An Invitation to 3-D Vision. Springer, Heidelberg (2006)

Integrated Digital Image Correlation for the Identification of Mechanical Properties

Hugo Leclerc, Jean-Noël Périé, Stéphane Roux, and François Hild

LMT-Cachan,
(ENS Cachan/CNRS/UPMC/PRES UniverSud Paris)
61 avenue du Président Wilson, F-94235 Cachan Cedex, France
{hugo.leclerc,jean-noel.perie,stephane.roux,francois.hild}@lmt.
ens-cachan.fr

Abstract. Digital Image Correlation (DIC) is a powerful technique to provide full-field displacement measurements for mechanical tests of materials and structures. The displacement fields may be further processed as an entry for identification procedures giving access to parameters of constitutive laws. A new implementation of a Finite Element based Integrated Digital Image Correlation (I-DIC) method is presented, where the two stages (image correlation and mechanical identification) are coupled. This coupling allows one to minimize information losses, even in case of low signal-to-noise ratios. A case study for elastic properties of a composite material illustrates the approach, and highlights the accuracy of the results. Implementations on GPUs (using CUDA) leads to high speed performance while preserving the versatility of the methodology.

Keywords: Digital Image Correlation, Finite Element Method, GPU, material property identification.

1 Introduction

Among full field measurement techniques used in Solid Mechanics [1], white-light correlation based methods are emerging because of their versatility and simplicity of use. Digital Image Correlation (DIC) softwares give access to dense displacement fields by matching digital images shot at distinct stages of loading in a mechanical test. Initiated in the early 1980s [2,3], DIC is an alternative to classical extensometry in many occasions, for instance, to study soft materials, measure large strain levels, analyze localized phenomena or heterogeneous tests. In addition, DIC underwent many rapid developments in different directions:

- Many softwares are currently available for performing stereo-correlation (to evaluate 3D displacements on the surface of samples or structures [4]).
- Because of the progress of high-speed digital cameras, these methods now tend to be applied to dynamic problems and transient phenomena [5].
- New full 3D-DIC developments are proposed. 3D imaging techniques (such as computed microtomography, μCT, and Magnetic Resonance Imaging, or

A. Gagalowicz and W. Philips (Eds.): MIRAGE 2009, LNCS 5496, pp. 161–171, 2009.

MRI) are mainly used for imaging purposes. Recent progress in scanning and reconstruction techniques allows one to get images of textured materials whose quality is sufficient to measure 3D displacements between two deformed states [6,7,8,9,12]. The challenge is now to process the large amount of image data in a reasonable time, and to exploit the measured displacement fields.

– Nowadays, DIC can also be used to drive (complex) experiments [13].

In Solid Mechanics, most of the used techniques are based on local matching procedures [4]. It consists in maximizing the cross-correlation function. Conversely, variational formulations may be used. They are mainly based on the brightness conservation equation [14,15,16]. A spatial regularization was introduced by Horn and Schunck [15] and consists in looking for smooth displacement solutions. However, this method is not appropriate for problems dealing with discontinuities in the apparent displacement [17]. In the latter case, the quadratic penalization is replaced by "smoother" ones based, for example, on robust statistics [18,19]. Furthermore, when dealing with deformable solids, other regularization techniques are introduced such as that based on the strain energy [20]. It can be noted that problems as complex as face tracking with three dimensional motions and deformations are handled by using *ad hoc* procedures [21].

However, the most attractive development of DIC in Solid Mechanics lies in its ability to identify parameters of constitutive laws [22] characterizing the mechanical behavior of materials or structures. Among the proposed identification methods [23], a widespread technique consists in updating the material properties in a Finite Element simulation (also referred to as Finite Element Model Updating, or FEMU) to reduce the difference between measured displacements and simulated ones [24]. The present paper focuses on this objective, and presents a novel method that conciliates the best of DIC and FEMU, while avoiding most of intermediate steps, together with a specific GPU implementation leading to considerable computation time savings.

The DIC method used herein is based on the Brightness Conservation (BC) assumption [14] written on a global level. However, instead of introducing spatial regularizations [15,18,17,19,20,21] with no or remote mechanical content, one rather takes advantage of meaningful (i.e., *mechanical*) bases to decompose the sought displacement field. The chosen DIC formulation [9,25] is such that any displacement basis is easily incorporated in the formulation. Among those, the simplest (without much mechanical content though), is the Finite Element (FE) Method. The structure is discretized using a set of finite elements and associated mesh. In that case, the DIC problem consists in looking for the nodal displacements minimizing a weak form of the BC functional. In each element, piecewise polynomial basis functions are used. The same formulation is used in 3D cases [12], where the challenge consists in performing time efficient computations, and also in dealing with a considerable amount of data (scans reach $2000 \times 2000 \times 2000$ voxels with 16-bit deep graylevel encoding).

General purpose graphics cards may then constitute a cost effective solution for performing massive parallel treatment using a standard PC. Graphics

Processing Units (GPUs) are for example more and more used for scientific purposes [26,27,28]. This new trend is referred to as General Purpose computation on Graphics Processing Unit (GPGPU). Considering the increasing need for fast and accurate Digital Image Correlation methods, usually highly parallelizable, a prototype of GPU-devoted DIC was developed.

The aim of the paper is to present the consequences of this kind of implementation in terms of applications and performances, in the field of mechanical analyses. The principle of the used FE-DIC method is first presented. In the second part, a specific form of the updating technique for identification (FEMU) is presented on a first application where kinematic fields are input data to identify material properties. In the third part, a new coupled DIC-FEMU procedure, designed to minimize the effect of a low Signal-to-Noise Ratio (SNR) is introduced. In the last part, implementation details are explained, including comments on speed performances.

2 Digital Image Correlation

2.1 Principle of the Proposed FE Scheme

The analysis of two gray level images f and g (f being the reference picture, and g the deformed one) is performed using the BC hypothesis that means that the image texture is passively advected by a displacement field $\mathbf{u}$, or

$$g(\mathbf{x}) = f(\mathbf{x} + \mathbf{u}(\mathbf{x})) \tag{1}$$

The problem consists in *identifying* the best displacement field by minimizing the correlation residual functional, Φ^2,

$$\Phi^2[\mathbf{u}] = \int_\Omega \varphi(\mathbf{x})^2 d\mathbf{x} \tag{2}$$

where

$$\varphi(\mathbf{x}) = |f(\mathbf{x} + \mathbf{u}(\mathbf{x})) - g(\mathbf{x})| \tag{3}$$

The minimization of Φ is intrinsically a non-linear and ill-posed problem. For these reasons, a discrete and weak format is preferred by adopting a general discretization scheme

$$\mathbf{u}(\mathbf{x}) = \sum_{n \in \mathcal{N}} u_n \boldsymbol{\psi}_n(\mathbf{x}) = [\boldsymbol{\psi}(\mathbf{x})]\{\mathbf{u}\} \tag{4}$$

where $\boldsymbol{\psi}_n$ are the vector shape functions, and u_n their associated degrees of freedom. In a matrix-vector format, $[\boldsymbol{\psi}]$ is a row vector containing the values of the shape functions $\boldsymbol{\psi}_n$ and $\{\mathbf{u}\}$ the column vector of the degrees of freedom. At this level of generality, one may choose to decompose the displacement field $\mathbf{u}(\mathbf{x})$ on a more or less "mechanically rich" basis. One can for example use classical FE shape functions $N_n(\mathbf{x})$. In an element Ω^e, the interpolated displacement $\mathbf{u}^e(\mathbf{x})$ then reads

$$\mathbf{u}^e(\mathbf{x}) = \sum_{n=1}^{n_e} \sum_\alpha a^e_{\alpha n} N_n(\mathbf{x}) \mathbf{e}_\alpha \tag{5}$$

where n_e is the number of nodes and $a^e_{\alpha n}$ the unknown nodal displacements. If one now minimizes the global residual Φ^2, one obtains a linear system $\mathbf{Ma} = \mathbf{b}$, where $\mathbf{a}$ is the vector of unknown nodal displacements. The matrix $\mathbf{M}$ and the right hand side vector $\mathbf{b}$ are respectively assembled using their elementary components $\mathbf{M}^e$ and $\mathbf{b}^e$

$$\mathbf{M}^e_{\alpha n \beta m} = \int_{\Omega_e} [N_m(\mathbf{x}) N_n(\mathbf{x}) \partial_\alpha f(\mathbf{x}) \partial_\beta f(\mathbf{x})] d\mathbf{x} \tag{6}$$

and

$$\mathbf{b}^e_{\alpha n} = \int_{\Omega_e} [f(\mathbf{x}) - g(\mathbf{x})] N_n(\mathbf{x}) \partial_\alpha f(\mathbf{x}) d\mathbf{x} \tag{7}$$

Up to now, classical bilinear shape functions associated with quadrilateral 4-node (Q4) or cubic 8-node elements (C8) were chosen for treating 2D (Q4-DIC [25]) and 3D images (C8-DIC [9]). In presence of discontinuous displacement fields (e.g., cracks), one may use enriched interpolation schemes such as those proposed in X-FEM approaches [10]. The method was tested with 4-node elements in 2D (XQ4-DIC [11]) and 8-node elements in 3D (XC8-DIC [12]). The method can naturally be generalized to any other kind of element in terms of type or interpolation degree. The reader is referred to the above mentioned references for further information concerning the DIC methodology, such as a multiscale procedure that is essential for robust convergence.

In the convenient framework of the in-house developed "LMT platform," one simply selects the type of elements used, and the associated interpolation schemes will be automatically generated [29].

3   Two-Stage Identification Procedure

The measured displacement fields can be used as input data for an identification procedure, e.g., to tune the parameters of a constitutive law. The elastic problem illustrated in Figure 1 is chosen as an example. A first snapshot of a Region Of Interest (ROI) is taken before the specimen is loaded biaxially, and a second picture is captured during loading. Digital Image Correlation allows one to measure the displacement field of the specimen.

The FEMU method [24] is a convenient way to identify mechanical properties, starting only from a picture sequence. It consists in using the displacement fields on the non free nodes of the mesh boundary as Dirichlet conditions for mechanical simulations. Simulations are run with varying parameter sets, until DIC and numerical displacements match at best. It is worth recalling that the accuracy of nodal displacements obtained from the DIC procedure is spatially varying as a result of the texture being non uniform. "Low contrasted" areas lead to high

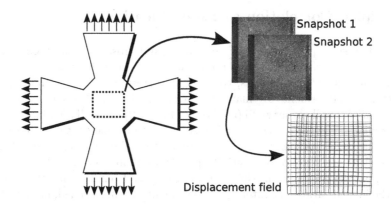

Fig. 1. First step of a two-stage material property identification procedure. Displacements are obtained on a mesh that is subsequently used for comparison with FE numerical simulations.

noise sensitivity, and should carry a smaller weight than "high contrasted" ones. One can prove that the best scalar product to quantify the matching between DIC and computed nodal displacements is

$$\langle \mathbf{a}, \mathbf{b} \rangle_\mathbf{M} = \langle \mathbf{a}, \mathbf{Mb} \rangle \tag{8}$$

where $\mathbf{M}$ is the assembled DIC matrix (see Equation (6)), which is symmetric and positive by construction.

In the case of fully differentiable energies, one can compute the derivatives of the simulated nodal displacement vector, $\mathbf{u}_s$, with respect to each material parameter, $\mathbf{p}_i$. Let $\mathbf{u}_d$ denote the nodal displacement vector obtained from DIC. The following functional

$$\mathcal{T}(\mathbf{p}_i) = \|\mathbf{u}_s(\mathbf{p}_i) - \mathbf{u}_d\|_\mathbf{M} \tag{9}$$

is to be minimized. This is achieved iteratively through successive linearizations

$$\mathbf{M} \left(\sum_i \frac{\partial \mathbf{u}_s}{\partial \mathbf{p}_i} (\mathbf{p}_i^{n+1} - \mathbf{p}_i^n) + \mathbf{u}_s(\mathbf{p}_i^n) - \mathbf{u}_d \right) = 0 \tag{10}$$

Figure 3 shows the change in $\mathbf{p}$ parameters, $\Delta \mathbf{p}^n$. In that case, only Poisson's ratio was searched for, using data shown in Figure 2. The convergence rate, $\log_{10}(\Delta \mathbf{p}^{n+1}/\Delta \mathbf{p}^n)$ is approximately 1.3. Only one local minimum is found.

One has to emphasize that material properties identification may become far more complex. Strong non-linearities may lead to several local minima and a degradation of the convergence rate. For illustrative examples and solutions, one may refer to Ref. [30]. In all cases, this two-step procedure leads to a trade-off to manage image noise. Coarse meshes are less subject to noise but cannot capture accurately the actual kinematics. Fine meshes allow one to represent complex displacement fields but the uncertainty level becomes detrimental. Figure 4 illustrates this point.

4   Coupled Correlation and Identification

Using the two-stage procedure (pure correlation and subsequent mechanical analysis), the noise sensitivity becomes dominant for small element sizes (Figure 4) yielding solutions, *if it converges*, that might correspond to local minima. Unfortunately, small elements are mandatory when dealing with "complex" geometries, e.g., with small corner radii (as those in Figure 2) or small angles, which are not uncommon. However, as one can do with any Finite Element representation, it is possible to use meshes refined only around "complex" borders. Nonetheless, for those elements, the noise sensitivity problem remains prominent.

The approach presented in this section originates from the idea that, at the end of the identification, the internal displacements are not needed, namely, they

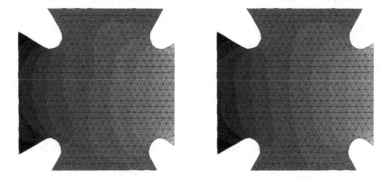

Fig. 2. Example of measured (left) and computed (right) displacement fields. The distance, defined in Equation (8), between these fields was minimized with respect to elastic parameters.

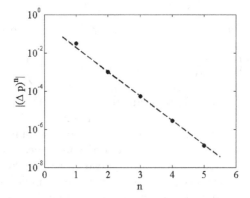

Fig. 3. Semi-log plot of the change in $\Delta \mathbf{p}^n$ parameters versus iteration number n. The data points are depicted by ($\bullet$), whereas the dotted line shows a regression onto an exponential decrease (for evaluation of Poisson's ratio, using data from the previous figure).

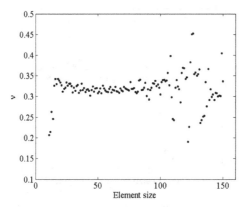

Fig. 4. Fitted Poisson's ratio vs. mesh size (in pixels). Coarse meshes lead to poor kinematic representations whereas fine meshes lead to high noise sensitivity.

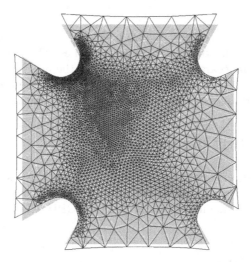

Fig. 5. Example of a mesh used for the coupled DIC-FEMU procedure. Element size is a decreasing function of $\|\partial \mathbf{u}_s / \partial \mathbf{p}_i\|$.

are only intermediate data, within a global procedure whose final goal is to find material properties. For mechanical simulations, one needs *at least* the Dirichlet boundary conditions (displacements) on non free borders. Thus, it is possible to contract the correlation problem such as the remaining unknowns are only the displacements on the non free borders, which are simple curves and do not need fine representations, and the material properties. For the experiment shown in Figure 2, one may for example use meshes like that of Figure 5.

Taking advantage of the freedom of using any displacement basis for the DIC analysis, one can compute with those fields *directly* from the FE modeling, instead of using a standard FE representation. In that case, the displacement basis

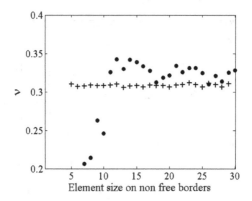

Fig. 6. Poisson's ratio versus element size (in pixels) around the free borders (elements on the non free border are of constant sizes) as obtained from the coupled DIC-FEMU procedure (+). The • symbols show the same quantity obtained with a uniform element size using the two-step procedure.

has a strong mechanical content. Figure 6 shows the change of Poisson's ratio ν *versus* the element size on high $\mathbf{p}_i$ sensitive zones. The non free borders are meshed with elements coarse enough to minimize the noise sensitivity. The stability of ν-estimates is quite spectacular as compared to the previous approach. Within the explored range of element size, $\nu = 0.308 \pm 0.003$.

5   CPU and GPU Implementation

Speed is a major concern. For this application, standard tools may lead to prohibitive running times. One had to look for specific implementations. We present here the main concepts behind the tools used for a GPU (CUDA compatible [31]) implementation.

5.1   Code Generation

A Finite Element representation implies an important set of assumptions but the user is free to choose the type of element. It would be possible to write each specific subroutine by hand, but hard-coding is a time consuming task. In the end, when one has to manage specific cases, actual running times are always a compromise between what a machine can do (e.g., in terms of floating point operations per second), and what a developer can do (in terms of available time).

Symbolic computing and automatic code generation is an attractive route to push farther the limits imposed by this trade-off. The main idea is to provide the computer with non degenerated information to let a pre-compiler manipulate the expressions, and choose the best options to generate an efficient code. For instance, in the LMT platform (an in-house built software platform dedicated to solid mechanics applications [29]), a triangular element is defined by a code like

```
class Triangle inherits NameElement
    static points   := [[0,0],[1,0],[0,1]] # ref space
    static def interval_var_inter( vi )
        return [ [0,1], [0,1-vi[0]] ]
```

This is sufficient to find, e.g., the shape functions and inverse functions, or at least ways to generated a code to find the values of the shape functions, starting from a pixel position. In the case of a triangle, the functions are linear, allowing for the generation of a code without any loops nor tests. For higher order elements, the code generator writes loops for all types of possible outputs. Up to now, C++, CUDA, x86 and x86_64 codes may be generated, allowing for an adaptation to different kinds of hardware.

Given a particular symbolic (representative) element e, one uses for example the following code to get some of the expressions to be generated (see Equations (6) and (7))

```
# e is of type SymbolicElement (e.g., triangle, ...)
var_inter := e.var_inter_for_pos( P )
mask   := e.var_inter_is_inside( var_inter )
phi   := e.shape_functions.subs( var_inter ) * mask
```

5.2 GPU Specific Optimizations

Working with NVIDIA boards, CUDA programming language is a natural choice. A multiprocessor manages one element at a time while threads manage one pixel inside the element of the corresponding multiprocessor. Registers and shared memory are fully used to store intermediate values. The `tex2D` instruction is of primary importance to get high efficiency, as cache and interpolations are automatically managed. With the current implementation of the procedure, the construction of matrix M for 1024^2-pixel pictures meshed by regular quads with sizes greater than 64 pixels requires about 2 ms. For a triangular mesh, one needs approximately 5 ms for the same picture at a medium precision, unless element sizes become small (leading to poor parallelism). The complete procedure from the pictures to the estimate of Poisson's ratio for the example reported in this study requires, 600, 30 and 650 ms, respectively for the DIC, FEMU, and coupled DIC-FEMU analyses. A new implementation for small element sizes is in progress.

6 Conclusions

A specific global Digital Image Correlation procedure was proposed. It is based on arbitrary Finite Element mesh and shape functions. This allows one to express the measured displacement field with the same description as that used for numerical computations. The identification of constitutive parameter based on DIC measurements was presented. The latter makes use of a specific scalar product issued from DIC that minimizes the sensitivity to noise.

Furthermore, the flexibility of the finite element representation offers a way to couple DIC and FEMU for enhanced accuracy and robustness of the identification. A simple elastic identification was presented as an illustration of the present procedure. However, the presentation of the methodology has been kept as general as possible to encompass within the same formalism non-linear constitutive laws.

Last, thanks to GPU computing and automatic code generation, the actual implementation leads to computation times within 1 second for Mbyte images.

Acknowledgements

The support of this research by the "Agence Nationale pour la Recherche" is gratefully acknowledged (VULCOMP project).

References

1. Rastogi, P.K. (ed.): Photomechanics, p. 77. Springer, Berlin (2000)
2. Sutton, M.A., Wolters, W.J., Peters, W.H., Ranson, W.F., McNeill, S.R.: Determination of Displacements Using an Improved Digital Correlation Method. Im. Vis. Comp. 1(3), 133–139 (1983)
3. Sutton, M.A., Cheng, M., Peters, W.H., Chao, Y.J., McNeill, S.R.: Application of an optimized digital correlation method to planar deformation analysis. Im. Vis. Comp. 4(3), 143–150 (1986)
4. Sutton, M.A., McNeill, S.R., Helm, J.D., Chao, Y.J.: Advances in Two-Dimensional and Three-Dimensional Computer Vision. In: Rastogi, P.K. (ed.) Photomechanics, pp. 323–372. Springer, Berlin (2000)
5. Elnasri, I., Pattofatto, S., Zhao, H., Tsitsiris, H., Hild, F., Girard, Y.: Shock enhancement of cellular structures under impact loading: Part I Experiments. J. Mech. Phys. Solids 55, 2652–2671 (2007)
6. Verhulp, E., van Rietbergen, B., Huiskes, R.: A three-dimensional digital image correlation technique for strain measurements in microstructures. J. Biomech. 37(9), 1313–1320 (2004)
7. Liu, L., Morgan, E.: Accuracy and precision of digital volume correlation in quantifying displacements and strains in trabecular bone. J. Biomech. 40, 3516–3520 (2007)
8. Lenoir, N., Bornert, M., Desrues, J., Bésuelle, P., Viggiani, G.: Volumetric digital image correlation applied to X-ray microtomography images from triaxial compression tests on argillaceous rock. Strain 43, 193–205 (2007)
9. Roux, S., Hild, F., Viot, P., Bernard, D.: Three dimensional image correlation from X-Ray computed tomography of solid foam. Comp. Part A 39(8), 1253–1265 (2008)
10. Moës, N., Dolbow, J., Belytschko, T.: A finite element method for crack growth without remeshing. Int. J. Num. Meth. Eng. 46(1), 133–150 (1999)
11. Réthoré, J., Hild, F., Roux, S.: Extended digital image correlation with crack shape optimization. Int. J. Num. Meth. Eng. 73(2), 248–272 (2008)
12. Réthoré, J., Tinnes, J.-P., Roux, S., Buffière, J.-Y., Hild, F.: Extended three-dimensional digital image correlation (X3D-DIC). C. R. Mécanique 336, 643–649 (2008)

13. Fayolle, X., Calloch, S., Hild, F.: Controlling testing machines with digital image correlation. Exp. Tech. 31(3), 57–63 (2007)
14. Fennema, C., Thompson, W.: Velocity determination in scenes containing several moving objects. Comput. Graph. Im. Proc. 9, 301–315 (1979)
15. Horn, B.K.P., Schunck, B.G.: Determining optical flow. Artificial Intelligence 17, 185–203 (1981)
16. Simoncelli, E.P.: Bayesian Multi-Scale Differential Optical Flow. In: Jähne, B., Haussecker, H., Geissler, P. (eds.) Handbook of Computer Vision and Applications, pp. 297–422. Academic Press, London (1999)
17. Mitiche, A., Bouthemy, P.: Computation and analysis of image motion: A synopsis of current problems and methods. Int. J. Comp. Vision 19, 29–55 (1996)
18. Black, M.: Robust Incremental Optical Flow, Ph.D dissertation, Yale University (1992)
19. Odobez, J.-M., Bouthemy, P.: Robust multiresolution estimation of parametric motion models. J. Visual Comm. Image Repres. 6, 348–365 (1995)
20. Bogen, D., Rahdert, D.: A strain energy approach to regularization in displacement field fits of elastically deforming bodies. IEEE Trans. Pattern Analysis and Machine Intelligence 18, 629–635 (1996)
21. DeCarlo, D., Metaxas, D.: Optical flow constraints on deformable models with application to face tracking. Int. J. Comp. Vision 38, 99–127 (2000)
22. Roux, S., Hild, F.: Digital Image Mechanical Identification (DIMI). Exp. Mech. 48(4), 495–508 (2008)
23. Avril, S., Bonnet, M., Bretelle, A.-S., Grédiac, M., Hild, F., Ienny, P., Latourte, F., Lemosse, D., Pagano, S., Pagnacco, E., Pierron, F.: Overview of identification methods of mechanical parameters based on full-field measurements. Exp. Mech. 48(4), 381–402 (2008)
24. Kavanagh, K.T., Clough, R.W.: Finite Element Applications in the Characterization of Elastic Solids. Int. J. Solids Struct. 7, 11–23 (1971)
25. Besnard, G., Hild, F., Roux, S.: "Finite-element" displacement fields analysis from digital images: Application to Portevin-Le Chatelier bands. Exp. Mech. 46, 789–803 (2006)
26. Belleman, R.G., Bédorf, J., Portegies Zwart, S.F.: High performance direct gravitational N-body simulations on graphics processing units II: An implementation in CUDA. New Astron. 13(2), 103–112 (2008)
27. Göddeke, D., Strzodka, R., Mohd-Yusof, J., McCormick, P., Buijssen, S.H.M., Grajewski, M., Turek, S.: Exploring weak scalability for FEM calculations on a GPU-enhanced cluster. Parallel Comput. 33, 685–699 (2007)
28. Gölddeke, D., Strzodka, R., Turek, S.: Performance and accuracy of hardware-oriented native-, emulated- and mixed-precision solvers in FEM simulations result. Int. J. Parallel, Emerg. Distrib. Syst. 22(4), 221–256 (2007)
29. Leclerc, H.: Plateforme metil : optimisations et facilités liées à la génération de code. In: Proc. 8e Colloque National en Calcul des Structures, Giens (2007)
30. Cooreman, S., Lecompte, D., Sol, H., Vantomme, J., Debruyne, D.: Elasto-plastic material parameter identification by inverse methods: Calculation of the sensitivity matrix. Int. J. Solids Struct. 44(13), 4329–4341 (2007)
31. NVIDIA Corporation, NVIDIA CUDA compute unified device architecture programming guide (2007), http://developer.nvidia.com/cuda

Recovery of 3D Solar Magnetic Field Model Parameter Using Image Structure Matching

Jong Kwan Lee[1] and G. Allen Gary[2]

[1] Department of Computer Science
Bowling Green State University, Bowling Green, OH 43403, USA
leej@bgsu.edu
[2] Center for Space Plasma and Aeronomic Research,
University of Alabama in Huntsville, Huntsville, AL 35899, USA
allen.gary@nasa.gov

Abstract. An approach to recover a 3D solar magnetic field model parameter using intensity images of the Sun's corona is introduced. The approach is a quantitative approach in which the 3D model parameter is determined via an image structure matching scheme. The image structure matching measures the positional divergence (i.e., pixel-by-pixel shortest Euclidean distance) between the real coronal loop structures in a 2D image to sets of modeled magnetic field structures to determine the best model parameter for a given region on the Sun. The approach's effectiveness is evaluated through experiments on synthetic images and a real image.

keywords: Structure Matching, Image-based Modeling, 3D Parameter Recovery.

1 Introduction and Background

Structure matching has been utilized in many problem domains, including pattern/object recognition (e.g., [4,8,23]), registration (e.g., [16]), image retrieval (e.g., [17]), etc. It may further be extended to derive underlying characteristics of an object or a physical model. In this paper, exploitation of an image structure matching scheme in recovering a 3D model parameter is described.

The target application of our work is the Sun's magnetism. Since the Sun is the most important source for life on Earth and its dynamic activities strongly impact our geo-space environment (e.g., solar storms can change the orbits of satellites and shorten satellite mission lifetimes [11]), study of the Sun is a high interest research topic. In particular, study of the solar magnetic field is the focus of important solar research since the Sun's dynamics are driven by its magnetic free energy.

3D magnetic field models are often used to aid the investigation of the solar magnetic field. One way these models are used is for hypothesizing properties of the solar magnetic field based on model properties (e.g., [7]). Thus, modeling of 3D magnetic field which well-characterizes the true solar magnetic field is crucial.

A. Gagalowicz and W. Philips (Eds.): MIRAGE 2009, LNCS 5496, pp. 172–181, 2009.
© Springer-Verlag Berlin Heidelberg 2009

Currently, solar scientists often employ a 3D magnetic field model with its key parameter (called *non-potentiality*) determined using a model parameter recovery method that considers only the Sun's 2D magnetic information near the surface (e.g., [5,6,7,15]). (This widely-used 3D solar magnetic field model and its parameter are discussed later in this section.) In particular, the parameter recovery method determines the magnetic field's non-potentiality using a *vector magnetogram* (i.e., a 2D photospheric image of local magnetic field direction and strength). However, since a vector magnetogram only contains 2D information near the solar surface, the measures from it may not be reliable bases to estimate the entire 3D field's non-potentiality.

In this paper, a model parameter recovery approach for the widely-used solar magnetic field model is investigated. The approach is based on a matching scheme which compares real 3D solar magnetic field structures in a 2D image with a set of possible magnetic field structures to recover the model parameter. Our approach extends the work of Carcedo et al. [3] and Wiegelmann et al. [21,22] to exploit the traces of 3D solar magnetic field in recovering the model parameter.

Next, some background of our focused solar magnetic field model is discussed.

1.1   3D Solar Magnetic Field Model

One of the most widely-used solar magnetic field models (which is the focus of our work) is the model introduced by Alissandrakis [1]. The model is called the *constant α force-free magnetic field model*. The model produces an estimate of the 3D magnetic field by applying the Fourier Transform to an image of the solar surface magnetic field (i.e., the *magnetogram*) which contains the magnetic flux density of the solar surface. The Fourier components of the Alissandrakis model's magnetic field are shown within Equation (1), which is a magnetic flux density expression:

$$B_x(u, v, z) = \frac{-i(uk - v\alpha)}{2\pi(u^2 + v^2)} e^{-kz} B_z(u, v, 0),$$

$$B_y(u, v, z) = \frac{-i(vk - u\alpha)}{2\pi(u^2 + v^2)} e^{-kz} B_z(u, v, 0), \tag{1}$$

$$B_z(u, v, z) = e^{-kz} B_z(u, v, 0),$$

where u and v are the variables in the Fourier domain (i.e., spatial frequency variables) for the x and y Cartesian coordinates, α is a magnetic field model parameter which represents the non-potentiality of the field, $k = \pm\sqrt{4\pi^2(u^2 + v^2) - \alpha}$, and B_x, B_y, and B_z are the transformed x, y, and z magnetic flux density components, respectively. As shown in Equation (1), α is the key free parameter when modeling a 3D solar magnetic field using the Alissandrakis model [1].

Non-potentiality (e.g., α in Equation (1)) of a magnetic field describes the amount of electric current flowing parallel to the magnetic field [19]. Non-potentia -lity can also be considered as a measure to describe the amount of twisting of the magnetic field, since magnetic current determines how much a magnetic field is twisted [19].

2   Related Work

In this section, the parameter recovery method [5] using the vector magnetogram is discussed briefly.

As mentioned previously, a vector magnetogram is a 2D image of local magnetic direction near the Sun's surface (i.e., in the innermost layer of the Sun's atmosphere). From a vector magnetogram, the net electric current (often denoted as I_N) flowing up (or down) from one of the magnetic bipolar regions (e.g., a positive pole or a negative pole) and the magnetic flux content (often denoted as Φ) of the active region of interest on the Sun can be determined. Using these measures, the non-potentiality (i.e., α) of the whole solar magnetic field is estimated using Equation (2):

$$\alpha = \frac{\mu I_N}{\Phi},$$
(2)

where μ is a constant (called the *permeability of free space*) of $4\pi \times 10^{-7}$.

However, the non-potentiality determined from Equation (2) may not be good one to represent the non-potentiality for the entire 3D solar magnetic field since the bases of the equation were measured only from near the Sun's surface.

3   Our Approach

Next, we describe our model parameter recovery approach for the constant α force-free magnetic field model [1].

As mentioned previously, our work extends the work of Carcedo et al. [3] and Wiegelmann et al. [22,21]. Carcedo et al. [3] attempted to determine the model parameter based on the normal directional divergences of magnetic field structures. Wiegelmann et al. [22] determined the model parameters for different loop sets of a solar region based on an iterative minimization process using a distance measure. Later, Wiegelmann et al. [21] compared the magnetic field structures generated from different magnetic field models using the same distance measure. (We note that Wiegelmann has also introduced a magnetic field reconstruction method [20] for other types of 3D solar magnetic field models.) Our approach extends these works by employing a new structure matching scheme using a high resolution solar image.

The approach introduced here differs from previous methods (that use only the vector magnetogram) by utilizing the real solar loop structures in the intensity images of the Sun's uppermost atmosphere, the corona. Figure 1 shows an

Fig. 1. Example of TRACE Coronal Intensity Image of the EUV Emission Above an Active Bipolar Region on the Sun

example coronal image of the Sun (taken from NASA's on going TRACE satellite mission [18]). In the figure, the tube-like thin arc structures are the coronal loop structures. The coronal loops structures are the visible traces of the 3D solar magnetic field [10]. Thus, fields with a higher magnitude of non-potentiality have coronal loop structures that appear more twisted than loops in a field with lower magnitude of non-potentiality.

The key part of our approach is to exploit the appearance (i.e., shapes) of the coronal loop structures in determining the non-potentiality using a new image structure matching scheme. Specifically, our structure matching scheme performs structure-by-structure comparisons between the loop structures in a coronal image and sets of possible magnetic field structures generated from the Alissandrakis model [1] to recover the α parameter.

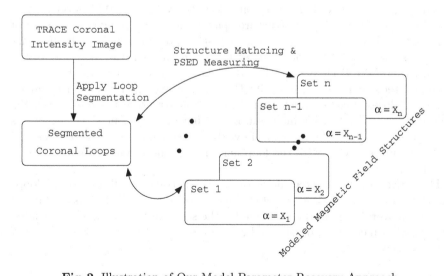

Fig. 2. Illustration of Our Model Parameter Recovery Approach

The model parameter recovery approach determines the α parameter of the magnetic field model by measuring the positional divergence between the real loop structures in the TRACE corona image and sets of possible magnetic field structures generated from the Alissandrakis model [1]. Specifically, this involves generating different sets of magnetic field structures using possible α values in the Alissandrakis model [1]. (The range of possible α values for modeling the solar magnetic field is discussed later in Section 4.) Each generated set are then compared with the real loop structures. The α value that generates a set of magnetic field structures which has the minimum divergence from the real loop structures is determined as the model parameter for the solar magnetic field. In particular, a set of modeled 3D magnetic field structures using a possible α value is projected onto a 2D image plane and the positional divergence is measured from the segmented real loop structures in the TRACE image to these projected magnetic field structures. We use pixel-by-pixel shortest Euclidean distances from the real loop structures to the modeled magnetic field structures as the positional divergence measure. We denote this distance measure the PSED measure.

Figure 2 shows an illustration of the new structure matching for our model parameter recovery approach. Each set of modeled magnetic field structures (generated with a different α value) is compared with the real coronal loop structures using structure matching scheme. Then, the α value of a magnetic field structure set that has the smallest overall PSED measure is determined as the model parameter.

The approach determines an overall PSED measure for the set of magnetic field structures associated with each possible α value. It is computed by first finding the average PSED for one particular structure of the set of modeled magnetic field structures associated with the α value. This one is the closest magnetic field structure (of each set) to the real coronal loop. It is selected by considering evenly-spaced small regions along the real loop. Whichever of each modeled set's magnetic field structures passes through the highest number of these regions is determined to be the closest one to the real loop. If there are multiple structures in the modeled set that pass through the same number of regions, average PSEDs for all of those magnetic field structures (to the real loop) is determined and the one with the smallest average PSED is determined to be the closest magnetic field structure. (If no structure passes through the regions, the small regions is increased in size until at least one structure passes through the regions.) The number of evenly-spaced small regions along the real loop structure can vary with the loop length (e.g., more small regions are used for a longer loop).

The PSEDs of the selected magnetic field structures for all real corona loops are then computed. The average of the selected structure's PSEDs is considered as its set's overall PSED. The α value for the set with the minimum associated overall PSED measure is determined as the best model parameter for the solar magnetic field.

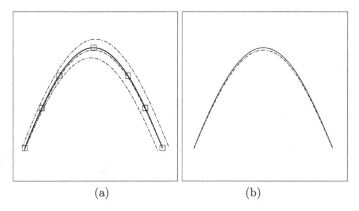

(a)   (b)

Fig. 3. Example of Closest Magnetic Field Structure Selection

The closest structure selection is illustrated using an example in Figure 3. In the figure, the solid curves represent the real coronal loops and the dotted curves represent the set of modeled magnetic field structures associated with one α value using the Alissandrakis model [1]. Figure 3 (a) shows the real loop and the modeled structures. The evenly-spaced small regions used in the structure matching process are shown as square boxes in Figure 3 (a). The real loop structure and the selected closest magnetic field structure for this case are shown in Figure 3 (b).

4   Experimental Results

We have evaluated the effectiveness of the new model parameter recovery approach using synthetic datasets and a real coronal image.

The first set of experiments using synthetic datasets involved creating 13 synthetic image datasets (each resulting from a different α value (± 0.012, ± 0.010, ± 0.008, ± 0.006, ± 0.004, ± 0.002, and 0.000 in units of inverse pixel lengths) in the Alissandrakis model [1]) and recovering the α parameter for each synthetic dataset. The α values used were reasonable since they are valid α values for the solar magnetic field (as described later). The synthetic images were created by first generating 3D magnetic field structures from the model and then projecting them onto a 2D image plane. These 2D projected modeled structures were used as the "real" image in our benchmarking.

The maximum magnitude of α, $|\alpha|$, for the solar magnetic field can be determined with respect to the size of the solar region viewed in a solar image [9]. Specifically, $|\alpha|$ should be less than $2\pi/L$, where L is the image size of the solar region viewed in pixel units. We have used a range of $[-0.012, 0.012]$ for the possible α values in the testing as this range is suitable for the size of the solar region viewed in the TRACE images (i.e., $L = 512$ for the TRACE images; thus, $|\alpha| < (2\pi/512) \approx 0.0122$).

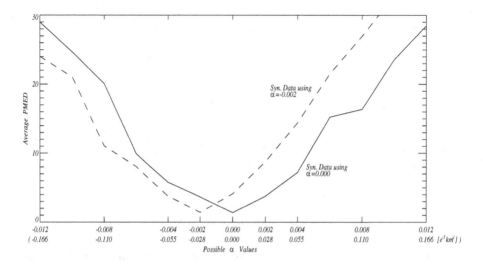

Fig. 4. Average PSED Measures for Two Synthetic Data

Table 1. Falsely Recovered α Values using Perturbed Synthetic Datasets

$\sigma = 10$		$\sigma = 20$	
True α	Recovered α	True α	Recovered α
-0.002	0.000	0.002	0.000
-	-	-0.002	-0.004
-	-	-0.006	-0.004
-	-	-0.008	-0.006
-	-	-0.010	-0.006

We have applied the new model parameter recovery approach to the 13 synthetic images. The loops in each synthetic image were compared to sets of magnetic field structures with different α values to determine the overall PSED measure. (Here, we note that the positions of the "modeled" magnetic field structures were slightly different from the "real" loops used for comparison since we assumed that the positions of "real" loops in 3D space were not known—the end points of the "modeled" structures may not coincide with the end points for the "real" loops.) Then, the best model parameter for each dataset was recovered by choosing the α value which has the smallest overall PSED measure.

Figure 4 shows the overall PSED measures determined by the new model parameter recovery approach for two of the 13 synthetic images (i.e., datasets generated using α values of -0.002 and 0.000). (Here, we note that the figure includes additional unit for solar scientists.) In the figure, a solid curve is used to show the overall PSED measures for the $\alpha = 0.000$ dataset. A dotted curve is used for the $\alpha = -0.002$ dataset. As shown in the figure, the overall PSED

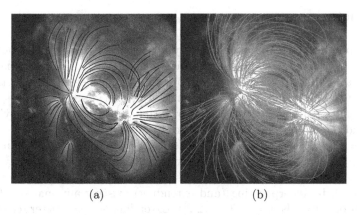

(a)   (b)

Fig. 5. Real TRACE Image Testing: (a) Segmented Coronal Loops (i.e., overlaid blue curves) for TRACE Image shown in Figure 1 and (b) Modeled Magnetic Field Structures (i.e., overlaid orange curves) with $\alpha = -0.006$

measures were the smallest for the true α values. Similar results were observed for the other 11 synthetic datasets; thus, our approach recovered all true α values for this set of testings.

The second set of experiments using the synthetic datasets included applying our approach to positionally-perturbed versions of the 13 synthetic datasets used previously. We perturbed the positions of real loops in each synthetic dataset by four levels of Gaussian noise. The added Gaussian noise has zero mean and σ of 1, 5, 10, and 20 (i.e., total number of 52 "perturbed" synthetic datasets were tested).

Using the perturbed synthetic datasets, our approach recovered the true α value for 46 cases. Table 1 shows the cases where our approach falsely recovered the α values. As shown in the table, our approach missed to recover the true α value for only one case of synthetic datasets with $\sigma = 10$ and only five cases of synthetic datasets with $\sigma = 20$.

We have also applied our model parameter recovery approach using one real coronal image and then visually-examined the correctness of recovered α value. Using TRACE corona image shown in Figure 1, we first manually-segmented the real coronal loops and then performed our structure matching scheme to determine PSED measures. (We used manual segmentation since existing automated coronal loop segmentation methods have some limitations.) In Figure 5 (a), the segmented loops are shown as blue overlays. For this real dataset, our approach determined -0.006 as the α value to model the solar magnetic field. The modeled magnetic field structures (with $\alpha = -0.006$) are shown as orange overlays in Figure 5 (b). As shown in the figure, the modeled magnetic field structure reasonably well-matches the coronal loop structures (which are the traces of the solar magnetic field). (Here, we note that the modeled structures shown in Figure 5 (b) include corresponding structures and other modeled structures to show how well all the modeled structures match with the coronal loop structures in the TRACE image.)

5   Conclusion

We have presented a parameter recovery approach for 3D solar magnetic field which employs an image structure matching scheme using a real solar image. The approach attempts to recover the non-potentiality parameter using the traces of 3D solar magnetic field. Through evaluation of the approach using synthetic datasets and a real dataset, we have shown that the approach can provide consistent and reasonable parameter recovery results.

For the future work, more complex cases can be considered. This can include result comparisons against the previous method's [5] results (which may include creating synthetic vector magnetograms). In addition, we may fully-automate our approach by incorporating (and extending) existing automated solar loop segmentation methods (e.g., [13,14,12,2]). Extending our parameter recovery approach to other solar magnetic field models may also be possible.

Acknowledgment

This work was partially supported by NASA's Office of Space Science–Solar and Heliospheric Physics Supporting Research and Technology Program.

References

1. Alissandrakis, C.E.: On the Computation of Constant α Force-free Magnetic Field. Astronomy and Astrophysics 100, 197–200 (1981)
2. Aschwanden, M.J., Lee, J.K., Gary, G.A., Smith, M., Inhester, B.: Comparison of Five Numerical Codes for Automated Tracing of Coronal Loops. Solar Physics 248, 359–377 (2008)
3. Carcedo, L., Brown, D.S., Hood, A.W., Neukirch, T., Wiegelmann, T.: A Quantitative Method to Optimise Magnetic Field Line Fitting of Observed Coronal Loops. Solar Physics 218, 29–40 (2003)
4. Cass, T.: Robust Affine Structure Matching for 3D Object Recognition. IEEE Trans. on Pattern Analysis and Machine Intelligence 20(11), 1265–1274 (1998)
5. Falconer, D.A.: A Prospective Method for Predicting Coronal Mass Ejections from Vector Magnetograms: Solar Eruptive Events. J. of Geophysical Research 106, 25185–25190 (2001)
6. Falconer, D.A., Moore, R.L., Gary, G.A.: Correlation of the Coronal Mass Ejection Productivity of Solar Active Regions with Measures of Their Global Nonpotentiality From Vector Magnetograms: Baseline Results. The Astrophysical J. 569, 1016–1025 (2002)
7. Falconer, D.A., Moore, R.L., Gary, G.A.: A Measure from Line-Of-Sight Magnetograms for Prediction of Coronal Mass Ejection. J. of Geophysical Research 108(A10), 1380–1386 (2003)
8. Jain, A., Hong, L., Bolle, R.: On-line fingerprint Verification. IEEE Trans. on Pattern Analysis and Machine Intelligence 19(4), 302–314 (1997)
9. Gary, G.A.: Linear Force-free Magnetic Fields for Solar Extrapolation and Interpretation. The Astrophysical J. Supplement Series 69, 323–348 (1989)

10. Gary, G.A.: Plasma Beta above a Solar Active Region: Rethinking the Paradigm. Solar Physics 203, 71–86 (2003)
11. NASA/Marshall Space Flight Center-Solar Physics, http://solarscience.msfc.nasa.gov/whysolar.shtml
12. Inhester, B., Feng, L., Wiegelmann, T.: Segmentation of Loops from Coronal Images. Solar Physics 248, 379–398 (2008)
13. Lee, J.K., Newman, T.S., Gary, G.A.: Oriented Connectivity-based Method for Segmenting Solar Loops. Pattern Recognition 39, 246–259 (2006)
14. Lee, J.K., Newman, T.S., Gary, G.A.: Dynamic Aperture-based Solar Loop Segmentation. In: 7th IEEE Southwest Symposium on Image Analysis and Interpretation, Denver, pp. 91–94 (2006)
15. Leka, K.D., Barnes, G.: Photospheric Magnetic Field Properties of Flaring Versus Flare-quite Active Regions. I. Data, General Approach, and Sample Results. The Astrophysical J. 595, 1277–1295 (2003)
16. Petti, P.L., Kessler, M.L., Fleming, T., Pitluck, S.: An Automated Image-registration Technique Based on Multiple Structure Matching. Medical Physics 21(9), 1419–1426 (1994)
17. Ravela, S., Manmatha, R., Riseman, E.M.: Image Retrieval Using Scale-Space Matching. In: Buxton, B.F., Cipolla, R. (eds.) ECCV 1996. LNCS, vol. 1064, pp. 273–282. Springer, Heidelberg (1996)
18. Schrijver, C.J., Title, A.M., Berger, T.E., Fletcher, L., Hurlburt, N.E., Nightingale, R.W., Shine, R.A., Tarbell, T.D., Wolfson, J., Golub, L., Bookbinder, J.A., Deluca, E.E., McMullen, R.A., Warren, H.P., Kankelborg, C.C., Handy, B.N., De Pontieu, B.: A New View of Solar Outer Atmosphere by the Transition Region and Coronal Explorer. Solar Physics 187, 261–302 (1999)
19. Schrijver, C.J., Zwaan, C.: Solar and Stellar Magnetic Activity. Cambridge University Press, New York (2000)
20. Wiegelmann, T.: Optimization Code with Weighting Function for the Reconstruction of Coronal Magnetic Fields. Solar Physics 219, 87–108 (2004)
21. Wiegelmann, T., Lagg, A., Solanki, S.K., Inhest, B., Woch, J.: Comparing Magnetic Field Extrapolations with Measurements of Magnetic Loops. Astronomy & Astrphys. 433, 701–705 (2005)
22. Wiegelmann, T., Neukirch, T.: Including Stereoscopic Information in the Reconstruction of Coronal Magnetic Fields. Solar Physics 208, 233–251 (2002)
23. Zhang, W., Wang, Y.: Core-based Structure Matching Algorithm of Fingerprint Verification. In: 16th Int'l Conf. on Pattern Recognition, Quebec City, Canada, pp. 70–74 (2002)

From Interactive Positioning to Automatic Try-On of Garments

Tung Le Thanh and André Gagalowicz

INRIA, The French National Institute for Research in Computer Science and Control
Domaine de Voluceau-Rocquencourt B.P. 105. 78153 Le Chesnay Cedex France
{tung.lethanh,andre.gagalowicz}@inria.fr
http://www-rocq.inria.fr/mirages/

Abstract. With the large development of computer graphics hardware and virtual cloth simulation techniques, virtual try-on of garments has become possible. An Internet-based shop will reduce significantly the cost of garment manufacturing as only paid garments will be manufactured. On top of that, a client will also be able to buy exactly what he wants and garments will be fitted to him/her. In this paper, we describe a virtual try-on system that allows a particular user to try garments easily. After loading his/her 3D digital model, he will be able to select the clothes he wants from a data base and see himself/herself wearing it virtually in 3D. The proposed system consists of two major parts : first, the 2D positioning part which requires a designer's interaction; second, a 3D part, fully automatic which will be installed in the user's computer, which allows tim to access a garments'catalog and see himself wearing virtually the chosen garment.

Keywords: Cloth simulation, garment, interactive positioning, virtual try on, database, web interface.

1 Introduction and Motivation

For more than two decades, the research area of virtual garment simulation received much attention and was applied in a wide domain, from movie industry to fashion and games.

In recent years, major advances have been achieved due to a deeper understanding of the physical behavior of cloth, major improvements in the control of the stability and accuracy of numerical computations. Much progress has been made towards visually pleasing and fast as well as correct simulations. With the quick evolution of computer graphics hardware nowadays, a garment simulation process can be realized within a short time. Efficient cooperations between cloth manufacturing industry and computer graphics have become possible. A major interesting research is dedicated to the realization of a 'virtual fitting room', where techniques and technologies for virtual garment management, cloth assembly, try-on evaluation and buying have to be developed. The goal is to reduce cloth manufacturing and stock costs by using internet facilities. We want to cover the whole process leading to this technology.

A. Gagalowicz and W. Philips (Eds.): MIRAGE 2009, LNCS 5496, pp. 182–194, 2009.

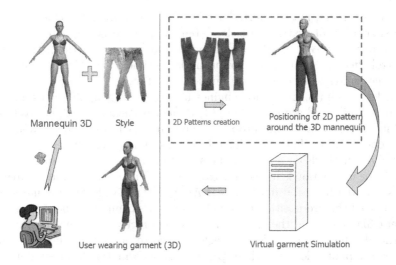

Mannequin 3D   Style   2D Patterns creation   Positioning of 2D pattern around the 3D mannequin

User wearing garment (3D)   Virtual garment Simulation

Fig. 1. Scenario of a virtual garment try-on system

In figure 1, we propose a natural computer-based scenario to try virtual clothes. This scenario does not require any complex 3D interaction of the user. The first step is to enter user's 3D geometry to the system. Such data can be obtained from actual 3D sensors. The only problem is that these sensors usually do not furnish REGULAR 3D surfaces (mainly due to the presence of holes) but regular surfaces are necessary to run 3D garment simulations. User's size measurements will be extracted automatically from this data and this information will be used to generate garment 2D patterns adapted to the user's body. The next step for the user will be to select a garment type and style from a data base furnished by the retailer , than choose a material. He will have then to see himself in 3D wearing the garment he chose.

Many systems, allowing to simulate virtual garments have been already developed. However, most of the efforts made in this area are related to the realism of the garment simulation but very few has been done at the level of automatic pre-positioning of garments on the client body model. simulate a garment specific well positioned around the 3D model. This pre-positioning, requires very good interaction skills usually of a specialized 3D graphics designer. It is inconceivable to ask such a performance from average clients.

In this paper, we describe a 'virtual garment try-on system that enables the user to try clothes without any 3D interaction skills. Pre-positioning of garment is done automatically and serves as initialization to the 3D garment simulator.

2 Related Works

In cloth modeling and animation, many researches have been made, from simplified cloth models [15], [11], to more accurate ones [1] [2] [7] [9]. Some surveys and comparisons of recent works are available in [21], [14].

In the last years, major efforts have been made in order to achieve fast and stable simulation of textile. Particle systems appear to be the most developed trend. Baraff and Witkin [1] have introduced an implicit method, that allows to control effectively the simulation using large time steps. Most of simulation engines nowadays work using their implicit-based model.

In order to use the simulation engines in everyday-life applications, garment have to be positioned around a digital body automatically. This pre-positioning is a difficult and challenging problem. Some approaches for dressmaking have been proposed. Some have been introduced in the literature [2], [13].

But these methods do not allow the automatic access to the garment simulation as some interaction is necessary. Towards the 'virtual fitting room' direction, majors research has been done by MIRALab [8] or MVM (My Virtual Model) [20]. To avoid the pre-positioning step in end-user's computer, their system uses a generic 3D model and they position clothes interactively on the generic model beforehand. The generic model is transformed corresponding to the user's shape or by using a photo-cloned method. Once the transformation has been done, garments, which were attached to generic model mesh, will be also transformed in order to produce the final result. In our opinion, this approach can give a result within a short time because only a simple 3D transformation is need. However, there is no flexibility regarding the type of garment shape desired by a client as pre-positioning is done beforehand.

Our system uses an automatic virtual dressing technique, which allows a particular user to use a realistic body scan model and to try virtual garments. This technique proposes a 2D manipulation method which will be coupled to a 3D mapping technique allowing to reach the final positioning.

The remainder of this paper is organized as follows: Section 3 describes the virtual garment positioning technique used in our system. Section 4 and Section 5 details our system components. We present briefly our simulation engine in Section 6, discuss our future work in section 7 before concluding in Section 8.

3   Overview of the Automatic Virtual Garment Positioning Method

The system proposed in this paper allows users to use their 3D digital body model as an input. This digital model contains detail information of user like sizes, textures, hair, eyes,... will help to produce a realistic scene. The challenge of such a system is the positioning of garment automatically around the 3D digital model. To perform the task, we divide the system in two different parts: the 2D part will help tailors to manipulate 2D pattern pieces, to define sewing information, and to specify pattern positions relatively in human body. The 3D part uses the output of 2D part as input. This part produces garment corresponding to user's sizes then place the garment around the 3D digital model. These two parts share the same data base over Internet.

Our idea is to use two generic silhouettes in 2D to describe the human body. In order to define the position of garment in the general human body, tailors only need to assign the garment to their position in two figurines. These two figurines, which correspond to two views (front and back) of the garment, present approximatively how the garment looks like finally. Once the 3D digital model of the user is analyzed, we reconstruct the user's silhouettes and calculate its transformation function (transforms the generic to the real silhouette). We then use this function to transform the figurines (with garment) to the user's silhouettes. Sewing is then performed automatically in 3D and the simulation engine will finish the job. More details can be found in [18].

4 Tailor Interactive Tool

We provided a set of basic functions in our 2D interaction tool. This tool allows tailors to import 2D pattern from industrial textile software. Tailors can also edit/delete the redundant patterns if needs.

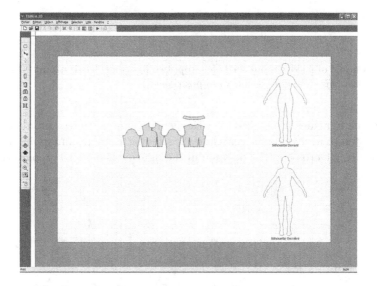

Fig. 2. Tailor Interaction Tool Interface

Implementing the 2D tool is a time consuming task because we aim to use the output data coming from various CAD/CAM textile software. Even through they use the same DXF format (Data eXchange Format), garments definition can vary from system to system.

Currently, our system supports the Nurb/Bezier/Polyline definition of 2D patterns curves. After a standardization phase, every garment is presented in Bezier form with control knots. This presentation allows editing or even creating a new garment easily.

4.1   Figurines and Cloth Positioning

2D figurines (front and back) are used as key element in our method 3. The figurines should be provided to the tailor or selected from a database. It's possible that a 2D pattern will be visible in the two views (front and back) of the cloth, so this pattern will be cut in two smaller pieces corresponding to each view. Note that, this action does not require a precise cut; the original pattern will always be reconstructed from its two small pieces used in the cloth simulation engine.

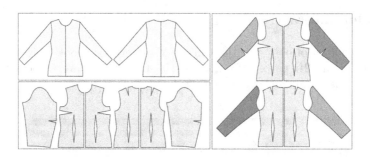

Fig. 3. Example of a skirt input data; left-top: two figurines; left-bottom: 2D patterns; right: 2D patterns were edited and were positioned

First, the figurines are placed over the generic silhouette (see figure 4). They can be edited to always be 'larger' than the silhouette. Next, a mapping between the garment patterns and the polygon in the figurine is done to prepare the pre-positioning.

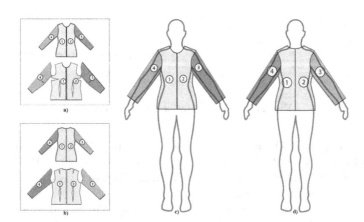

Fig. 4. Example of mapping between the garment patterns and the polygons of the figurines, each pattern and its correspondent are assigned with the same number

Note that the tailor has to provide two figurines so that each pattern piece and its correspondent polygon must have the same number of line segments (except segments containing clamp). With the 2D interaction tool, the tailor can add or modify a figurine easily to match his requirements.

4.2   Sewing Definition

In order to simplify the tailor's work, cloth clamps are detected automatically thanks to their specific structure : the end-angle between two successive line segments is always smaller than a predefined threshold and the two lines have the same length.

In addition, a color is assigned to each line segment which refers to its sewing attribute:

- Red color for segments sewn with an other segment;
- Green color if it corresponds to a pattern divided in two smaller pieces;
- Gray color if the segment is a free border (see an example of the coloring in figure 4).

With this intuitive coloring , our system can detect the sewing information automatically and check for the consistency of the garment construction.thanks to the figurine structures. If two polygons in the figurine (corresponding to two 2D patterns) share a same Red or Green segment, they are sewn over the segment. The sewing cross-over plan between front and back patterns can be determined based on Red/Green segments.

Of course, tailors can correct the sewing information if needed but it's very rare except when tailors want to test the behavior of special complicated garments. Furthermore, Tailors can furnish or adjust gradation information in order to produce a correct garment for a specific size. Textures of garments could be added for the realistic rendering of the final scene. Retailers will have to furnish the textures using scanned images of all pieces of cloth that are susceptible to offer or from a predefined database.

Next, garments, figurines and seam information are sent to a server which acts like a clothes database center. Tailors need to assign an icon presenting the garment. This icon allows end-user to figure out easily how the garment looks like.

Until this step, our system does not require any 3D interaction of tailor. He is used to work in 2D CAD/CAM environments, thus any tailor will use our 2D interaction tool without difficulty.

5   3D Interaction Tool

5.1   Interface

We provide to the end-user an interaction tool that combines a Web panel interface, a pre-positioning automatic module and a cloth simulation engine. Figure 5 presents the interface of our 3D interaction tool installed in an end-user's computer. This tool allows the user to select a garment and the material used to

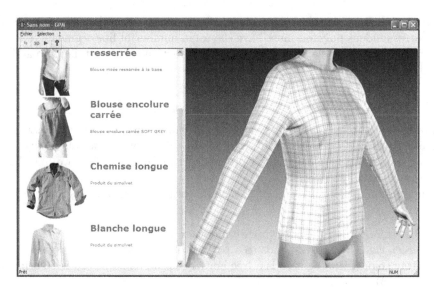

Fig. 5. 3D Interaction Tools Interface. The garment icons are shown in the left panel. Once the user clicks on an icon, its virtual garment will be shown in the right panel.

design it by clicking on simple icons and see **himself/herself** in 3D wearing the garment corresponding the user's size and choice.

- In the first step the user has to enter his own 3D digital shape(+texture). This data can be obtained is from a body scanner or simply from a customizable body modeling tool.
- Next, the user clicks on an icon of a garment he would like to buy. Through this simple click,the 2D patterns corresponding to the chosen garment will be fetched from a database server installed in the seller's site.
- Finally, he clicks on the icon corresponding to the material (color and type) he selects for his garment. Another, more informative type of interaction consists of the furniture by the seller of o booklet of samples of the material that are available for the garment production. The client will be able to touch it and transmit the code written on it to the system. When doing so, the client will, once again, call other databases, one containing the digital image of the chosen material, the other one will contain the Kawabata characteristic curves (tension, shear and flexion) which will be used to design the mechanical non linear and hysteretic mass/spring model of the chosen material.

In the next sessions, we will describe our method allowing to extract the user's size and to position the garment around the 3D digital character automatically.

5.2   Manipulations with the 3D Digital Mannequin of the User

Our system uses the 3D digital model (geometry + texture)of the client as input of the 3D interaction tool; this model is fundamental to simulate realistically the

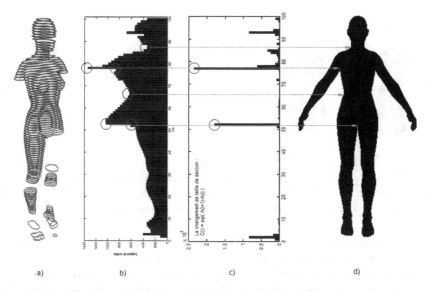

Fig. 6. Study of the biggest sections. a) the biggest horizontal sections, b) lengths of the sections, c) variations of the section lengths : $C_i = (l_i - l_{i-1})^2$. d) the feature point positions of the input model shown in its silhouette.

client. An acceptable model must have two extended arms. The angles between these extended arms and the body should be large enough so that there is no contact between them and the body.

First, we need to understand this 3D shape and find its size (and other measurements). We compute the inertia matrix of the shape and the eigenvalues and associated eigenvectors give us the main directions of the shape (vertical direction, and then the two lateral directions of the body). than it becomes easy to determine the front part and the back part of the body. We then slice the body in horizontal sections and compute the lengths of all these sections. The inspection of the maxima of these sections give us the major characteristic lines (including the size) as shown in figure 6.

These 3D measurements lines are equivalent to the measuring tool that the dressmaker places around the body of a person to take his measurements (such as neck, bust, pelvis, hip, etc.). Here, we performed the same task (automatically) as the dressmaker.

The detected feature lines will be used to determine the operations to put the garment around the 3D character, which will be described in the next paragraphs.

5.3 Transformation of the Tailor-Made Figurine to the User Silhouette

With the set of feature lines detected, we build a graph containing a fix number of polygons as shown in figure 7a, 7b. We then determine the transformation

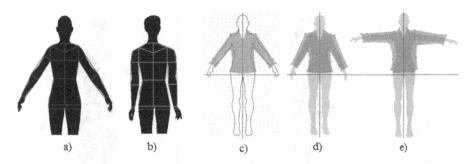

Fig. 7. Feature point graphs. a) feature point graph in the generic silhouette; b) feature point graph on a real silhouette; c) Figurines of a shirt positioned in the generic silhouette; d, e) Figurines transformed in the client silhouettes.

between the generic silhouette and the real one. This transformation will help to reconstruct the figurines on the real silhouette as well.

This problem is easy to solve since we know the generic feature points positions and their matches in the real silhouette. We can use the projective plan method or the moving least-square method to get the transformation function. We then simply extrapolate the transformation computed on the silhouettes to the figurines

Note that, with the projective plan transformation method, only polygon and its correspondent is used to calculate the local transformation function regardless of its neighbors, this can cause some undesired results for the transformation of figurines.

We then divide the silhouette contour (of each polygon) in a fixed number of small segments. These segments are used as inputs to determine the transformation function. The moving least-square method gives us valid results.

5.4 Transformation of Garment Onto the Real Figurine

In our proposed method, each garment 2D pattern has one correspondent polygon in the figurines. Since the figurines are transformed over the real silhouettes (with their polygons), the 2D patterns can be fitted over their correspondents in the client silhouettes in order to build the garment in 3D.

Firstly, the figurine 2D pattern garments are meshed along horizontal and vertical directions. Next, each original 2D pattern is meshed along the warp/weft directions. Moving least-square method will be used to determine the position of each practical mass (of the garment 2D) in the polygon. To perform the task, line segments (of the original 2D pattern) and their matches (of the figurine patterns) were divided in a fix number of small segments. These pairs of small segments are then used as input of the moving least-square method. Figure 8 shows the transformation between a 2D figurine pattern and its corresponding 3D pattern. (once again, more details can be found in [18]).

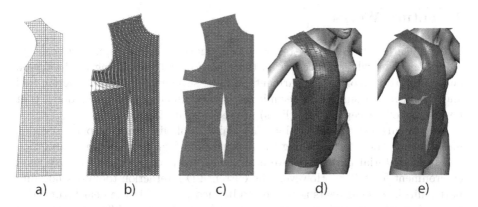

a)   b)   c)   d)   e)

Fig. 8. Transformation between 2D patterns of the garment and their matches. a) each 2D pattern of the tailor figurines (drawn by him) is meshed; b) meshing of the original garment 2D pattern obtained by the mapping of a); c meshing of the original 2D pattern (with the clamps) along the WARP/WEFT directions. d) use of a, b) and c) to determine the 3D location of each mass from c)on the 3D mannequin. e) clamping of the 2D pattern obtained in d).

Once all the 2D patterns of the garment are transformed from their figurine positions, only a simple sewing process is needed to produce the final result.

6   Cloth Simulation Engine EMILION

In this paragraph, we briefly describe the cloth simulation engine EMILION that is used to compute the final shape of the garment (output of our 3D tool).

EMILION is developed in the MIRAGES project, part of the National Research Institute in Computer Science and Automatic Control (INRIA) to simulate the exact draping of cloth and its correct fit on digitized human bodies. The emphasis in the development of EMILION was to produce physically correct mechanical models suitable for the textile industry and fast numerical algorithms to solve the equations of dynamics derived. The basis of the simulation system is a non linear hysteretic mass/spring network. The stiff differential equations are solved by an implicit-modified ordinary differential equation solver in order to enable large time steps [1]. Collision detection in EMILION is based on k-DOP hierarchies [25], collision response is realized by constraints as described in [1]. EMILION includes the integration of physical material parameters gained from Kawabata experiments [17,19] in order to insure the realism of the simulation result including the implementation of hysteresis effects and buckling [5].

Because of its physical basis and the possibility to use large time steps to solve the equation of dynamics, the EMILION garment simulation system fits for virtual try-on applications. Even though we use this particular simulator, our application is independent of the choice of the simulation engine, and it would be easy to integrate any other simulator because of its flexible architecture.

7   Future Works

The application described in this work is intended to be a complete virtual try-on system. The whole process from the design over the tailoring to the try-on and customization of virtual garments shall be supported by 2D interaction tool and 3D automatic simulation techniques. The specificity of this technique is that no interaction (except simple clicks) are required from the user.

We are currently implementing basic tailor tools like drawing pens, scissors, measurement tools for designing the patterns and figurines. Sewing information and even material parameters were also implemented providing an easy work environment for tailors. However, the current 2D interaction tool does not support multi-layers garments that need to be incorporated in the near future. But we are lack of any mechanical friction parameters between different material if such simulation has to be realized.

An important topic of future research will be the implementation of techniques for speeding up the simulation engine; GPU computing is a good choice since robust graphics cards are highly developed and this technology became very popular today.

8   Conclusion

In this paper we have discussed the implementation of a system allowing virtual try-on. We have presented first a 2D interaction tool that is meant to be used by tailors and a 3D module, providing the drape of garments on the 3D digital model of the user.

The 2D pattern garment manipulation tool allows to select, cut, create new figurines and to define garment position relatively on human body. In this tool, sewing information was extracted automatically based on the structure definition of garment and figurines. We provide two generic silhouettes representing the two views (front and back) of a generic human body. Since the figurines are positioned in the generic silhouettes, no interaction with the real client body are necessary. This way garments are automatically positioned around the 3D digital mannequin of any client without difficulty.

Our 3D interaction tool provides a web interface combined with a prepositioning module and a garment simulation engine. The web interface allows communicating with a Database server in order to retrieve the 2D patterns of a chosen garment. The pre-positioning module creates automatically the 2D patterns corresponding to user's size and sews these patterns to produce the final garment around the 3D mannequin. Finally, the cloth simulation engine will drape the garment to produce a realistic result of the client wearing virtually the garment he/she has not yet bought.

Acknowledgments. We would like to thank our project partner Nadine Corado for providing example input data for our application. This project is financed by French Research Agency (ANR).

References

1. Baraff, D., Witkin, A.: Large Steps in cloth simulation. In: Cohen, M. (ed.) SIGGRAPH 1998 Conference Proceedings, Annual Conference Series, Orlando, Florida, USA, July 1998, pp. 43–54. ACM SIGGRAPH (1998)
2. Benoit, L., Nadia, M.T., Daniel, T.: Cloth Animation with Self Collision Detection. In: Proceedings IFIP Conference on Modeling in Computer Graphics, Tokyo, pp. 179–187. Springer, Heidelberg (1991)
3. Bridson, R., Fedkiw, R., Anderson, J.: Robust treatment of collisions, contact and friction for cloth animation. In: ACM Transactions on Graphics, SIGGRAPH 2002, vol. 21 (2002)
4. Breen, D.E., House, D.H., Wozny, M.J.: Predicting the drape of woven cloth using interacting particles. In: SIGGRAPH 1994, pp. 365–372 (1994)
5. Charfi, H., Gagalowicz, A.: A new technique to measure Rayleigh's Viscosity parameters of textile. In: Industrial Simulation Conference, ISC 2005 (June 2005)
6. Chittaro, L., Corvaglia, D.: 3D Virtual Clothing: from Garment Design to Web3D Visualization and Simulation. In: Web3D 2003: Proceeding of the 8th international conference on 3D Web technology, Saint Malo, France, p. 73 (2003)
7. Choi, K.-J., Ko, H.-S.: Stable but responsive cloth. In: ACM Transactions on Graphics, SIGGRAPH 2002, vol. 21 (2002)
8. Cordier, F., Lee, W., Seo, H.: Nadia Magnenat-Thalmann. Virtual-Try-On on the Web. In: VRIC, Virtual Reality International Conference, Laval Virtual 2001, May 16–18 (2001)
9. Decaudin, P., Julius, D., Wither, J., Boissieux, L., Sheffer, A., Cani, M.-P.: Virtual Garments: A Fully Geometric Approach for Clothing Design. EUROGRAPHICS 2006 25(3) (2006)
10. Eberhardt, B., Etzmu, O., Hauth, M.: Implicit-explicit schemes for fast animation with particle systems. In: Eurographics Computer Animation and Simulation Workshop 2000 (2000)
11. Feynman, C.: Modeling the Appearance of Cloth. Master's thesis, Dept. of EECS, Massachusetts Inst. of Technology, Cambridge, Mass. (1986)
12. Fuhrmann, A., Gross, C., Luckas, V.: Interactive animation of cloth including self collision detection. Journal of WSCG 11(1), 141–148 (2003)
13. Gross, C., Fuhrmann, A., Luckas, V.: Automatic pre-positioning of virtual clothing. In: Preedings of the 19th spring conference on Computer Graphics, Budmerice, Slovakia, pp. 99–108 (2003)
14. House, D.H., Breen, D.E. (eds.): Cloth Modeling and Animation. A. K. Peters (2000)
15. Jerry, W.: The Synthesis of Cloth Objects. Proceeding SIGGRAPH 1986, Computer Graphics 20(4), 49–54 (1986)
16. Ju, X., Werghi, N., Siebert, P.: Automatic segmentation of 3d human body scans. In: IASTED International Conference on Computer Graphics and Imaging, CGIM 2000, pp. 239–244 (2000)
17. Kawabata, S.: The Standardization and Analysis of Hand Evaluation, 2nd edn. Text. Mach. Soc., Japan (1980)
18. Le Thanh, T., Gagalowicz, A.: Virtual Garment Pre-positioning. In: Gagalowicz, A., Philips, W. (eds.) CAIP 2005. LNCS, vol. 3691, pp. 837–845. Springer, Heidelberg (2005)
19. Matsudaira, M., Kawabata, S., Niwa, M.: Measurement of Mechanical Properties of Thin-Dress Fabrics for Hand Evaluation. Journal of the Textile Machinery Society of Japan 31(3), 53–60 (1985)

20. http://www.mvm.com
21. Ng, H.N., Grimsdale, R.L.: Computer techniques for modeling cloth. Computer Graphics and Applications 16(5), 28–41 (1996)
22. Provot, X.: Deformation Constraints in a Mass-Spring Model to Describe Rigid Cloth Behavior. In: Proc. Graphics Interface 1995, pp. 147–154 (1995)
23. Provot, X.: Collision and self-collision handling in cloth model dedicated to design garments. In: Graphics Interface 1997, pp. 177–189 (1997)
24. Terzopoulos, D., Platt, J., Barr, A., Fleischer, K.: Elastically Deformable Models. Computer Graphics, 205–214 (July 1987)
25. Klosowski, J.T., Held, M., Mitchell, J.S.B., Sowizral, H., Zikan, K.: Efficient Collision Detection Using Bounding Volume Hierarchies of k-DOPs. IEEE Transactions on Visualization and Computer Graphics 4(1), 21–36 (1998)

Level Set Segmentation of Knee Bones Using Normal Profile Models

Gaetano Impoco

University of Catania,
Viale A. Doria 6, 95100 Catania, Italy
impoco@dmi.unict.it

Abstract. We address the problem of segmenting bone structures from CT scans of the knee joint, in the level set framework. Our method is based on intensity profiles along the normals to the evolving contour. The evolution is guided by the similarity of image intensity profiles to profile models. The evolution stops when the intensity profiles closely match the model. The profile models are built using a manually labelled training sample.

Keywords: Image segmentation, Medical images, Level Sets.

1 Introduction

Digital medical imaging devices are becoming more and more valuable for diagnosis, pre-operative planning, and post-operative outcome evaluation of a surgical intervention. Acquiring data of internal organs by means of medical imaging devices is common clinical practice both for diagnosis and pre-operative planning. Although this information could be fruitfully exploited for quantitative measurements, in current clinical practice it is mainly employed for qualitative observation. Computer-aided procedures for planning and evaluation are desired to allow quantitative evaluation and information interchange among surgeons.

Digital models of the involved organs or tissues can be of great value for this purpose. In particular, planning and evaluation often involve the creation of a model from patient-specific data, acquired by means of CT, MRI, and so on. We have recently employed models of human body parts to select the best-fitting prosthesis for knee joint replacement [1] and to compute ad-hoc geometric measurements for breast plastic and reconstructive surgery [2].

A systematic use of digital models for quantitative analysis is far from being reached. This is mostly due to the lack of robust and reliable systems to build accurate models of the structures of interest. A large variety of segmentation methods have been developed for medical image processing and segmentation. Nonetheless, ad-hoc solutions are often preferred to properly detect complex structures, such as vessels, organs, or skeletal structures.

In this paper, we address the problem of segmenting bone structures from CT scans of the knee joint. A CT scan is a stack of 2D sections of the imaged

A. Gagalowicz and W. Philips (Eds.): MIRAGE 2009, LNCS 5496, pp. 195–206, 2009.

body part, called slices. Each slice can be considered as a greylevel image, where intensity values (measured in *Hounsfield Units*, HU) are roughly proportional to the density of the material in the sampled volume associated to the pixel. This is of great help for segmentation of tissues on the basis of their density. However, the HU ranges for different tissues often overlap, and can vary between different patients and between different slices of the same study (e.g., the density of a femur is remarkably lower close to the articulation than in its central area). This is especially true at articulations and for soft tissues and trabecular bone in aged patients, where osteoporosis reduces the density of bones.

As a result, the boundaries of bones, as well as other anatomical structures, may not be clearly distinguishable from neighbouring tissues. Most standard segmentation algorithms do not take into account the geometry of the segmented structures, leading to broken boundaries that are difficult to cope with e.g., for simulations [1].

Level Sets [3] have been fruitfully employed in the segmentation of medical images, thanks to their ability to easily enforce smoothness constraints on the boundaries between regions. Binary segmentation is obtained evolving the boundary towards a rest position that minimises an energy functional. The boundary is implicitly represented by an embedding function, Φ, usually chosen as the signed distance function from the contour. The *level set equation* is

$$\frac{\partial \Phi}{\partial t} = -\nabla\Phi \cdot \frac{d\boldsymbol{x}}{dt} = -\nabla\Phi \cdot \mathbf{F} \tag{1}$$

where $\mathbf{F}$ is a function encompassing the partial derivatives of Φ evaluated at $\boldsymbol{x}$, and drives the evolution of the contour. Equation 1 is evaluated at the nodes of a finite grid. Resolution anisotropy is easily dealt with by using an appropriate grid spacing.

The level set framework can be easily employed for segmentation by defining a function $\mathbf{F}$ such that Equation 1 goes to zero close to the boundaries of the interest objects in the image. A regularisation term is often contained in $\mathbf{F}$, such as local curvature. Most of the level set segmentation algorithms in the literature are based on two different approaches: *edge-based*, and *region-based*. The edge-based method, introduced by Caselles et al. [4], employs an image gradient term to stop the evolution of the contour close to image intensity edges. The region-based method, proposed by Chan and Vese [5], employs global statistics about the inner and outer regions of the evolving contour to segment the image into two homogenous regions. These basic methods are intrinsically bimodal. Multi-region segmentation can be obtained by simultaneously evolving a number of different contours.

Medical image segmentation is probably one of the most successful applications of level sets. Due to regularisation terms, level sets tend to show a nice smooth structure. The curvature term drives the contour towards a circular shape. Hence, level set methods are particularly suited for segmentation of convex objects. Effective results are reported for the segmentation of organs with (almost) convex sections, such as colon, liver, heart, and the like (see for example [6,7]). Level sets have also been used to segment complex shapes, such as brain sections [8,9] and vascular structures [10,11,12,13] in noisy images.

The difficulty of segmenting bone structures is often underestimated since bones are usually well distinguished from other tissues. This is often true, especially for young individuals and for stiff bones. Most interesting clinical cases, however, look like that in Figure 2. A CT study is shown of an aged patient affected by osteoporosis, before undergoing a surgical operation for total knee replacement. Due to age and osteoporosis, bone stiffness is much lower compared to that of young people. The slice represents a section of the articulation, where the bone tissue is less dense than at the middle part of long bones. Clearly, the boundary of the bone is difficult to outline even for expert medical staff. If the bone model is intended for simulation (e.g., [1]) the accurate segmentation of those blurred regions is of much importance.

A common problem with CT images is resolution. It is often the case that distinct bones are too close with respect to the imaging resolution, leading to merging of distinct contours. A topology-preserving level set method is presented in [17]. The Narrow Band algorithm is modified in order to forbid unwanted topology changes. In [18], a sort of inter-contour skeleton is defined between the initial contours, and evolved together with the level set to keep the contours separated. A comparison of generic methods applied to segmentation of bones in CT images is presented in [19].

One of the first attempts to include a specific image term for bone data is due to Lorigo et al. [14]. It is as simple as local variance to describe trabecular bone in MRI images, but it shows an early interest in effective bone segmentation. In [15], probability density functions (PDFs) are generated at various distances from the contour using the implicit representation Φ. Each PDF encodes the probability to find a certain pixel value at a given distance from the contour. The same authors propose a similar concept in [16], where a typical intensity profile along the normal to the contour is matched against the true profiles.

We propose a segmentation method funded on the edge-based level set framework, in which the evolution is guided by the similarity of contour normal profiles to a profile model. Our method resembles in spirit the methods proposed in [15] and [16]. However, the former compares curvatures along the contour rather than profiles normal to it, while the latter models the ideal profile as a Gaussian, without specifying the true shape for a specific application. The main contribution of this paper with respect to these methods is to employ profile models learned from the medical data at hand, rather than generic profiles e.g., with Gaussian shape. Bone probability distribution is also exploited by means of an area-based stopping term, based on single pixel classification.

2   Segmentation by Matching Normal Profile Models

The basic idea of our method is as follows. Given an input image, we compare a profile model with the intensity profile along the normals to the evolving contour.

The evolution stops when the image intensity profiles match the model. In order to build a profile model which is consistent with the medical data at hand, a labelled training set is used to compute the typical profile across the boundaries of cortical bones.

The evolution equation used in our method is

$$\frac{\partial \Phi}{\partial t} = \delta_\varepsilon(\Phi)[(\nu - \mu \cdot \kappa + \lambda \Upsilon_{I,\Omega}) \cdot \Psi_{I,P}(\Phi)] |\nabla \Phi| \qquad (2)$$

where κ is the local curvature, I is the input image, and ν, μ and λ are constants. The term $(\nu - \mu \cdot \kappa)$ encompasses a common regularisation force, the curvature κ (weighted by μ), and an inflation force ν. The function $\Upsilon_{I,\Omega}$ is related to the probability of a HU value to belong to bone

$$\Upsilon_{I,\Omega}(\boldsymbol{x}) = (p(\boldsymbol{x}) - p_{in})^2 - (p(\boldsymbol{x}) - p_{out})^2 \qquad (3)$$

where $p(\boldsymbol{x}) = p(I(\boldsymbol{x}) \in c_{bone}|\Omega)$ is the probability of $\boldsymbol{x}$ to be a bone pixel, conditioned by its neighbourhood Ω. This probability is weighted by the mean probability values, p_{in} and p_{out}, respectively inside and outside the contour. This equation recalls Chan and Vese's region-based method [20] with pixel probabilities used in place of grey levels.

The term $\Psi_{I,P}(\Phi)$ relates the intensity profile of the image I along the contour normal to a profile model P, and is defined as

$$\Psi_{I,P}(\Phi) = 1 - |cc(X_I, P)| \qquad (4)$$

Here, X_I is the intensity profile of radius r of the image I along the normal to the contour, P if a profile model of the same length, and the function $cc(X_I, P)$ is the correlation coefficient of the two vectors X_I and P. For a perfect correlation (i.e., the vectors are identical) the correlation coefficient $cc(X_I, P) = \pm 1$, while it goes to zero for extremely different distributions. Hence, $\Psi_{I,P}(\Phi) \in [0, 1]$ and is minimum for highly correlated profiles. Finally, the function $\delta_\varepsilon(\Phi)$ is a smoothed Dirac delta such that $\delta_\varepsilon(\Phi) = 1$ at $\Phi(\boldsymbol{x}) = 0$, which prevents topology changes to occur far from the contour.

From Equations 2 and 4 it should be clear that the evolution stops when the image intensity profiles perfectly match the profile model. In applications, however, $\Psi_{I,P}(\Phi)$ rarely comes close to zero because of several reasons, such as acquisition noise and bone density (i.e., intensity) variability due to age, gender, and so on. This may draw the contour across the desired bone boundaries. A simple solution is to set a threshold τ to introduce a tolerance to profile similarity strength. Equation 4 is rewritten as

$$\Psi_{I,P}(\Phi) = max((1 - \tau) - |cc(X_I, P)|, 0) \qquad (5)$$

where the maximum prevents $\Psi_{I,P}(\Phi)$ from taking negative values.

Notice that the correlation coefficient $cc(X_I, P)$, being normalised with respect to the vector means, compares the deviations of profile points from the mean values. Thus, the function $\Psi_{I,P}(\Phi)$ does not take into account the absolute intensity values of the profile. This can be an advantage when working with datasets showing a wide range of variation of bone densities. However, pixels having a profile similar to the model, but different absolute intensity, can be misclassified as bone.

3   Profile Models

Profile models are computed from a training set of CT scans. The radiologists of the Radiology Department of the Vittorio Emanuele Hospital of Catania were asked to manually label each scan of the training set. An interface was provided containing thresholding controls as well as the possibility to label single image areas. For labelling purposes, pixels are grouped using a region growing segmentation method, in order to let the users easily select image areas.

The labelling procedure has three main steps. First, the radiologists set two thresholds on HU (intensity) values to label soft tissue, trabecular bone, and cortical bone. Background is automatically excluded from labelling, using a standard automatic thresholding technique. These thresholds are selected by looking at a single slice and then are propagated to the whole dataset. They are used as a basis for further manual refinement in the second step, where misclassified regions can be selected separately and singularly re-labelled. In the third step, cortical bone is reduced to a single contour line.

The labelling software also provided a HU tolerance when selecting regions. However, the users found it of little use and its concept tricky to understand. Hence, this feature was not used. Finally, in order to make the labelling environment more familiar to the radiologists, we show three panels containing the original slice, its segmented version (used to select regions), and the current annotation.

The contour normals are computed for all contour points and then averaged with the weighted contribution of neighbouring normals, in order to obtain a more robust estimation. Intensity profiles of radius r are computed by interpolating pixel values along each contour normal. Finally, the mean profile is used as a representative profile model.

Two models were generated: one for slices close to the knee joint, P_d, and one for the others, P_m. We made this distinction because the bone density distributions of these two classes are widely different (see Figures 1 and 2). The model P_d is obtained as the mean profile of all intensity profiles in slices of distal femur and proximal tibia. P_m is generated from the remaining slices.

Finally, a pixel probability model was generated by computing the PDF of bone pixels. In our current model, we approximate the probability term $\Upsilon_{I,\Omega}$ in Equation 3 by assuming that pixel values are statistically independent. The neighbourhood Ω is thus empty. Bayesian classification is used to compute the posterior probability.

4   Results and Discussion

Nine knee datasets were imaged at the Radiology Department of the Vittorio Emanuele Hospital of Catania, giving a total of 849 slices, a sample rich enough

Table 1. Parameter settings used for the acquisition of the knee CT data employed in our experiments

Parameter	Value	
Exposure	200	Sv
kVp	140	kiloVolt
Slice thickness	1.3	mm
Slice spacing	0.6	mm
Slice resolution	512×512	pixels
Number of slices	70–140	

for testing 2D segmentation. The data was captured by means of a Multidetector CT Scanning (MDCT) device in spiral mode, using the acquisition parameters reported in Table 1. The age of the patients ranged from 70 to 80 years and all of them suffer from osteoporosis and arthrosis. This choice is motivated by the fact that this is one of the most difficult cases (due both to age and to severe osteoporosis and arthrosis) and the most widespread in clinical cases of knee replacement surgery [1]. The acquired datasets were manually labelled by expert radiologists of the Vittorio Emanuele Hospital. 50% of the labelled datasets were used for learning, and the remaining were employed for testing.

In our experiments, we place the initial contour around the interest structure and let it shrink. The inflation force ν in Equation 2 could be accordingly set to zero. We set it to a small value, $\nu = 0.05$, to push the contour beyond small isolated, misclassified features. A small value, $\mu = 0.1$, is assigned to the curvature term, as well. The probability term has a predominant weight, $\lambda = 1$. Choosing a convenient value of τ in Equation 5 is less obvious, since it strongly affects the convergence of the level set to the bone contours. We experimentally found that $\tau = 0.4$ works well for most of our data. Finally, the profile radius was set to $r = 10$. These settings are used in all our experiments.

Figure 1 shows the segmentation of a femur of an aged patient. The slice lies close to the condyles (i.e., the two projections on the lower extremity of femur). Here the bone density is higher than the density of soft tissues but the intensity gradient is low. Most segmentation algorithms fail with this kind of data, because they do not model the real intensity variations around the edges of the interest object.

A more challenging case is shown in Figure 2. This slice shows a section of the condyles of a femur at the knee joint. Due to osteoporosis, the density of cortical bone is as low as that of trabecular bone, and very close to the density of soft tissues. Here we use a different profile model, P_d, that fits better this data (see Section 3).

The segmentation of two slices is compared to the ground truth in Figure 3. In these examples, the level set flows over the real bone boundary (to be precise, the level set stops before reaching the true contour). We observed this trend in most of the classified slices. This is probably due to the choice of the parameter τ in Equation 5, since it relates the evolution speed to the probability of getting close to the bone.

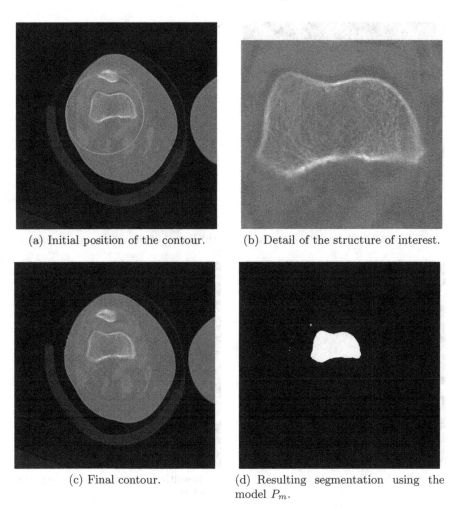

(a) Initial position of the contour.   (b) Detail of the structure of interest.

(c) Final contour.   (d) Resulting segmentation using the model P_m.

Fig. 1. Segmentation of a slice close to the distal end of a femur of a 70 years old patient. The detail shows that the cortical bone appears blurred, especially at the base of the condyles.

As already discussed in Section 2, the function $\Psi_{I,P}(\Phi)$ in Equation 5 does not take into account the absolute intensity values of the profile. This can be considered a sort of invariance with respect to absolute intensity. The algorithm benefits from this invariance being more robust to variations of bone densities between slices. On the other hand, pixels can be found which match the profile model but their absolute density is much lower than bone pixels. Due to this invariance, this term alone has no means to discriminate those pixels from true bone pixels.

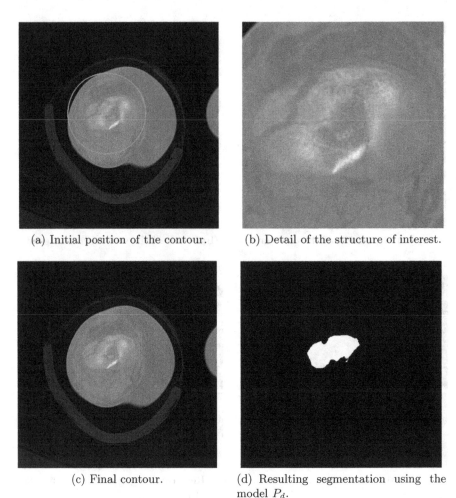

(a) Initial position of the contour.   (b) Detail of the structure of interest.

(c) Final contour.   (d) Resulting segmentation using the model P_d.

Fig. 2. Segmentation of a slice crossing the condyles of a femur of a 70 years old patient. The detail shows that the cortical bone is almost undistinguishable from trabecular bone and soft tissues.

Figure 4(b) shows an example of this effect. An area classified as bone is clearly part of soft tissues. We overcome this problem using the term $\Upsilon_{I,\Omega}$ defined in Equation 3. Here, the intensity level (HU value) is directly used to compute the probability of being bone.

Figure 5 shows another example of the performance of this term. Poor results are obtained using the profile term alone (Figure 5(b)). In particular, the contour leaks into the bone at some blurred areas, since the profile here does not match the model, being almost flat. The probability term greatly improves the segmentation quality.

Pixel misclassification could also be avoided by modifying the matching function in Equation 5. An explicit term could be added to weight the absolute

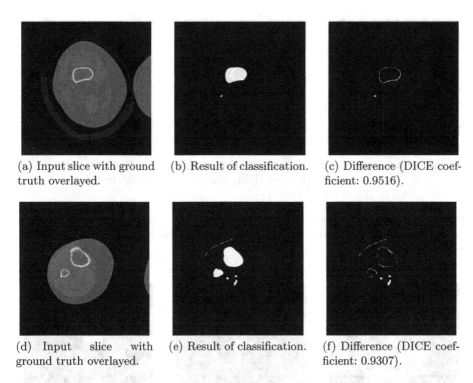

(a) Input slice with ground truth overlayed.

(b) Result of classification.

(c) Difference (DICE coefficient: 0.9516).

(d) Input slice with ground truth overlayed.

(e) Result of classification.

(f) Difference (DICE coefficient: 0.9307).

Fig. 3. Qualitative comparison of the automatic classification of two slices against the manually-labelled ground truth

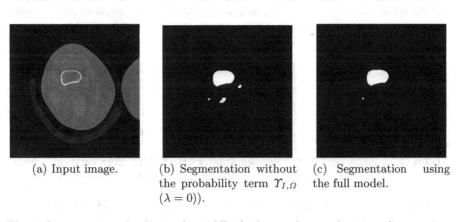

(a) Input image.

(b) Segmentation without the probability term $\Upsilon_{I,\Omega}$ ($\lambda = 0$)).

(c) Segmentation using the full model.

Fig. 4. Segmentation of a slice in the middle the femur of an aged patient. Some regions of soft tissue are misclassified as bone. Adding the probability term $\Upsilon_{I,\Omega}$ helps removing most of them.

intensity of the profile. Other matching functions can also be used in place of cross correlation, computing the absolute intensity difference with respect to the profile model. One such function is the *sum of squared error* (SSE), which is

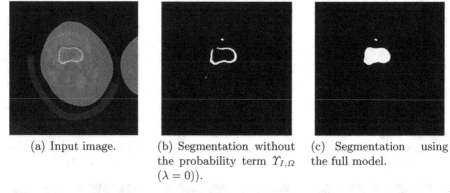

(a) Input image.

(b) Segmentation without the probability term $\Upsilon_{I,\Omega}$ ($\lambda = 0$)).

(c) Segmentation using the full model.

Fig. 5. Segmentation of a slice in the middle the femur of an aged patient. Without the probability term $\Upsilon_{I,\Omega}$ the contour leaks into the bone. Notice that the small isolated region correctly represents the patella.

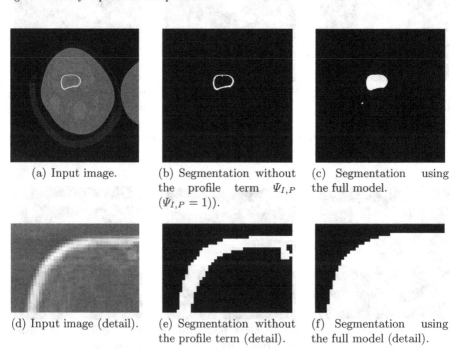

(a) Input image.

(b) Segmentation without the profile term $\Psi_{I,P}$ ($\Psi_{I,P} = 1$)).

(c) Segmentation using the full model.

(d) Input image (detail).

(e) Segmentation without the profile term (detail).

(f) Segmentation using the full model (detail).

Fig. 6. The same slice as in Figure 4. Contribution of the profile term $\Psi_{I,P}$.

minimised for perfectly matching profiles and is high for similar profiles with different mean value. Preliminary experiments with SSE, however, showed a worse performance with respect to cross correlation. Probably, a better option is to use the labelled data to learn a statistical model of intensity profiles. The contour evolution would be driven in order to maximise the probability of observing certain intensity profiles. This seems a promising direction for future work.

Figure 6(b) shows the segmentation of an image without using the profile term $\Psi_{I,P}$. For the sake of comparison, the same slice as in Figure 4 is used. Clearly, the contour leaks into the bone, even if it stops close to the cortical part. This is due to the fact that the statistical distribution of cortical bone HU values is highly different from the statistics of trabecular bone. Profile models are thus a useful stopping criterion to avoid contour leaks. As an important side effect, profile matching enforces smoothness of the segmented contour (compare the results in Figure 6(e) and Figure 6(f)).

We deem important one final remark about the scale of bone structures in the image. Our method is not scale invariant as it is, since the shape of intensity profiles is dependent on scale. However, the thickness of bones is strongly related to the size of their section. If the scale is known, the profiles can be easily scaled to the data. The scale can be set by the user simply by drawing an axis connecting opposite points in the bone contour. This is a common measurement procedure in the clinical practice.

References

1. Andrä, H., Battiato, S., Bilotta, G., Farinella, G.M., Impoco, G., Orlik, J., Russo, G., Zemitis, A.: Structural simulation of a bone-prosthesis system of the knee joint. Sensors, Special Issue on Medical Imaging 8, 5897–5926 (2008)
2. Farinella, G.M., Impoco, G., Gallo, G., Spoto, S., Catanuto, G., Nava, M.: Objective outcome evaluation of breast surgery. In: Larsen, R., Nielsen, M., Sporring, J. (eds.) MICCAI 2006. LNCS, vol. 4190, pp. 776–783. Springer, Heidelberg (2006)
3. Osher, S., Fedkiw, R.P.: Level Set Methods and Dynamic Implicit Surfaces. Springer, Heidelberg (2002)
4. Caselles, V., Kimmel, R., Sapiro, G.: Geodesic active contours. International Journal on Computer Vision 22, 61–79 (1997)
5. Chan, T., Vese, L.: An active contour model without edges. In: Nielsen, M., Johansen, P., Fogh Olsen, O., Weickert, J. (eds.) Scale-Space 1999. LNCS, vol. 1682, pp. 141–151. Springer, Heidelberg (1999)
6. Hodge, A.C., Fenster, A., Downey, D.B., Ladak, H.M.: Prostate boundary segmentation from ultrasound images using 2D active shape models: Optimisation and extension to 3D. Computer Methods and Programs in Biomedicine 84, 99–113 (2006)
7. Lynch, M., Ghita, O., Whelan, P.: Segmentation of the left ventricle of the heart in 3-D+t MRI data using an optimized nonrigid temporal model. IEEE Transactions on Medical Imaging 27, 195–203 (2008)
8. Jonasson, L., Hagmann, P., Pollo, C., Bresson, X., Richero Wilson, C., Meuli, R., Thiran, J.: A level set method for segmentation of the thalamus and its nuclei in DT-MRI. Tensor Signal Processing 87, 217–352 (2007)
9. Angelini, E.D., Song, T., Mensh, B.D., Laine, A.F.: Brain MRI segmentation with multiphase minimal partitioning: A comparative study. International Journal of Biomedical Imaging 2007 (2007)
10. Lorigo, L.M., Faugeras, O.D., Grimson, W.E.L., Keriven, R., Kikinis, R., Nabavi, A., Westin, C.F.: CURVES: Curve evolution for vessel segmentation. Medical Image Analysis 5, 195–206 (2001)

11. Nain, D., Yezzi, A.J., Turk, G.: Vessel segmentation using a shape driven flow. In: Barillot, C., Haynor, D.R., Hellier, P. (eds.) MICCAI 2004. LNCS, vol. 3216, pp. 51–59. Springer, Heidelberg (2004)

12. Manniesing, R., Velthuis, B.K., van Leeuwen, M.S., van der Schaaf, I.C., van Laar, P.J., Niessen, W.J.: Level set based cerebral vasculature segmentation and diameter quantification in CT angiography. Medical Image Analysis 10, 200–214 (2006)

13. Sum, K., Cheung, P.: Vessel extraction under non-uniform illumination: a level set approach. IEEE Transaction on Biomedical Engineering 55, 358–360 (2008)

14. Lorigo, L.M., Faugeras, O.D., Grimson, W.E.L., Keriven, R., Kikinis, R.: Segmentation of bone in clinical knee MRI using texture-based geodesic active contours. In: Wells, W.M., Colchester, A.C.F., Delp, S.L. (eds.) MICCAI 1998. LNCS, vol. 1496, pp. 1195–1204. Springer, Heidelberg (1998)

15. Leventon, M.E., Faugeras, O., Grimson, W.E.L., Wells, W.M.: Level set based segmentation with intensity and curvature priors. In: IEEE Workshop on Mathematical Methods in Biomedical Image Analysis (MMBIA 2000), Washington, DC, USA, pp. 4–11. IEEE Computer Society, Los Alamitos (2000)

16. Leventon, M.E., Grimson, W.E.L., Faugeras, O.: Statistical shape influence in geodesic active contours. In: IEEE Conference on Computer Vision and Pattern Recognition (CVPR 2000), pp. 316–323 (2000)

17. Han, X., Xu, C., Prince, J.L.: A topology preserving level set method for geometric deformable models. IEEE Transactions on Pattern Analysis and Machine Intelligence 25, 755–768 (2003)

18. Sebastian, T.B., Tek, H., Crisco, J.J., Kimia, B.B.: Segmentation of carpal bones from CT images using skeletally coupled deformable models. Medical Image Analysis 7, 21–45 (2003)

19. Truc, P.T.H., Kim, Y.H., Lee, Y.K., Lee, S.Y., Kim, T.S.: Evaluation of active contour-based techniques toward bone segmentation from CT images. In: World Congress on Medical Physics and Biomedical Engineering, pp. 3121–3125. Springer, Heidelberg (2006)

20. Chan, T., Vese, L.: Active contours without edges. IEEE Transactions on Image Processing 10, 266–277 (2001)

Detection of Overlapped Ellipses by Combining Region and Edge Data

Lin Zheng and Quan Liu

School of Information engineering, Wuhan University of Technology,
430070, Wuhan, China
linzheng@whut.edu.cn, qliu@public.wh.hb.cn

Abstract. This paper describes an approach for detecting overlapped ellipses by combining region and edge data. The Principle Component Analysis method is used to give the shape and position of an ellipse. A region based EM iterative algorithm is proposed to calculate the number of ellipses and their initial shapes in the overlapped region. As a result, every edge point is assigned to a certain ellipse by statistics decision. Then an edge fitting algorithm is employed to refine the ellipses' geometric parameters based on the edge data. Above coarse-to-fine algorithm is applied to detect the overlapped fruits and the moving targets. The result is stable and accurate.

1 Introduction

Overlapped ellipses detection is frequently met in computer vision and can be applied to fruits segmentation, articulated object detection, overlapped object distinguishing, etc.

There are two broad methods proposed to extract the overlapped ellipses. The first is region based method, which use Gaussian mixture model to fit overlapped regions supposed that the number of ellipses is known [1][2]. Region based method is stable, but has no apparent relation with the region's edge. Hence the result is coarse. The difficulty to use the Gaussian mixture method is how to calculate the number of ellipses.

The second is the edge based method, which employs some ellipses to fit the region's edge. The edge based method gives an accurate description of the target's edge, but the result is sensitive to the noise and outliers because edge pixels are rare. R. A. McLaughlin proposes a Randomized Hough Transform (RHT) edge based method to find overlapped ellipses [3]. T.C. Chen presents a Randomized Circle Detection (RCD) method to detect overlapped circles [4]. These two methods are all time wasting. A. Fitzgibbon proposed a direct least square method to fit partial data, which has a good fitting result and run rapid [5].

In this paper, the region and edge features are used together and a coarse-to-fine algorithm is proposed to detect irregular overlapped ellipses adaptively. At the coarse stage, an Expectation-Maximization algorithm is designed to calculate the number of ellipses in this region, initial shapes of the ellipses and the weights of every edge point with all ellipses. Then every edge point is assigned to an ellipse. As a result, all

A. Gagalowicz and W. Philips (Eds.): MIRAGE 2009, LNCS 5496, pp. 207–216, 2009.

ellipses get an edge point set respectively. At the fine stage, an edge fitting technology is employed to refine the ellipses by fitting above edge point sets.

The proposed method has two advantages. First it can calculate the number of ellipses correctly and adaptively. Second it has an accurate fitting to the image.

This paper is arranged as following. In section 2, an ellipse description method based on Principal Component Analysis (PCA) is introduced. In section 3, a region based Expectation-Maximization (EM) algorithm is designed to calculate the number of ellipses. Edge feature is used to get an accurate result of the ellipses in section 4. In section 5, some experiments are presented to show the efficiency of this method.

2 Description of an Ellipse with PCA

Since an ellipse coincides with a 2D Gaussian distribution on shape, Gaussian distribution is employed to describe an ellipse in this paper. Let $X_j = [x_j, y_j]^T$, $j = 1,2,\cdots,L$ be a pixel in a region, and P_{ij} be the probability of this pixel belonging to the i-th ellipse, $i = 1,2,\cdots,N$. By statistical analysis, the mean of the i-th ellipse is calculated by equation (1):

$$\mu_i = \frac{\left[\sum_{j=1}^{L} P_{ij} x_j \quad , \quad \sum_{j=1}^{L} P_{ij} y_j \right]^T}{\sum_{j=1}^{L} P_{ij}} \tag{1}$$

and the covariance matrix is calculated by equation (2):

$$\Sigma_i = \begin{bmatrix} \sigma_{11} & \sigma_{12} \\ \sigma_{12} & \sigma_{22} \end{bmatrix} = \frac{\sum_{j=1}^{L} P_{ij} [X_j - \mu_i][X_j - \mu_i]^T}{\sum_{j=1}^{L} P_{ij}} \tag{2}$$

According to paper [6], we can compute the two eigen-values of Σ_i :

$$\lambda_1 = \frac{1}{2}\left(\sigma_{11} + \sigma_{22} + \sqrt{(\sigma_{11} - \sigma_{22})^2 + 4\sigma_{12}^2} \right) \tag{3}$$

$$\lambda_2 = \frac{1}{2}\left(\sigma_{11} + \sigma_{22} - \sqrt{(\sigma_{11} - \sigma_{22})^2 + 4\sigma_{12}^2} \right) \tag{4}$$

and the two eigenvectors, i.e. the principle components: the long one $[v_1, v_2]^T$ and the short one $[v_2, -v_1]^T$. where

$$v_1 = \frac{\sigma_{12}}{\sqrt{(\lambda_1 - \sigma_{11})^2 + \sigma_{12}^2}} \qquad (5)$$

$$v_2 = \frac{\lambda_1 - \sigma_{11}}{\sqrt{(\lambda_1 - \sigma_{11})^2 + \sigma_{12}^2}} \qquad (6)$$

The angle between the long major axis and x-axis is

$$\phi = \arctan(v_2 / v_1) \qquad (7)$$

As introduced in paper [1], the ellipse to envelop the region has the following form:

$$x + jy = (a_1 \cos t + j a_2 \sin t)e^{j\phi} + (\bar{x} + j\bar{y}) \qquad (8)$$

Where (x, y) is an edge point of the ellipse. $a_i = 2\sqrt{\lambda_i}$ is the half axis. t is the rotation angle from the long major axis to the point (x, y) anti-clockwise. Referring to figure 1, there are some examples of single ellipse detection with PCA.

Fig. 1. Ellipse description with PCA. (a) Coins; (b) Lemon; (c) Watermelon.

3   Overlapped Ellipses Detection by EM

To calculate the number of the ellipses in the image, a region based EM method is proposed. As a result, every edge point is assigned to an ellipse, belonging to which this edge point has the highest probability.

3.1   EM Algorithm for Parameter Estimation

Let U be the observed samples, $\Im$ be the number of classes, Θ be the parameters of all classes, J be the relation between samples and classes. U and J form the complete data. The estimation of Θ based on Maximum a Posterior (MAP) is:

$$\Theta^* = \arg\max_{\Theta} \sum_{J \in \mathfrak{J}} P(\Theta, J \mid U) \qquad (9)$$

Which is a marginal probability over J.

If we know the relation J between the samples and classes, every class can be calculated by its sample set directly. This is a complete data problem.

In fact, incomplete data problem always exists i.e. J is unknown. For example, a camera can be employed to detect fruits on the fruit packing line. One of the detected foreground regions may contain several fruits because of the occlusion. This arise a problem to determine how many fruits in the region and which fruit any pixel in the region belongs to.

It is difficult to solve the equation (9) with analytic method because of so many unknowns. Hence some methods are proposed to simplify it. Supposed the initial Θ is correct, it is easy to compute which class a sample belongs to i.e. get the initial estimation of J based on the equation (9). This step is called Expectation. Then supposed J is correct, and the parameter model Θ is refined by using the equation (9), which is called Maximization. By cycling back and forth between the Expectation and the Maximization, Θ and J will converge. This idea is called EM [7].

3.2   Design of EM for Overlapped Ellipses Detection

An extracted isolated foreground region may contain only one target, such as in figure 1. In this situation, one ellipse is enough to describe it with little error. But sometimes the region consists of several targets which occlude each other. The error between the region and the ellipse is large, and more ellipses are demanded.

To solve the problem, an EM iterative algorithm based on region statistical character is proposed to calculate the number of ellipses and estimate the coarse geometric parameters for all the ellipses:

Step1: At the initial, one ellipse is used to fit the foreground. The number of ellipses is $M = 1$. PCA method is employed to calculate the statistics, the centre, major axis and half axis of the ellipse.

Step2: Compute the error between the foreground region S_{region} and ellipse region $S_{Ellipse}$:

$$\varepsilon = \frac{diff(S_{Ellipse} - S_{region}) + diff(S_{Region} - S_{Ellipse})}{S_{Region}}$$

where $diff(S_{Ellipse} - S_{Region})$ is the areas in ellipses but not in foreground. $diff(S_{Region} - S_{Ellipse})$ is the areas in foreground but not in ellipses.

Step3: If the error is less than a threshold T_1, end the program; Otherwise, the ellipse which has the largest error among all ellipses is divided into two parts by its short axis. Set $M = M + 1$.

Step4: Compute the weights in Expectation step:

Supposed the statistics and geometric parameters of ellipses are correct, compute the normalized weight of a foreground pixel P_j with i-th ellipse, $i = 1, 2, \cdots, M$:

$$P_{ij} = \frac{P(X_j \mid C = i)}{\sum\limits_{k=1}^{M} P(X_j \mid C = k)} \qquad (10)$$

where,

$$P(X_j \mid C = i) = \left[\frac{\exp(-\frac{1}{2}(X_j - \mu_i)^T \Sigma_i^{-1}(X - \mu_i))}{|2\pi\Sigma_i|^{0.5}} \right] \qquad (11)$$

Step5: Refine parameters in Maximization Step:

According to the probabilities of all pixels with an ellipse, recalculate the geometric parameters of the ellipse by equation (1)~(7).

Step6: If there is no change for all geometric parameters, go to step 2 to decide if a new ellipse is needed. Otherwise go to step 4 for another EM iteration.

The threshold T_1 relates with the likelihood to be an ellipse. Normally we can set $T_1 = 0.1 \sim 0.2$ as a prior knowledge when the target has a high likelihood to an ellipse.

3.2.1 Update Weights in Overlapped Areas

As we can see, the weights are normalized in equation (10) and summed to 1. This introduces a problem: the ellipses, which overlap each other, have a weight far less than 1 in the overlapped region. Hence the ellipses will undervalue these overlapped regions in computing the mean and variance. This problem becomes more serious when the overlapped region occupies a high ratio of the object, such as the ellipse at the top-left in figure 2(a), (c). There is an obvious error between the estimated ellipse and the real object in figure 2(c).

To solve this problem, the weights are set as 1 for all concerned targets in the overlapped region, shown in figure 2(b), and recalculate the statistics in two or three final iterations. The reconstructed ellipses are show in figure 2(d), where the top left and bottom right ellipses are closer to the edge than in figure 2(c).

3.2.2 Combine Overlapped Ellipses

During the EM optimization process, the ellipses may be captured by some local minimum and some new ellipses are added to help the algorithm reach the global

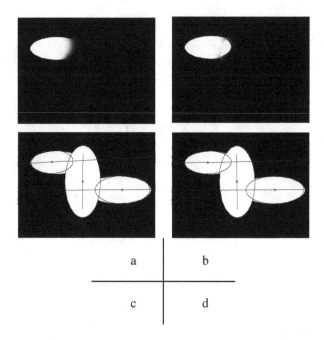

a | b

c | d

Fig. 2. Weight Update in Overlapped areas. (a) The weigh for Top-Left ellipse before update; (b) The weigh for Top-Left ellipse after update; (c) The fitting before update; (d) The fitting after update.

minimum with little error. As a result, the ellipses may be over complete. It is reasonable to combine some of these ellipses. For example, the two ellipses at the bottom of figure 6(b) have a high ratio of overlap and similar directions. It is reasonable to unit them as a complete canoe.

In this paper, a probability P is designed to decide whether or not the two overlapped ellipses will be merged, which is an error function as in figure 3(a):

$$P = erf(W(\theta) \cdot P_O(S_{A \cap B}))$$ (12)

where, $P_O(S_{A \cap B}) = S_{A \cap B} \big/ \min\{S_A, S_B\}$ is the probability that A overlaps B. S_A and S_B is the area of ellipse A and B respectively. $S_{A \cap B}$ is their overlapped area.

$W(\theta)$ is the direction coincidence degree of the major axis of ellipse A and B, which has the following form:

$$W(\theta) = 1 + \beta \cdot \max(0, \frac{\theta_C - \theta}{\theta_C})$$ (13)

where, θ is the angle between the major axis of the two ellipses. $\theta_C \in [0, \pi/2]$ sets the valid angle which excludes larger values. Usually set $\theta_C = \pi/4$. β gives a 0-1 decision according to the ratio of long half axis to short half axis:

$$\beta = \begin{cases} 2 \sim 3, & if \quad \dfrac{a_{A1}}{a_{A2}} \geq 1.1 \quad \& \quad \dfrac{a_{B1}}{a_{B2}} \geq 1.1 \\ 0, & Others \end{cases} \tag{14}$$

Which means that the influence of direction coincidence can be neglected as either of the two ellipses approximates to a circle. $W(\theta)$ has the shape as figure 3b.

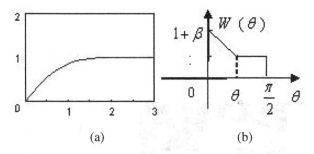

(a)   (b)

Fig. 3. (a) Error function; (b) Direction coincidence degree

Setting a threshold T_M. If $P_M \geq T_M$, the ellipse A will be merged into the ellipse B. Then $S_B = S_B + S_A$, and $P_{Bj} = P_{Bj} + P_{Aj}$ for any pixel P_j in the overlapped area.

Since equation (12) is nonlinear, T_M is difficult to calculate analytically. By experience method, T_M is selected in the range [0.6, 0.8].

4 Refine Ellipses by Edge Fitting

The ellipses are closer to the edge after the weight update, as in figure 2(d). But some error remains because the region based method has no direct relation with the object edge.

An edge based ellipse refine method is proposed to remove above error after region based method. Because ellipses are very close to the edge after the region based EM, most edge points have a heavy weight with its own ellipse. Hence an edge fitting method is designed as following:

Step1:For any edge pixel $P_j = (x_j, y_j)$, $j = 1, 2, \cdots, L$, decide which ellipse it is from:

$$i^* = \arg \max_i W_{ij} \tag{15}$$

The result is shown in figure 4(a), (b), (c). Every ellipse has a distinguished edge point set.

Step2: direct least square fitting method (DLS) is employed to refine every ellipse by fitting the corresponding edge sample set [5]. The result is shown in figure 4(d), (e), (f) and 4(g), which has a high accuracy than in figure 2.

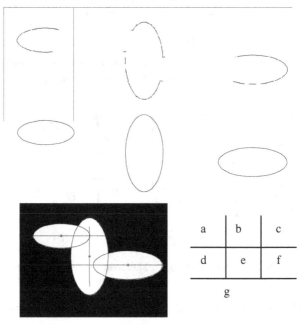

Fig. 4. Refine Ellipse by Edge Fitting. (a), (b), (c) edge point set for the ellipses respectively; (d),(e),(f) edge fitting result respectively; (g) fitting result in original image

5 Experiments

(1) Detect overlapped fruits
In figure 5, several fruits occlude each other. In figure 5(a), 3 ellipses are used, which leads to some error. In figure 5(b), 5 ellipses are used and the error becomes very small.

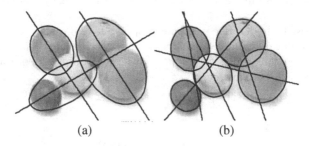

(a)   (b)

Fig. 5. Detect overlapped fruits. (a) Fitting with 3 ellipses; (b) Fitting with 5 ellipses

(2) Detect canoe and player

In figure 6(a), the canoe and player are extracted with a statistical method introduced in paper [8]. The segmented result is rough because of splashed water. In this situation, region based EM gives a good initial estimation for canoe and operator. In figure 6(b), three ellipses are used. By overlapped ellipses combining technology, two ellipses at bottom are merged.

In the edge fitting refine stage, direct least square fitting method is employed to detect the two ellipses. The result is shown in figure 6(c).

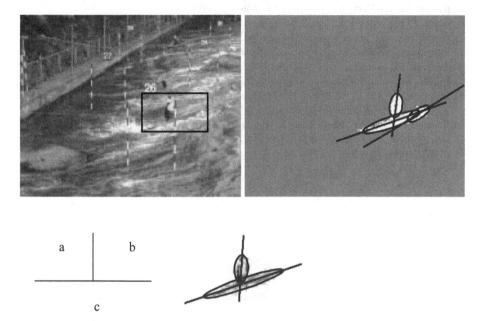

Fig. 6. Extract canoe and player. (a) Canoe and operator pointed out by a rectangle; (b) fitting with 3 ellipses; (c) edge fitting result with 2 ellipses.

6 Conclusion

In this paper, a coarse-to-fine multi-ellipse target detection algorithm is proposed by using region and edge data. In the coarse stage, a region based EM algorithm is designed to calculate the number of ellipses in a foreground and provide a good initial shape for every ellipse. Then an edge fitting technology is employed to refine ellipses.

This algorithm is used in several situations. The experiments show it is stable and accurate.

Acknowledgement

This work was supported by National Natural Science Foundation of China (50775167/E051303).

References

1. Wijewickrema, S.N.R., Paplinsk, A.P.: Principal Component Analysis for the Approximation of an Image as an Ellipse. In: Proc. Int. Conf. in Central Europe on Computer Graphics and Visualization, Plzen, Czech Republic, February 2005, pp. 69–70 (2005)
2. Yang, X.Y., Krishnan, S.M.: Image segmentation using finite mixtures and spatial information. Image and Vision Computing 22, 735–745 (2004)
3. Mclaughlim, R.A.: Randomized Hough transform: improved ellipse detection with comparison. Pattern Recognition Letters 19, 299–305 (1998)
4. Chen, T.C., Chung, K.L.: An Efficient Randomized Algorithm for Detecting Circles. Computer Vision and Image Understanding 83, 172–191 (2001)
5. Fitzgibbon, A., Pilu, M.: Direct Least Square Fitting of Ellipses. IEEE Trans. On Pattern Analysis and Machine. 21, 477–480 (1999)
6. Jolliffe, I.: Principal Component Analysis. Springer, Heidelberg (2002)
7. Dempster, A., Laird, N., Rubin, D.: Maximum likelihood from incomplete data via the EM algorithm. Journal of the Royal Statistical Society, Series B 39, 1–38 (1977)
8. Zheng, L., Beetz, M., Gedikli, S.: Visual Tracking System for Water Surface Moving Target. In: Proceedings of the SPIE. MIPPR 2007, Wuhan, China, November 2007, vol. 6786, 67863K, pp. 1–10 (2007)

Flash Lighting Space Sampling

Matteo Dellepiane, Marco Callieri, Massimiliano Corsini, Paolo Cignoni,
and Roberto Scopigno

Visual Computing Lab, ISTI-CNR, Pisa, Italy
name.surname@isti.cnr.it
http://vcg.isti.cnr.it

Abstract. Flash light of digital cameras is a very useful way to picture
scenes with low quality illumination. Nevertheless, especially low-end
cameras integrated flash lights are considered as not reliable for high
quality images, due to known artifacts (sharp shadows, highlights, uneven lighting) generated in images. Moreover, a mathematical model of
this kind of light seems difficult to create. In this paper we present a color
correction space which, given some information about the geometry of
the pictured scene, is able to provide a space-dependent correction of
each pixel of the image. The correction space can be calculated once in a
lifetime using a quite fast acquisition procedure; after 3D spatial calibration, obtained color correction function can be applied to every image
where flash is the dominant illuminant. The correction space presents
several advantages: it is independent from the kind of light used (provided that it is bound to the camera), it gives the possibly to correct
only determinate artifacts (for example color deviation) introduced by
flash light, and it has a wide range of possible applications, from image
enhancement to material color estimation.

Keywords: Color, shading, shadowing, and texture.

1 Introduction

The vast diffusion of digital cameras started a revolution in the way photography is used to measure the appearance of real-world objects. However, despite
the significant improvements of this kind of devices, there are several issues,
regarding the correctness of the acquired data, that need a careful evaluation.
Particularly, the setup and calibration of the lighting environment is often one
of the most limiting aspects for simplified appearance acquisition techniques.

All cameras are equipped with a flash: this tool can provide a practical, easy,
cheap controlled lighting environment. Unfortunately, built-in flashes usually
produce a variety of undesirable effects, like uneven lighting (overexposed in the
near field and dark in the distance), highlights and distracting sharp shadows.
Moreover, "white balance" setting for flash is not effective because it applies the
same correction throughout the image. More interesting results could be obtained
by knowing the geometry of the scene, and using a model of the behavior of flash
light. However, a mathematical modeling of flash light spatial behavior can be

A. Gagalowicz and W. Philips (Eds.): MIRAGE 2009, LNCS 5496, pp. 217–229, 2009.
© Springer-Verlag Berlin Heidelberg 2009

hard to be obtained, because of the large variety of camera/flash models, and the irregular spatial light distribution produced by the flash reflectors and lenses.

For these reasons we propose Flash Lighting Space Sampling (FLiSS), a correction space where a correction matrix is associated to each point in the camera field of view. Once that basic information about the geometry of the scene is known, the proposed structure permits to correct each pixel according to the position of the corresponding point in space. This structure proves to be simple and effective, and it has several advantages, which will be presented and discussed during the analysis of results and the concluding remarks.

2   Related Work

The work proposed in this paper shares some aspects with various subjects, such as computational photography, image enhancement and lighting modeling and estimation. Some of the most interesting and related works will be presented.

Flash/No-Flash Digital Photography. The use of flash/no-flash pairs to enhance the appearance of photographs is a relatively recent research topic where several interesting works appeared. The *continuous flash* [17] was a seminal work, where flash and no-flash pairs were combined to create adjustable images. Two almost contemporaneous papers [9,23] proposed techniques to enhance details and reduce noise in ambient images, by using flash/no-flash pairs. These works provide features for detail transfer, color and noise correction, shadows and highlights removal. While the systems are not completely automatic, very interesting results can be easily obtained. The goal of a more recent work [1] is to enhance flash photography by introducing a flash imaging model and a gradient projection scheme, to reduce the visual effects of noise. Flash/no-flash pairs are used by [21] to detect and remove ambient shadows.

Color constancy and white balance. White balance is a key issue in the context of the *color constancy problem*, that studies the constancy of perceived colors of surfaces under changes in the intensity and spectral composition of the illumination. Several works in this field rely on the assumption that a single illuminant is present: the enhancement of photos can be based on geometric models of color spaces [10], statistical analysis of lights and colors [11] or natural images [12], study of the edges of the image [25].

Another group of papers deals with mixed lighting conditions. Methods can be semi-automatic [20] or automatic. Automatic methods usually work well under quite strong assumptions, like hard shadows and black-body radiators lights [19] or localized gray-world model [8]. A very recent work [18] proposes a white balance technique which renders visually pleasing images by recovering a set of dominant material colors using the technique proposed by [22]. One of the assumptions is that no more than two light types (specified by the user) illuminate the scene. Most of the cited works share some of the main hypotheses of our method. Nevertheless, the knowledge of some information about the geometry of the scene eliminates the need for other restricting assumptions (such as smooth illumination, gray-world theory, need of user interaction).

Illumination estimation and light models. The works in the field of illumination estimation have two principal aims: the estimation of the lighting of an environment [7,24] or the measure of the characteristics of a luminary. Our work is more related to the second group.

One of the first attempts to model both the distant and the near behavior of a light source is the *near-field photometry* approach of Ashdown [2]. Near-field photometry regards the acquisition of a luminary by positioning a number of pinhole cameras (or moving a single camera) around it, and measuring the incident irradiance on a CCD sensor. The results are mapped onto an hemicube that represents the final model of the luminary acquired. Heidrich et al used a similar method [16] by moving the camera on a virtual plane and representing the light sources with a Lumigraph [14]. This representation was named *canned light source*. More recently, Goesele et al. [13] improved the near-field photometric approach using a correction filter to compensate the fact that a digital camera is not a real pinhole camera. Our approach recalls near-field photometry; the main difference is that we estimate the data to "correct" the effect of the light source on known colors, instead of building a model of the light source of interest.

3   Definition of FLiSS

The aim of our work is to build a spatial color correction function that associates a specific color correction procedure to each point in the camera frustum space. We call this particular data structure *color correction space*. Such an approach allows to override the limitations assumed in most of the color correction approaches [3,4], that is that the illumination is constant, or easily model-able, across the scene. Our main assumptions are: flash light can be considered the dominant illumination in the scene; the light interaction can be described using just sRGB space (we do not account full spectra data); surfaces are non-emitting. These hypotheses are common among existing techniques which deal with single illumination, and they cover most of the real cases. Typically, the color calibration of digital photographs consists in taking a snapshot of a pre-defined color reference target, such as a Macbeth ColorChecker^{TM} or an AGFA IT8^{TM}, placed near the subject of interest, and estimating the parameters of a transformation that maps the colors of the reference target in the image into its real colors.

A quite simple to model the correction is a linear affine transformation $c' = Ac + B$. Obviously, due to the nonlinear nature of image color formation, this kind of correction is a rough approximation and many other approaches could be used [3,4]. Moreover, the correction is effective for the image parts that are close (and with a similar illumination) to the reference target. On the other hand, in practice, this simple and compact approach works reasonably well in most cases.

The linear transformation can be written as a 4×3 matrix:

$$\begin{pmatrix} R' \\ G' \\ B' \end{pmatrix} = \begin{bmatrix} c_{11} & c_{12} & c_{13} & c_{14} \\ c_{21} & c_{22} & c_{23} & c_{24} \\ c_{31} & c_{32} & c_{33} & c_{34} \end{bmatrix} \begin{pmatrix} R \\ G \\ B \\ 1 \end{pmatrix} \tag{1}$$

In the following we refer to this matrix as the color correction matrix $\mathcal{C}$ and to its elements as the correction parameters. We explain how we estimate $\mathcal{C}$ in Section 4.

Roughly speaking, the parameters of $\mathcal{C}$ have the following meaning: (c_{11}, c_{22}, c_{33}) are related to the change in contrast of the color; $(c_{12}, c_{13}, c_{21}, c_{23}, c_{31}, c_{32})$ are related to the color deviation caused by the color of the flash light (if the flash is purely white light, these components tend to zero); (c_{14}, c_{24}, c_{34}) are related to the intensity offset. We use the term contrast in the sense that the multiplication for the coefficients expands the range of values of the channels.

Given the assumptions above, we can finally define FLiSS. Flash Lighting Space Sampling is a color correction space where a color correction matrix is associated to each point in a camera frustum. The correction space will be calculated starting from several sampled points in the camera space. The process of correcting an image will, for each pixel in the image, use the appropriate correction matrix according its corresponding position in the camera space. Due the continuous nature of the correction space, the correction will prove to be reliable even without a precise digital model of the scene, so that an approximate reconstruction such as the ones generated, for example, by stereo matching could be used as well.

4   Acquisition Procedure and Data Processing

The aim of our work was to try to build a procedure which could be used in a general case, without using prototypal or expensive devices. The computation of our color correction space is necessary only once (or very few times) in a camera lifetime. But even with this assumption, it was important to define some as simple and fast as possible acquisition procedures. Hence, we decided to sample the camera space view frustum by taking flash lighted photos of a small color target, calculating the correction matrix in those points and subsequently building the entire space by interpolation. We performed the acquisition with three different models of digital cameras, shown in Figure 1(left). These models are representative of three categories of non-professional cameras: compact, digital SLR (single-lens reflex) and digital SLR with external flash.

As a color target we used a Mini Macbeth ColorChecker; its size (about $3.5'' \times 2.5''$) allows the assumption that the light variation across it is negligible. To sample the view frustum of the camera we sliced it with several planes of acquisition at different distances, moving the Mini Macbeth in different positions for each plane. We divided a distance range between 50 and 220 cm in 7 planes, with 25 positions for each plane, as shown in Figure 1(right). The color

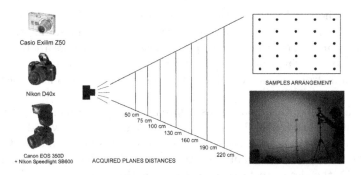

Fig. 1. Left: Digital cameras used for light space sampling. Right: the scheme of acquisition for flash space sampling, with a snapshot of the acquisition setup.

target was placed on a tripod, and always faced towards the camera. For each position multiple snapshots were taken, in order to deal with the known variability of flash behavior, and keeping a fixed exposure time and aperture during the entire procedure.

The snapshots were acquired in sRGB RAW format to avoid any other internal processing by the digital camera, except for the Casio compact camera, with which we were forced to use JPEG images. The acquisition procedure of all the needed images for each model took a couple of hours. The processing of acquired data was subdivided in two main phases. In the first one we calibrated all the acquired images, using the color target reference. In the second phase we built the color correction space through parameters interpolation. Figure 2 shows a schematization of the entire data processing.

Color Calibration

The color calibration was done using the calibration model explained in Section 3. The parameters of the matrix C were estimated by solving a linear system with the c_{ij} as unknowns. Our parameter estimation algorithm takes inspiration from the RANSAC approach.

Since four colors could be sufficient to write a system with 12 equations and 12 unknowns, several random combinations of colors are used to find the best solution in terms of quality. The quality of the color correction is evaluated

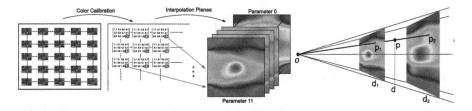

Fig. 2. Data processing. Left: The acquired color targets are calibrated generating a set of color correction matrices, and then interpolated on each plane. Right: Each correction parameter is interpolated to fill the whole camera space.

considering the CIELab distance [6] between the real colors and the corrected ones. Hence, after extracting the Mini Macbeth in a semi-automatic way and segmenting each color patch, several permutations of the colors are tried in order to find the best estimation in terms of quality. This robust approach produces better results from a perceptual point of view than a standard least square approach.

After calibration, each image is associated to: the 12 parameters of the matrix C, the position of the Mini Macbeth relative to the camera view and the distance at which the image has been taken. Since several shots of the same position of the target are taken, obtained values are the means of all the obtained estimations. Before this, the data are further processed with a statistical analysis [15] to remove the outliers from the acquired data.

Data Interpolation

Starting from a set of color correction matrices for several points in the space, we could try either to fit a mathematical function or to calculate the intermediate color correction matrix through interpolation. Here, we opt for interpolation, leaving the first approach as an interesting direction of future research. Given a point p in camera space, we calculate the corresponding color correction matrix as the linear interpolation in the squared distance between the camera center O and the point p in the following way:

$$C_{ij}(p) = C_{ij}(p_1) + (d^2 - d_1^2)\frac{C_{ij}(p_2) - C_{ij}(p_1)}{d_2^2 - d_1^2} \qquad (2)$$

where $C(x)$ indicates the color correction matrix at the point x; p_1 and p_2 are the intersection points between the line starting from O and passing through p. These points lie on the acquisition plane immediately before and after p (Figure 2right); d_1, d and d_2 are the distances between the point O and p_1, p and p_2.

Since $C(p_1)$ and $C(p_2)$ are not known in advance, they have to be estimated to evaluate (2). In fact, only few positions on the acquisition plane are measured: hence another interpolation is required. For this planar interpolation we use radial basis method, with gaussians centered on each sample: standard deviation σ defines how much each sample influences the neighbors. In formulas:

$$C_{ij}(p) = \frac{\sum\limits_{i=1}^{N} C_{ij}(P_i) \exp\left[-\frac{(p - P_i)^2}{\sigma^2}\right]}{\sum\limits_{i=1}^{N} \exp - \left[\frac{(p - P_i)^2}{\sigma^2}\right]} \qquad (3)$$

where N is the total number of samples for the plane and P_i are the positions of samples.

In conclusion, in order to calculate a color correction matrix for a point in the camera space, we first need to calculate two linear interpolations in the acquisition planes, then a linear squared interpolation for the final result.

Regarding the practical implementation, the pre-computed interpolation planes are stored as floating point textures plus, for each texture, some additional information (the distance from the camera center and a scale factor). With this representation, the correction algorithm can be entirely implemented on the GPU, and the pre-computation of the interpolation planes reduces the evaluation of $C_{ij}(p_1)$ and $C_{ij}(p_2)$ in (2) to two texture lookups.

5   Data Analysis

The analysis of the data obtained from the acquisition was performed before the validation of results. A first analysis was on the value ranges obtained for the single coefficients of the matrix. Table 1 shows the statistics relative to all the coefficients calculated for the Nikon camera. The single mean values of

Table 1. Statistics of single coefficients for Nikon camera

	Contrast			Intensity offset			Color deviation					
Coefficient	c_{11}	c_{22}	c_{33}	c_{14}	c_{24}	c_{34}	c_{12}	c_{13}	c_{21}	c_{23}	c_{31}	c_{32}
Mean value	1.20	1.25	1.33	52.07	50.90	50.47	.07	-.06	.005	-.12	.07	-.31
Variance	0.06	0.09	0.09	81.57	123.5	141.3	.001	.003	.002	.009	.001	.012
Min value	0.83	0.84	0.87	27.5	15.4	14.2	-.06	-.43	-.29	-.67	-.21	-.85
Max value	2.59	2.84	2.86	72.0	75.0	76.8	.23	.01	.08	.01	.12	-.11

each group are very similar, and variance describes a general stability of data. Moreover, color deviation coefficients describe the flash as very near to a white light (with a slight deviation in green channel). Contrast and Intensity offset groups show quite low values in variance, but, as it could be expected, the needed modification of color values increases a lot with the distance. Further information about the properties of flash light can be inferred from the analysis of the azimuthal and normal sections of the correction space. In Figure 3 the planes associated to coefficient c_{11} are shown: isolines help in understanding the shape of the light. The light wavefront is quite similar for all the models, but several differences arise as well. The most regular profile is the one associated to the external flash (Canon camera): this is probably the most reliable kind of illumination. Nevertheless, the isolines show that the shape of the light is not similar to a sphere, but it could be better approximated with series of lobes. Moreover, it can be noted that the maximum value of the coefficient is higher respect to the other two examples: this is because the external flash was set with the largest possible field of view, resulting in a very bright illumination for near points, with the need of more correction for the far ones. The Casio associated space is the less regular one, and this is probably due to the fact that the camera model was the least expensive one, and the images were processed and stored directly in JPEG format. Anyway, an interesting observation is that the correction space seems to be slightly shifted on the right, like the position of the flash with respect to the lens (Figure 1).

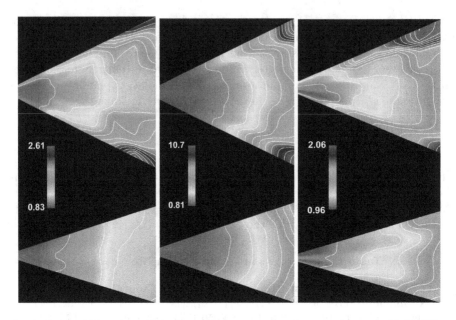

Fig. 3. Azimuthal and normal plane for parameter c_{11} in camera space. (Left) Nikon camera. (Center) Canon camera. (Right) Casio camera.

6   Results

In order to evaluate the results of the correction introduced by FLiSS, we created a setup where the color values in selected regions were known in advance: the scene was formed by a series of identical objects, set at different distances from the camera. The reference objects were simple structures (formed by bricks of different colors) of LEGO DUPLO©, whose size and color are known to be practically identical for all components. The reference RGB color value of each brick was calculated by calibration with the Mini Macbeth. Seven blocks of LEGO© bricks were put in a setting shown in Figure 4(left), and the corresponding 3D model of this configuration was created. This scene has the advantage that, in every single photos, identical elements are present in different parts of the image and at different distances in space. Images were aligned to the 3D model using a registration tool: thus color correction was performed on each pixel which was framing a structure.

One possible criticism to our work could be: would a simple model, for example a point light, be enough to provide a good correction? For this reason, we modeled the flash as a point light and we corrected the same images using this model. We obtained an acceptable estimation of the intensity of the light, by modeling the light degradation with a quadratic law and taking into account the knowledge of the distance of each pixel in the scene, and its reference color: we used this light estimation to correct the color value. Results of corrections on one of the images are shown in Figure 4(left): the original, the Fliss corrected, and the point light corrected images are shown. While in both cases the colors across the scene seem similar, obtained color values are clearly different between the two results. In order to check the accuracy of color correction we measured the perceptual distance between produced colors and the previously calculated reference colors. We chose the CIE76 Delta E measure, which is the Euclidean distance between the CIELAB values associated to two colors. If Delta E is smaller than one, colors are perceived as identical; for values smaller than eightten, the colors can be considered as very similar.

Figure 4 shows the results of the calculation of these distance values. The top boxes display reference colors, then for each piece (each line is associated to the piece number indicated in Figure 4 left) the three columns represent the average color value for original, Fliss corrected and point light corrected image. The Delta E value for each color shows the distance with respect to the reference. It is quite clear that, although producing similar colors, the point light correction returns results which are different respect to the reference: only red color is accurately corrected. Slight improvements are introduced only for distant objects. In order to achieve better results, probably a different modeling for each channel would be necessary. On the contrary, Fliss correction proves to be very reliable, with an average Delta E value which is always smaller than ten. Only in the case of block number *1* the correction is less effective: this can be due to the fact that the object is out of focus, or the color in the original image may be saturated.

A second validation experiment was performed in order to show that the correction introduced by Fliss ca be reliable also with low quality geometric information of the scene. We reconstructed the geometry of a common scene starting from a few photos, using the Arc3D web service [26]. Figure 5 shows the

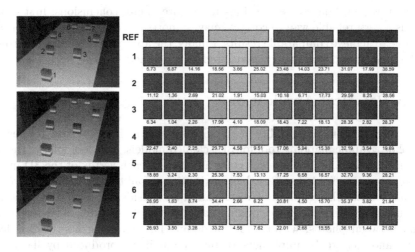

Fig. 4. Left: example of images used for validation: top, original; center, Fliss corrected; bottom, point light model corrected. Only the Lego blocks part of the image was corrected. Right: correction accuracy estimation: for each Lego block, average color and Delta E value (respect to reference) in original (left column), Fliss corrected (center column) and Point Light corrected (right column) images are shown.

Fig. 5. Example of color correction using low quality geometry: Top left, original flash image; Top right, a snapshot of the extracted geometry; Bottom, Fliss corrected image

starting image, together with the geometry obtained, and the corrected image (the correction was applied only where geometry data was present). While the obtained 3D model is only an approximated representation of the geometry in the image, the result of FLiSS use is satisfying: for example the color of the tablecloth after correction is the same throughout the scene.

Analysis of validation tests brings us to three main conclusions: first of all, Fliss is a very reliable way to correct images; moreover, simpler light models are not able to achieve comparable results. Finally, the approach is reliable regardless of the quality of the geometry associated to the image.

7   A Practical Application

To further show the potentiality of FLiSS, we tested its impact in the framework of a photographic mapping tool for 3D scanned models. The tool follows the approach described in [5] where the color assigned to each vertex of the 3D model is computed as weighted sum of the contributions of all the photos which project on that vertex. These weights are calculated by considering several metrics and their use guarantees the better quality data to prevail, obtaining a smooth transition between photos, without loss of detail. This mapping approach is fast, robust and easy to be implemented, but the artifacts produced by flash, like highlights, hard shadows and non uniform illumination cannot be automatically corrected.

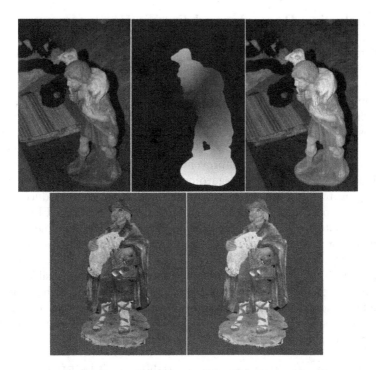

Fig. 6. Top: an example of color correction on a flash photograph: the original image (left); a normalized map of the offset correction on the image (center); the corrected image (right). Bottom: a colored 3D model, visualized with no illumination (only color values assigned to vertices). Color correction OFF in left image, ON in right one.

We selected a test set of objects (colored Nativity statues of different heights and materials), to assess the quality and the impact of the flash light correction. We generated the 3D models of the statues with a Minolta 3D scanner and the photos with the Nikon D40x and flash light as the principal illuminant. Images were aligned on the models, and each image of the set was corrected using FLiSS: an example of color correction on a single image is shown in Figure 6.

With an accurate estimation of light position (obtained with a simple procedure involving a simple LEGO© calibration rig), it was also easy to detect the zones where shadows and highlights were present, and remove them by using the redundancy of data provided by the images of the set.

The effects of color correction on a final 3D Model can be seen in Figure 6: the model on the right, obtained with color correction, appears much more "flat" with respect to the model produced without color correction and artifact removal (left). In conclusion, these images show one of the possible applications of our color correction space, where the effectiveness of results is strengthened by the fact that acquisition of color information was fast and easy.

8   Concluding Remarks and Future Work

We presented FLiSS, a novel structure to apply a spatial color correction on images taken using flash light. The method requires basic information of the geometry of the scene, and it's able to automatically correct the color values in the image. FLiSS needs to be estimated only once in a camera lifetime, and this calibration procedure is adequately fast and easy. FLiSS has several advantages: it is robust and flexible, and although its use best fits with the case of flash light, the structure can be used with any kind of light which is bound to the camera. Future work can include both improvements in the proposed structure and exploitation of its possible uses. Regarding the first issue, further effort could be put in trying to find even simpler acquisition procedures, or reduce the number of samples positions. Regarding possible future applications, FLiSS could be used in the context of image enhancement, by applying for example only part of the elements of the correction matrix to obtain visual pleasantness or enhanced readability. Moreover, by working on different flash-lighted images and extracting the local illumination data from the FLiSS, it should be possible to recover simple models of the optical properties of a surface. Finally, while the simple point light model for flash exhibits strong limitations, FLiSS data could be used to obtain a more complex model that could approximate in a very accurate way the behavior of the flash light.

Acknowledgements. This work was partially funded by EU IST IP 3DCO-FORM and Tuscany Regional Project STArT. We would like to thank Michael Goesele for his kind support, Giuliano Kraft and Francesco Dellepiane for providing us the hardware for experimental studies.

References

1. Agrawal, A., Raskar, R., Nayar, S.K., Li, Y.: Removing photography artifacts using gradient projection and flash-exposure sampling. ACM Tr. on Graphics 24(3), 828–835 (2005)
2. Ashdown, I.: Near-Field Photometry: Measuring and Modeling Complex 3-D Light Sources. In: ACM SIGGRAPH 1995 Course Notes, pp. 1–15 (1995)
3. Barnard, K., Cardei, V., Funt, B.: A comparison of computational color constancy algorithms. I: Methodology and experiments with synthesized data. Image Processing, IEEE Trans. 11(9), 972–984 (2002)
4. Barnard, K., Martin, L., Coath, A., Funt, B.: A Comparison of Computational Color Constancy Algorithms Part II: Experiments With Image Data. IEEE Trans. on Image processing 11(9), 985 (2002)
5. Callieri, M., Cignoni, P., Corsini, M., Scopigno, R.: Masked photo blending: mapping dense photographic dataset on high-resolution 3d models. Computer & Graphics 32(4), 464–473 (2008)
6. Commission Internationale de l'Eclairage (CIE). Colorimetry CIE 15 (2004)
7. Debevec, P.: Rendering synthetic objects into real scenes: bridging traditional and image-based graphics with global illumination and high dynamic range photography. In: SIGGRAPH 1998, pp. 189–198. ACM Press, New York (1998)

8. Ebner, M.: Color constancy using local color shifts. In: Pajdla, T., Matas, J(G.) (eds.) ECCV 2004. LNCS, vol. 3023, pp. 276–287. Springer, Heidelberg (2004)
9. Eisemann, E., Durand, F.: Flash photography enhancement via intrinsic relighting. In: ACM Trans. on Graphics, vol. 23, ACM Press, New York (2004)
10. Finlayson, G.D., Hordley, S.D.: Improving gamut mapping color constancy. IEEE Trans. on Image Processing 9(10), 1774–1783 (2000)
11. Finlayson, G.D., Hordley, S.D., Hubel, P.M.: Color by correlation: A simple, unifying framework for color constancy. IEEE Trans. on Pattern Analysis and Machine Intelligence 23(11), 1209–1221 (2001)
12. Gijsenij, A., Gevers, T.: Color constancy using natural image statistics. In: Int. Conf. on Comp. Vision and Pat. Recogn., Minneapolis, USA, June 2007, pp. 1–8 (2007)
13. Goesele, M., Granier, X., Heidrich, W., Seidel, H.-P.: Accurate light source acquisition and rendering. ACM Tr. Gr. 22(3), 621–630 (2003)
14. Gortler, S.J., Grzeszczuk, R., Szeliski, R., Cohen, M.F.: The lumigraph. In: SIGGRAPH 1996, pp. 43–54 (1996)
15. Grubbs, F.: Procedures for detecting outlying observations in samples. Technometrics 11, 1–21 (1969)
16. Heidrich, W., Kautz, J., Slusallek, P., Seidel, H.-P.: Canned lightsources. In: Rend. Tech., pp. 293–300 (1998)
17. Hoppe, H., Toyama, K.: Continuous flash. Technical Report MSR-TR-2003-63, Microsoft Research (2003)
18. Hsu, E., Mertens, T., Paris, S., Avidan, S., Durand, F.: Light mixture estimation for spatially varying white balance. In: SIGGRAPH 2008, ACM Press, New York (2008)
19. Kawakami, R., Ikeuchi, K., Tan, R.T.: Consistent surface color for texturing large objects in outdoor scenes. In: ICCV 2005: Int. Conf. on Computer Vision, Washington, DC, USA, pp. 1200–1207 (2005)
20. Lischinski, D., Farbman, Z., Uyttendaele, M., Szeliski, R.: Interactive local adjustment of tonal values. ACM Tr. Gr. 25(3), 646–653 (2006)
21. Lu, C., Drew, M.S., Finlayson, G.D.: Shadow removal via flash/noflash illumination. In: W. on Mult. Signal Processing, pp. 198–201 (2006)
22. Omer, I., Werman, M.: Color lines: Image specific color representation. In: CVPR 2004, June 2004, vol. II, pp. 946–953. IEEE, Los Alamitos (2004)
23. Petschnigg, G., Szeliski, R., Agrawala, M., Cohen, M., Hoppe, H., Toyama, K.: Digital photography with flash and no-flash image pairs. In: SIGGRAPH 2004, pp. 664–672 (2004)
24. Unger, J., Wenger, A., Hawkins, T., Gardner, A., Debevec, P.: Capturing and rendering with incident light fields. In: EGRW 2003, pp. 141–149 (2003)
25. van de Weijer, J., Gevers, T.: Color constancy based on the grey-edge hypothesis. In: ICIP, pp. 722–725 (2005)
26. Vergauwen, M., Van Gool, L.: Web-based 3d reconstruction service. Mach. Vision Appl. 17(6), 411–426 (2006)

Error Analysis of Stereo Calibration and Reconstruction

Agnieszka Bier[1] and Leszek Luchowski[2]

[1] Institute of Mathematics, Silesian University of Technology, Gliwice, Poland
`agnieszka.bier@polsl.pl`
[2] Institute of Theoretical and Applied Informatics, Polish Academy of Sciences,
Gliwice, Poland
`leszek.luchowski@iitis.gliwice.pl`

Abstract. This paper addresses the problem of the propagation of input data errors in the stereovision process and its influence on the quality of reconstructed 3D points. We consider only those particular camera calibration and 3D reconstruction algorithms which employ singular value decomposition (SVD) methods. Using the SVD Jacobian estimation method developed by Papadopoulo and Lourakis, we determine all the partial derivatives of outputs with respect to the inputs and present a set of tests applying them in various stereovision conditions in order to evaluate their impact on the quality of 3D reconstruction.

Keywords: Stereovision, sensitivity, projection matrix, 3D reconstruction.

1 Introduction

Many applications require stereovision algorithms to work within a given tolerance of error. The problem of precision in 3D reconstruction has been studied by many authors. For example, in [1] the authors investigate the error of a 3D point reconstructed by triangulation in the case of parallel image planes and derive the probability that the results are within a specified error margin. The precision analysis of 3D reconstruction from image sequences including the covariance matrix method and the evaluation of 3D reconstruction error have been thoroughly discussed in [6]. The latter is also considered in [4] in the sense of confidence intervals for the coordinates of the 3D point reconstructed from the cameras set in normal configuration and using disparity maps. Another approach is presented in [9], where the method of bounding boxes is used in the uncertainty analysis of 3D reconstruction. The sensitivity of 3D reconstruction of a specific kind of scene is analyzed in [2]. Hartley and Zisserman [3] analyzed the uncertainties in identifying a homography between two 2D images based on given points, using the best approximation in terms of the Mahalanobis distance between given and reconstructed coordinates. While not directly representing stereo vision, their

A. Gagalowicz and W. Philips (Eds.): MIRAGE 2009, LNCS 5496, pp. 230–241, 2009.
© Springer-Verlag Berlin Heidelberg 2009

findings resemble ours (see Section 3) in that error increases with the distance of points from the origin of the coordinate system.

In this paper we present yet another approach to the problem by performing sensitivity analysis of SVD-based stereovision algorithms.

The paper is organized as follows. Section 2 introduces the calculation models for camera calibration and 3D reconstruction, and gives a detailed sensitivity analysis of both processes. In Section 3 we evaluate the sensitivity of the algorithm for different configurations of scene images varying with respect to camera positioning. The paper concludes with a discussion of the results.

2 Sensitivity Analysis of the Stereovision Process

2.1 Calculation Model

A 3D scene and its 2D image are mathematically represented as a set of pairs of corresponding 2D and 3D points, one being a *projection* of the other. Throughout this work, we will denote 2D coordinates with lowercase letters, and the corresponding 3D coordinates with capital letters. The homogeneous coordinates of $\mathbf{x} = [sx, sy, s]$ and $\mathbf{X} = [X, Y, Z, 1]$, where $\mathbf{x}$ is the projection of $\mathbf{X}$, are related by the *projection matrix* $\mathbf{P}$:

$$s \begin{bmatrix} x \\ y \\ 1 \end{bmatrix} = \mathbf{P} \begin{bmatrix} X \\ Y \\ Z \\ 1 \end{bmatrix}, \tag{1}$$

Calibration uses a set of known 3D points and their 2D images to determine the projection matrix. The equation (1) then leads to a system of equations which are linear with respect to entries of the projection matrix $\mathbf{P}$ and hence can be written in the form:

$$\mathbf{MR} = 0. \tag{2}$$

$\mathbf{M}$ is a $2p \times 12$ matrix, where $p \geq 6$ is the number of pairs of points (x_i, y_i) and $(X_i, Y_i, Z_i, 1)$, $i = 1, 2, ..., p$ used for calibration, and $\mathbf{R}$ is a column vector composed of the entries of $\mathbf{P}$ written row by row:

$$\mathbf{M} = \begin{bmatrix} \cdots & \cdots & \cdots \\ \mathbf{X_i} & 0\ 0\ 0\ 0 & -x_i \mathbf{X_i} \\ 0\ 0\ 0\ 0 & \mathbf{X_i} & -y_i \mathbf{X_i} \\ \cdots & \cdots & \cdots \end{bmatrix}, \tag{3}$$

$$\mathbf{R} = [\mathbf{P_1}, \mathbf{P_2}, \mathbf{P_3}]^T. \tag{4}$$

The equation (2) usually does not have an exact nonzero solution, as in most cases it is over-determined. An approximate (least-squares) solution is found by using singular value decomposition.

The approximate solution $\mathbf{R_a}$ to (2) is determined as

$$\mathbf{M} = \mathbf{UDV^T} \quad \text{(SVD)}, \qquad \mathbf{R_a} = [\mathbf{V_{1,12}}, \mathbf{V_{2,12}}, ..., \mathbf{V_{11,12}}, \mathbf{V_{12,12}}]^T \quad (5)$$

$\mathbf{R_a}$ minimizes the norm

$$\|\mathbf{MR}\| = \sum_{i=1}^{2p} (\mathbf{MR})_i^2$$

among all vectors $\mathbf{R}$ of unit length.

To restore the matrix $\mathbf{P}$ we need only to rearrange the entries of $\mathbf{R_a}$.

The reconstruction of the 3D scene is the process of recovering unknown 3D coordinates from $K \geq 2$ 2D images of the scene made from different points of view. We assume at least two calibrated cameras are used to take pictures of the scene, and the calibration stage performed for each of the cameras has yielded the projection matrices $\mathbf{P}^{(1)}, \mathbf{P}^{(2)}, ..., \mathbf{P}^{(K)}$. Let $[x^{(i)}, y^{(i)}]$ be the Euclidean coordinates of the 2D projections of the unknown 3D point $[X, Y, Z]$, in the image produced by the camera with projection matrix $\mathbf{P}^{(i)}$ $(i = 1, 2, ..., K)$.

Recalling equation (1) for each camera and the pair of 2D and 3D coordinate vectors, we obtain a system of equations

$$s^{(i)} \begin{bmatrix} x^{(i)} \\ y^{(i)} \\ 1 \end{bmatrix} = \mathbf{P}^{(i)} \begin{bmatrix} X \\ Y \\ Z \\ 1 \end{bmatrix}, i = 1, 2, ..., K \quad (6)$$

Usually, having $K \geq 2$ views ensures that the system of equations (6) is (over)determined with respect to the unknown 3D coordinates, and therefore it has a unique (possibly least-squares approximate) solution. An under-determined system of equations can result from inappropriate camera positioning and it will not be covered here. We will, however, deal with "almost-under-determined" systems, which appear to be quite frequent in real life 3D reconstruction processes.

After rearranging the system of equations (6), we obtain the following matrix equation:

$$\mathbf{L} \begin{bmatrix} X \\ Y \\ Z \end{bmatrix} = \mathbf{B} \quad (7)$$

for $\mathbf{L}$ being a matrix of dimension $2K \times 3$:

$$L = \begin{pmatrix} \mathbf{P}_{1,1}^{(1)} - x^{(1)}\mathbf{P}_{3,1}^{(1)} & \mathbf{P}_{1,2}^{(1)} - x^{(1)}\mathbf{P}_{3,2}^{(1)} & \mathbf{P}_{1,3}^{(1)} - x^{(1)}\mathbf{P}_{3,3}^{(1)} \\ \mathbf{P}_{2,1}^{(1)} - y^{(1)}\mathbf{P}_{3,1}^{(1)} & \mathbf{P}_{2,2}^{(1)} - y^{(1)}\mathbf{P}_{3,2}^{(1)} & \mathbf{P}_{2,3}^{(1)} - y^{(1)}\mathbf{P}_{3,3}^{(1)} \\ \mathbf{P}_{1,1}^{(2)} - x^{(2)}\mathbf{P}_{3,1}^{(2)} & \mathbf{P}_{1,2}^{(2)} - x^{(2)}\mathbf{P}_{3,2}^{(2)} & \mathbf{P}_{1,3}^{(2)} - x^{(2)}\mathbf{P}_{3,3}^{(2)} \\ \mathbf{P}_{2,1}^{(2)} - y^{(2)}\mathbf{P}_{3,1}^{(2)} & \mathbf{P}_{2,2}^{(2)} - y^{(2)}\mathbf{P}_{3,2}^{(2)} & \mathbf{P}_{2,3}^{(2)} - y^{(2)}\mathbf{P}_{3,3}^{(2)} \\ \cdots & \cdots & \cdots \\ \mathbf{P}_{2,1}^{(K)} - y^{(K)}\mathbf{P}_{3,1}^{(K)} & \mathbf{P}_{2,2}^{(K)} - y^{(K)}\mathbf{P}_{3,2}^{(K)} & \mathbf{P}_{2,3}^{(K)} - y^{(K)}\mathbf{P}_{3,3}^{(K)} \end{pmatrix},$$

and **B** the column vector of size $2K$:

$$
\mathbf{B} = \begin{bmatrix}
x^{(1)}\mathbf{P}_{3,4}^{(1)} - \mathbf{P}_{1,4}^{(1)} \\
y^{(1)}\mathbf{P}_{3,4}^{(1)} - \mathbf{P}_{2,4}^{(1)} \\
x^{(2)}\mathbf{P}_{3,4}^{(2)} - \mathbf{P}_{1,4}^{(2)} \\
y^{(2)}\mathbf{P}_{3,4}^{(2)} - \mathbf{P}_{2,4}^{(2)} \\
\vdots \\
y^{(K)}\mathbf{P}_{3,4}^{(K)} - \mathbf{P}_{2,4}^{(K)}
\end{bmatrix}.
$$

Equation (7) can be solved for the unknown Euclidean vector $[X, Y, Z]$ using the pseudo-inverse matrix, which leads to the least-squares solution:

$$
\begin{bmatrix} X \\ Y \\ Z \end{bmatrix}_{ls} = \mathbf{L}^+\mathbf{B}, \tag{8}
$$

where $\mathbf{L}^+$ denotes the Moore-Penrose pseudo-inverse of matrix $\mathbf{L}$. The vector $[X, Y, Z]_{ls}^T$ minimizes the least-squares norm:

$$
\sum_{j=1}^{2K} \left((\mathbf{L}[X, Y, Z]^T)_j - \mathbf{B_j} \right)^2.
$$

2.2 Sensitivity of Camera Calibration

At the calibration stage we assume that the 3D coordinates of the points $\mathbf{X}_i$ are known accurately, while the 2D coordinates of their projection $\mathbf{x_i}$ are read from the images with some error. We give here the analysis of propagation of the measurement errors to the entries of the resulting projection matrix.

In Section 2.1 we calibrated the cameras using the SVD method to solve a system of linear equations (2). Now we will examine the sensitivity of this method to input errors. The dependencies of SVD outputs (the entries of the three matrix components) on the entries of the input matrix were thoroughly discussed in [7]. The method proposed by Papadopoulo and Lourakis allows the Jacobian of the SVD components to be determined with respect to the entries of the matrix being decomposed, considering the SVD as a transformation of the matrix entries. We apply this method to estimate calibration error. Assume that $M = [m_{ij}]$ defined in Section 2.1 contains both accurate inputs and error-burdened ones. Let M have the SV decomposition defined in (5). Then, following Equation (9) in [7], we have:

$$
\frac{\partial \mathbf{V}}{\partial m_{i,j}} = -\mathbf{V}\Omega_{\mathbf{V}}^{ij}, \qquad \frac{\partial \mathbf{U}}{\partial m_{i,j}} = \mathbf{U}\Omega_{\mathbf{U}}^{ij},
$$

where $\Omega_{\mathbf{U}}^{ij}$ and $\Omega_{\mathbf{V}}^{ij}$ are matrices of size $2p \times 2p$ and 12×12, respectively, and their entries can be determined as solutions of the following systems:

$$
\begin{cases}
d_l \Omega_{\mathbf{U}\,kl}^{ij} + d_k \Omega_{\mathbf{V}\,kl}^{ij} = u_{ik}v_{jl}, \\
d_k \Omega_{\mathbf{U}\,kl}^{ij} + d_l \Omega_{\mathbf{V}\,kl}^{ij} = -u_{il}v_{jk}.
\end{cases} \tag{9}
$$

In this notation, d_i is the i-th diagonal entry of the D component in SVD, while $u_{i,j}$ and $v_{i,j}$ are the entries of matrices $\mathbf{U}$ and $\mathbf{V}$.

If $d_k \neq d_l$, then the obtained entries of matrix $\mathbf{\Omega_V^{ij}}$ have the form:

$$\mathbf{\Omega_V^{ij}}_{kl} = \frac{d_l u_{il} v_{jk} + d_k u_{ik} v_{jl}}{d_k^2 - d_l^2}$$

If two or more singular values are equal, then - as suggested in [7] - the equations (9) related to those values have to be solved with a least squares method in order to obtain the Jacobian with the smallest possible norm. Then the partial derivatives can be determined from the equation:

$$\frac{\partial v_{kl}}{\partial m_{ij}} = -\sum_{s=1}^{12} v_{ks} \mathbf{\Omega_V^{ij}}_{sl}. \tag{10}$$

Returning to the notation of the projection matrix we have:

$$\frac{\partial \mathbf{P_{k,l}}}{\partial m_{i,j}} = \frac{\partial v_{4(k-1)+l,12}}{\partial m_{i,j}}, \tag{11}$$

hence, if (x_s, y_s) is the 2D image of the s-th 3D point $(X_s, Y_s, Z_s, 1)$,

$$
\begin{aligned}
\frac{\partial \mathbf{P_{k,l}}}{\partial x_s} &= \sum_{(i,j)} \frac{\partial \mathbf{P_{k,l}}}{\partial m_{i,j}} \cdot \frac{\partial m_{i,j}}{\partial x_s} = \\
&= -X_s \cdot \frac{\partial \mathbf{P_{k,l}}}{\partial m_{2s-1,9}} - Y_s \cdot \frac{\partial \mathbf{P_{k,l}}}{\partial m_{2s-1,10}} - Z_s \cdot \frac{\partial \mathbf{P_{k,l}}}{\partial m_{2s-1,11}} - \frac{\partial \mathbf{P_{k,l}}}{\partial m_{2s-1,12}} = \\
&= -X_s \cdot \frac{\partial v_{4(k-1)+l,12}}{\partial m_{2s-1,9}} - Y_s \cdot \frac{\partial v_{4(k-1)+l,12}}{\partial m_{2s-1,10}} - Z_s \cdot \frac{\partial v_{4(k-1)+l,12}}{\partial m_{2s-1,11}} - \\
&\quad - \frac{\partial v_{4(k-1)+l,12}}{\partial m_{2s-1,12}}
\end{aligned} \tag{12}
$$

and

$$
\begin{aligned}
\frac{\partial \mathbf{P_{k,l}}}{\partial y_s} &= \sum_{(i,j)} \frac{\partial \mathbf{P_{k,l}}}{\partial m_{i,j}} \cdot \frac{\partial m_{i,j}}{\partial y_s} = \\
&= -X_s \cdot \frac{\partial v_{4(k-1)+l,12}}{\partial m_{2s,9}} - Y_s \cdot \frac{\partial v_{4(k-1)+l,12}}{\partial m_{2s,10}} - Z_s \cdot \frac{\partial v_{4(k-1)+l,12}}{\partial m_{2s,11}} - \\
&\quad - \frac{\partial v_{4(k-1)+l,12}}{\partial m_{2s,12}},
\end{aligned} \tag{13}
$$

where the partial derivatives $\frac{\partial v_{4(k-1)+l,12}}{\partial m_{2s-1,9}}$, $\frac{\partial v_{4(k-1)+l,12}}{\partial m_{2s-1,10}}$, $\frac{\partial v_{4(k-1)+l,12}}{\partial m_{2s-1,11}}$, $\frac{\partial v_{4(k-1)+l,12}}{\partial m_{2s-1,12}}$, $\frac{\partial v_{4(k-1)+l,12}}{\partial m_{2s,9}}$, $\frac{\partial v_{4(k-1)+l,12}}{\partial m_{2s,10}}$, $\frac{\partial v_{4(k-1)+l,12}}{\partial m_{2s,11}}$ and $\frac{\partial v_{4(k-1)+l,12}}{\partial m_{2s,12}}$ are determined from equations (10).

2.3   Sensitivity of 3D Reconstruction

Having determined the projection matrices of the cameras and given the 2D coordinates, in the images, of a feature point, we can use formula (8) to recover its 3D coordinates. Note that projection matrices as well as 2D coordinates are burdened with errors. The uncertainty of the entries of the projection matrices were derived in the previous section. For now, we assume that the projection matrices are given with their uncertainties as $\mathbf{P}^{(i)} \pm \triangle\mathbf{P}^{(i)}$ and we are not interested in the sourceof these errors. At the same time, we consider the errors of reading the 2D coordinates $(x^{(i)}, y^{(i)})$ of a feature point $\overline{\mathbf{X}} = [X, Y, Z]^T$.

Recall the solution (8) to the 3D reconstruction problem given in section 2.1. We are now interested in the partial derivatives of the reconstructed coordinates with respect to any of the inputs that can be burdened with errors.

$$\frac{\partial \overline{\mathbf{X}}}{\partial r} = \frac{\partial (\mathbf{L}^+ \mathbf{B})}{\partial r} = \frac{\partial \mathbf{L}^+}{\partial r} \mathbf{B} + \mathbf{L}^+ \frac{\partial \mathbf{B}}{\partial r}, \tag{14}$$

where r represents any of the parameters $\mathbf{P}_{k,l}$, $x^{(i)}$, $y^{(i)}$. Clearly, vector $\frac{\partial \mathbf{B}}{\partial r}$ can be determined directly from the vector $\mathbf{B}$, as for $i = 1, 2, ..., K$ we have

$$\frac{\partial \mathbf{B}_{2i-1}}{\partial \mathbf{P}_{3,4}^{(i)}} = x^{(i)}, \qquad \frac{\partial \mathbf{B}_{2i-1}}{\partial \mathbf{P}_{1,4}^{(i)}} = -1, \qquad \frac{\partial \mathbf{B}_{2i}}{\partial \mathbf{P}_{3,4}^{(i)}} = y^{(i)}, \qquad \frac{\partial \mathbf{B}_{2i}}{\partial \mathbf{P}_{1,4}^{(i)}} = -1,$$

$$\frac{\partial \mathbf{B}_{2i-1}}{\partial x^{(i)}} = \mathbf{P}_{3,4}^{(i)}, \qquad \frac{\partial \mathbf{B}_{2i}}{\partial y^{(i)}} = \mathbf{P}_{3,4}^{(i)},$$

and

$$\frac{\partial \mathbf{B}_j}{\partial \mathbf{P}_{k,l}^{(i)}} = \frac{\partial \mathbf{B}_j}{\partial x^{(i)}} = \frac{\partial \mathbf{B}_j}{\partial y^{(i)}} = 0$$

elsewhere. Likewise, the derivatives for the entries of the matrix $\mathbf{L}$ are:

$$\frac{\partial \mathbf{L}_{2i-1,j}}{\partial \mathbf{P}_{1,j}^{(i)}} = 1, \qquad \frac{\partial \mathbf{L}_{2i,j}}{\partial \mathbf{P}_{2,j}^{(i)}} = 1, \qquad \frac{\partial \mathbf{L}_{2i-1,j}}{\partial \mathbf{P}_{3,j}^{(i)}} = -x^{(i)}, \qquad \frac{\partial \mathbf{L}_{2i,j}}{\partial \mathbf{P}_{3,j}^{(i)}} = -y^{(i)}, \quad j = 1, 2, 3,$$

$$\frac{\partial \mathbf{L}_{2i-1,1}}{\partial x^{(i)}} = -\mathbf{P}_{3,1}^{(i)}, \qquad \frac{\partial \mathbf{L}_{2i-1,2}}{\partial x^{(i)}} = -\mathbf{P}_{3,2}^{(i)}, \qquad \frac{\partial \mathbf{L}_{2i-1,3}}{\partial x^{(i)}} = -\mathbf{P}_{3,3}^{(i)},$$

$$\frac{\partial \mathbf{L}_{2i,1}}{\partial y^{(i)}} = -\mathbf{P}_{3,1}^{(i)}, \qquad \frac{\partial \mathbf{L}_{2i,2}}{\partial y^{(i)}} = -\mathbf{P}_{3,2}^{(i)}, \qquad \frac{\partial \mathbf{L}_{2i,3}}{\partial y^{(i)}} = -\mathbf{P}_{3,3}^{(i)},$$

while all other derivatives $\frac{\partial \mathbf{L}_{s,t}}{\partial \mathbf{P}_{k,l}^{(i)}}$, $\frac{\partial \mathbf{L}_{s,t}}{\partial x^{(i)}}$, $\frac{\partial \mathbf{L}_{s,t}}{\partial y^{(i)}}$ are zeros.

The study of $\frac{\partial \mathbf{L}^+}{\partial r}$ needs more complicated analysis. Recall from section 2.1 that $\mathbf{L}^+ = \mathbf{V}\mathbf{\Sigma}^+\mathbf{U}^T$, hence for every pair of indices (i, j):

$$\frac{\partial \mathbf{L}^+}{\partial l_{i,j}} = \frac{\partial \mathbf{V}}{\partial l_{i,j}} \mathbf{\Sigma}^+ \mathbf{U}^T + \mathbf{V} \frac{\partial \mathbf{\Sigma}^+}{\partial l_{i,j}} \mathbf{U}^T + \mathbf{V}\mathbf{\Sigma}^+ \frac{\partial \mathbf{U}^T}{\partial l_{i,j}}. \tag{15}$$

Since all entries of the diagonal matrix $\mathbf{\Sigma}^+$ are the reciprocals of the entries of matrix $\mathbf{\Sigma}$, except for those equal to zero, which remain unchanged, we obtain:

$$\frac{\partial \mathbf{\Sigma}^+_{k,k}}{\partial l_{i,j}} = -\frac{1}{(\Sigma_{k,k})^2} \cdot \frac{\partial \Sigma_{k,k}}{\partial l_{i,j}},$$

if $\Sigma_{k,k} \neq 0$, and

$$\frac{\partial \mathbf{\Sigma}^+_{k,k}}{\partial l_{i,j}} = 0$$

otherwise. Additionally, for the case of $\Sigma_{k,k} \neq 0$, following the relations in [7] we have:

$$\frac{\partial \mathbf{\Sigma}^+_{k,k}}{\partial l_{i,j}} = -\frac{1}{(\Sigma_{k,k})^2} \cdot \mathbf{U}_{i,k} \mathbf{V}_{j,k}.$$

The rest we get from [7]:

$$\frac{\partial \mathbf{U}}{\partial l_{i,j}} = \mathbf{U}\Omega_{\mathbf{U}}^{ij}, \qquad \frac{\partial \mathbf{V}}{\partial l_{i,j}} = -\mathbf{V}\Omega_{\mathbf{V}}^{ij},$$

where $\Omega_{\mathbf{U}}^{ij}$ and $\Omega_{\mathbf{V}}^{ij}$ are matrices defined for the decomposition $\mathbf{L} = U\Sigma V^T$ and their entries can be obtained from the systems of equations analogous to (9).

Applying all these relations to the formula (15), we determine the derivatives of the entries of $\mathbf{L}^+$ with respect to particular entries of matrix $\mathbf{L}$. Now, for parameter r, which represents one of parameters $\mathbf{P}_{\mathbf{k,l}}$, $x^{(i)}$, $y^{(i)}$, we obtain:

$$\frac{\partial \mathbf{L}^+}{\partial r} = \frac{\partial \mathbf{L}^+}{\partial \mathbf{L}} \cdot \frac{\partial \mathbf{L}}{\partial r}, \tag{16}$$

and more precisely

$$\frac{\partial \mathbf{L}^+}{\partial r} = \sum_{(i,j)} \frac{\partial \mathbf{L}^+}{\partial \mathbf{L}_{i,j}} \cdot \frac{\partial \mathbf{L}_{i,j}}{\partial r}.$$

Finally, we have all the derivatives needed to compute $\frac{\partial \overline{\mathbf{X}}}{\partial r}$ using equation (14):

$$\frac{\partial \overline{\mathbf{X}}}{\partial r} = \left(\sum_{(i,j)} \frac{\partial \mathbf{L}^+}{\partial \mathbf{L}_{i,j}} \cdot \frac{\partial \mathbf{L}_{i,j}}{\partial r} \right) \mathbf{B} + \mathbf{L}^+ \frac{\partial \mathbf{B}}{\partial r}. \tag{17}$$

3   Practical Applications

The purpose of the tests performed on the simulated 3D scene and its images is to confront the theory presented in the previous section with an actual stereovision process. We used POV-Ray (Persistence of Vision Ray-Tracer) to generate photo-realistic images from descriptions of scenes and camera positions. An important advantage of a virtual scene is that accurate 3D coordinates are known and can be used as a reference when evaluating the results of 3D reconstruction.

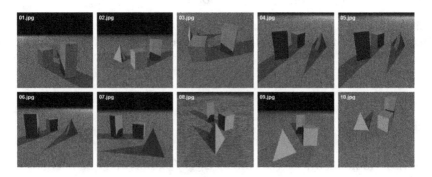

Fig. 1. Images obtained in POV-Ray

Our test scene consisted of a tetrahedron, a cuboid and a cube, all of a similar size, standing on a plane. For testing purposes, 11 pictures of the scene were created in POVRay, each from a different point of view and camera direction. Ten of them are presented in Figure 1. The eleventh is shown in Figure 2 along with the labeling of the vertices.

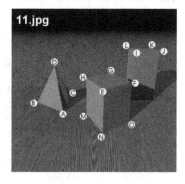

Fig. 2. Image 11.jpg and vertex labels

All the resulting images were next subjected to calibration. The seven visible vertices of the cube served as calibration points. Using formulas 12 and 13 derived in the previous section, we determined the derivatives $\frac{\partial \mathbf{P}^{(i)}}{\partial x_j}$ and $\frac{\partial \mathbf{P}^{(i)}}{\partial y_j}$, which are the derivative matrices of the i-th projection matrix with respect to the x and y coordinate of the j-th calibration point in the i-th image.

In the following analysis, the measure of the sensitivity $\mathcal{S}_j(P^{(i)})$ of the projection matrix $\mathbf{P}^{(i)}$ to error in the j-th calibration point is defined as the sum of squares of all entries of matrices $\frac{\partial \mathbf{P}^{(i)}}{\partial x_j}$ and $\frac{\partial \mathbf{P}^{(i)}}{\partial y_j}$:

$$\mathcal{S}_j(P^{(i)}) := \sum_{k,l} \left[\left(\frac{\partial \mathbf{P}_{k,l}^{(i)}}{\partial x_j} \right)^2 + \left(\frac{\partial \mathbf{P}_{k,l}^{(i)}}{\partial y_j} \right)^2 \right].$$

Such definition seems reasonable as it gathers the influence of both coordinates on the whole projection matrix. It has, however, some drawbacks. As the components in the sum are taken with equal weights, their impact on further processing is not taken into account. This sum cannot therefore be regarded as a measure of calibration quality, which should be considered from the point of view of the quality of the whole stereovision process.

Table 1 presents the sensitivities of projection matrices to all calibration points separately and combined sensitivity to errors in coordinates of all calibration points, calculated as square root of the sum of squares of sensitivities to particular points.

The implementation of the differential method presented in the previous section allowed us to determine those of all eleven projection matrices which are the most sensitive to input errors, and those calibration points which influence the most and the least the precision of calibration. One can observe that some calibration points have significantly smaller impact on projection matrices than others. Moreover, the three worst calibration sensitivity measures are achieved for the images for which the calibration process is performed with all cube vertices except the one which is least distant to the origin of the scene.

Considering the distances of those particular calibration points from the origin of the 3D scene coordinate system, a general tendency can be observed. The points lying furthest from the origin have the greatest impact on the projection matrix entries, in other words, calibration is more sensitive to the points lying

Table 1. Calibration sensitivity to individual calibration points and combined sensitivity to all calibration points (multiplied by 10^4)

Image \ Point	E	F	G	H	M	N	O	P	Combined
01.jpg	2	3	21	1	-	5	1	5	22.5
02.jpg	31	92	19	25	17	50	17	-	116.15
03.jpg	21	3	6	14	-	10	17	2	32.8
04.jpg	16	2	0	5	4	-	3	1	17.63
05.jpg	12	1	0	4	4	-	5	1	14.25
06.jpg	2	1	0	1	1	-	0	0	2.65
07.jpg	7	6	1	1	2	6	-	3	12.04
08.jpg	1	2	2	2	0	0	-	2	4.12
09.jpg	0	13	1	8	9	5	-	1	18.47
10.jpg	18	10	3	50	5	24	10	-	60.28
11.jpg	288	115	8	27	2	112	35	-	332.8

further from the origin. A possible reason for this is that the absolute values of coordinates are multipliers in the formulas for projection matrix derivatives. Therefore, if they are smaller, they lead to smaller derivatives. Indeed, translating the system of 3D coordinates so that the center of the base of the cube was close to the origin resulted in a calibration error that was a fraction of its previous value, i.e. the projection matrix entries appeared to be less sensitive to input errors. Conversely, setting the system of 3D coordinates so that the cube was standing further from the origin resulted in a calibration error that was a multiple of its original value.

A more detailed examination of the results revealed that for almost all images and points the entries $P_{1,4}$ and $P_{2,4}$ are the most sensitive to errors in the $x^{(k)}$ and $y^{(k)}$ coordinate.

The tests can however only serve as an illustration of the observed tendency and cannot be treated as a proof for the hypothesis. Even if the tendency is considered as a rule, one can ask whether a small sensitivity of the projection matrix to calibration inputs is desirable or not. Moreover, if the calibration points are located far from the camera, a small difference in 2D image coordinates results in a large difference in 3D scene coordinates. This especially applies to picture 06.jpg, where the distance between the cube vertices used for calibration and the camera is definitely the greatest among all the images. ¿From the presented point of view, a small sensitivity to input errors works against the quality of the process in the sense of precise reconstruction. This feature of camera calibration should then be taken into account, when choosing the camera position for best reconstruction results.

3D reconstruction was performed for every pair among the 11 views, i.e. for a total of 55 pairs. In general, 8 feature points - the vertices of the tetrahedron and cuboid - were subject to 3D reconstruction. However, due to the fact that not all of these points are visible in all pictures, the number of reconstructed points varied from five to eight. Each pair of images was used for reconstruction and yielded data which we used to evaluate error propagation.

We begin the analysis of results by comparing the reconstructed 3D coordinates with the ideal ones taken from POVRay. In order to measure the quality of each 3D reconstruction, we calculated the mean-square error of all 3D coordinates reconstructed for a given pair. The results are shown in Table 2.

Table 2. The quality of 3D reconstruction

Image	01.jpg	02.jpg	03.jpg	04.jpg	05.jpg	06.jpg	07.jpg	08.jpg	0.9.jpg	10.jpg	11.jpg
01.jpg		0.0018	0.0017	0.0101	0.0022	0.0202	0.0025	0.0062	0.0025	0.0018	0.0016
02.jpg	0.0018		0.0004	0.0021	0.0013	0.0286	0.0020	0.0007	0.0019	0.0011	0.0104
03.jpg	0.0017	0.0004		0.0017	0.0005	0.0098	0.0018	0.0005	0.0013	0.0002	0.0009
04.jpg	0.0101	0.0021	0.0017		1.2742	0.2662	0.0041	0.0011	0.0015	0.0010	0.0013
05.jpg	0.0022	0.0013	0.0005	1.2742		0.0867	0.0031	0.0005	0.0018	0.0003	0.0005
06.jpg	0.0202	0.0286	0.0098	0.2662	0.0867		0.0219	0.0133	0.0160	0.0114	0.0198
07.jpg	0.0025	0.0020	0.0018	0.0041	0.0031	0.0219		0.0115	0.0026	0.0021	0.0020
08.jpg	0.0062	0.0007	0.0005	0.0011	0.0005	0.0133	0.0115		0.0127	0.0001	0.0007
09.jpg	0.0025	0.0019	0.0013	0.0015	0.0018	0.0160	0.0026	0.0127		0.0058	0.0016
10.jpg	0.0018	0.0011	0.0002	0.0010	0.0003	0.0114	0.0021	0.0001	0.0058		0.0017
11.jpg	0.0016	0.0104	0.0009	0.0013	0.0005	0.0198	0.0020	0.0007	0.0016	0.0017	

Two facts can be observed. First, the pair of images taken from points of view that differed least (04.jpg and 05.jpg) resulted in the least precise reconstruction. The reason for this is that narrowly spaced viewpoints lead to small angles between gaze directions. Therefore, a slight inaccuracy in image coordinates results in a large change of reconstructed 3D location, especially the depth. In the extreme case of coincident cameras, 3D reconstruction is impossible.

The theoretically calculated sensitivity of reconstructed points to errors in the input 2D coordinates - according to formula (16) - is also highest for this image pair.

The second observation regarding the quality of reconstruction with use of picture 06.jpg, the one with the projection matrix least sensitive to input data errors, is quite surprising. This image yields the worst performance (in terms of overall error in pairs with every other image). This shows that a small sensitivity of the estimated projection matrix to input data errors does not guarantee a good reconstruction. We should emphasize that while talking about calibration precision, we mean the precision of the estimation of the projection matrix and disregard the impact of its entries on the reconstruction quality.

All the 11 projection matrices obtained in the calibration stage were examined regarding the sensitivity of their entries to input 2D coordinates of the calibration points. The most sensitive entry was identified for every image, feature point, and coordinate. In 133 cases out of 154, this was either $P_{1,4}$ or $P_{2,4}$. Most of the rest have single deviations from that rule and there are only two projection matrices having more than two. A question arises how this influences the quality of reconstruction.

We also considered the accuracy of 3D reconstruction of individual points. We found that the points most distant from origin were reconstructed less accurately. Using the differential method as above, the sensitivity of these reconstructed 3D points (points J and K on Fig. 1) to errors in the 2D inputs was determined for all reconstruction image pairs. Pairs including picture 06.jpg turned out to be slightly more sensitive to the errors than others. The same was true, to a lesser extent, about reconstructed points other than J and K.

Our next experiment used an analytical model of camera pairs and a scene.

The scene consisted of a cube and tetrahedron, each with an edge length of 13 cm, both centered at the origin of the 3D coordinate system. This meant they overlapped, but in a simulated environment this was not a problem.

The intrinsic parameters of the cameras represented a focal distance (35mm equivalent) of 52mm and an image sensor with 3072 x 2048 pixels (for an aspect ratio of 3:2), placed symmetrically with respect to the optical axis.

The extrinsic parameters positioned the camera to look directly at the origin of the 3D coordinates from a distance of 100 cm. 72 such virtual cameras were placed on a horizontal circle (the Y axis being vertical) around the origin.

Each camera was virtually calibrated using seven of the vertices of the cube, imitating a real scene where it is impossible to see all 8. The first camera was then paired with every other one, resulting in 71 pairs. For each pair, the 3D position of the vertices of the tetrahedron were reconstructed from their coordinates projected by the two cameras. The camera parameters and the simulation results were substituted to the formulas of Sections 2.2 and 2.3 to compute the sensitivities of reconstructed 3D points to errors in the 2D coordinates used for either calibration or reconstruction. Results are presented in Figure 3.

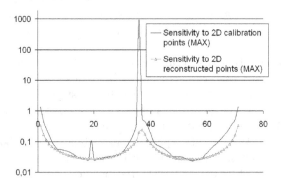

Fig. 3. Sensitivity of reconstruction as function of choice of second camera

The horizontal axis is the number (2 through 72) of the camera forming a pair with Camera 1. The central point (camera 37) represents a pair of cameras facing each other across the scene. The sensitivity on the vertical axis is expressed in centimeters of reconstruction error per pixel of error in image coordinate.

It can be seen that the sensitivity increases dramatically when the optical axes of the two cameras are close to each other, whether gazing in the same or opposite directions. There is also an unexpected, smaller increase in sensitivity to 2D coordinate error in calibration, occurring for perpendicular cameras. In this position, the optical axes of both cameras are parallel to edges of the calibration cube, which may have affected the stability of our matrix computations. Further experiments will be needed to clarify this.

To summarize: the experiments and theoretical analysis have brought some valuable observations, which may serve as material for further and more detailed discussion of the factors that condition the quality of the whole stereovision process, as well as its particular stages. The main observations are:

- The choice of calibration points influences the precision of the recovered projection matrix. Points closer to the origin of the assumed system of 3D coordinates have a smaller impact on the projection matrix entries.
- Increasing the distance between the camera and the scene increases the impact of 2D reconstruction input errors on the quality of the process.
- Bringing the two cameras closer together, or facing each other, decreases the 3D reconstruction quality and increases the impact of 2D reconstruction input errors on the quality of the process.

Acknowledgments. This work was supported in part by the Polish Ministry of Science and Higher Education, under the research grant N N516 1862 33.

References

1. Blostein, S., Huang, T.: Error analysis in stereo determination of 3-d point positions. IEEE Trans. Pat. An. Mach. Int. PAMI-9(6) (1987)
2. Grossmann, E., Santos-Victor, J.: Least-squares 3D reconstruction from one or more views and geometric clues. Comp. Vis. Im. Und. 99, 151–174 (2005)
3. Hartley, R., Zisserman, A.: Multiple view geometry in computer vision. Cambridge University Press, Cambridge (2003)
4. Kamberova, G., Bajcsy, R.: Precision in 3-D points reconstructed from stereo (1997), http://www.cis.upenn.edu/~kamberov/doc/eccv.ps.gz
5. Ma, Y., Soatto, S., Košecká, J., Sastry, S.: An invitation to 3-D vision. In: IAM, vol. 26. Springer, NY (2004)
6. Min, S., Rixin, H., Daojun, W.: Precision analysis to 3D reconstruction from image sequences
7. Papadopoulo, T., Lourakis, M.: Estimating the Jacobian of the singular value decomposition: Theory and applications, INRIA RR 3961 (2000)
8. Solina, F.: Errors in stereo due to quantization, Univ. Pensylvania, Tech. Rep. MS-CIS-85-34 (1985)
9. Telle, B., Stasse, O., Yokoi, K., Ueshiba, T., Tomita, F.: Three characterisations of 3D reconstruction uncertainty with bounded error. Proc. IEEE, 3894–3899 (2005)
10. Verri, A., Torre, V.: Absolute depth estimates in stereopsis. J. Opt. Soc. Amer. 3(3), 297–299 (1986)

Spatio-Temporal Tracking of Faces by Stereo Vision

Markus Steffens[1,2], Werner Krybus[1], Christine Kohring[1], and Danny Morton[2]

[1] South Westphalia University of Applied Sciences, Luebecker Ring 2,
59494 Soest, Germany
{steffens,krybus,kohring}@fh-swf.de
[2] University of Bolton, Deane Road, Bolton BL3 5AB UK
d.morton@bolton.ac.uk

Abstract. This report contributes a new approach for the robust tracking of humans' heads and faces based on a spatio-temporal scene analysis. The framework comprises aspects of structure and motion problems, as there are feature extraction, spatial and temporal matching, re-calibration, tracking, and reconstruction. The scene is acquired through a calibrated stereo sensor. A cue processor extracts invariant features in both views, which are spatially matched by geometric relations. The temporal matching takes place via prediction from the tracking module and a sixmilarity transformation of the features' 2D locations between both views. The head is reconstructed and tracked in 3D. The reprojection of the predicted structure limits the search space of both the cue processor as well as the re-construction procedure. Due to the focused application, the instability of calibration of the stereo sensor is limited to the relative extrinsic parameters that are re-calibrated during the re-construction process. The framework is practically applied and proven. First experimental results will be discussed and further steps of development within the project are presented.

1 Introduction and Motivation

Scene analysis in the current context comprises the process of modeling objects observed in a scene. This is generally based on the recognition and localization of pictorial and iconic image features indicating sought objects. These can be low-level or mid-level features such as shapes or areal patches in a certain feature space. The aim is to fit a certain object model from the extracted features to infer certain knowledge about the objects. The states inferred from said fitted models, for example position and orientation in object space, cover large errors, especially with generic models applied to a wide range of scenes.

Most methods for scene analysis are based on exemplar features [14, 15]. Those approaches extract low-level features and compare found candidates with previously learned exemplars. Scene analysis based on low-level feature models has the characteristic that its performance is highly influenced by the states of extracted features such as location and orientation in case of geometric states. The visual information in unconstrained environments might be uncertain due to large camera angles causing strong perspectivity, bad lighting conditions, or motion blur. In such cases an exemplar

A. Gagalowicz and W. Philips (Eds.): MIRAGE 2009, LNCS 5496, pp. 242–253, 2009.

method will generally fail in accuracy. However, in other cases such as pure recognition tasks based on coded templates the data might be sufficient.

When analyzing the process of fitting an object model to indicator features, it becomes apparent that all indicator features are inherently linked to a semantic being an assumed characteristic of focused objects. For example geometries explicitly related to eyes or nostrils in facial images. The recognition and localization step is always restricted to said indicator features. Knowledge from other sources, which are not semantically related to the objects under observation, is neglected.

To increase the accuracy of inferred states the amount of independent information has to be increased. This will lead to a more robust scene analysis in case the curse of dimensionality is respected, that is, no redundant information is used. Assuming that the set of indicator features is complete, one can think of utilizing generic features from the overall scene besides indicator features and higher contextual information. One of several questions arising is how to combine information from different sources so that the objects can be modeled more accurate.

Contextual information is studied in the field of scene understanding. It is a high-level process inferring knowledge about the scene in a semantic form. Early approaches are based on exemplar methods only. Due to inherent limitations of exemplar methods regarding accuracy and robustness, early studies in the area of image understanding were developed for constrained environments such as medical image and document analysis. With the help of high-level contextual information, such as the correlation between location and object or between activity and object, current approaches try to solve the problems of understanding complex scenes in the presence of insufficient and inaccurate information [1, 2, 3]. Although current methods inferring knowledge from unconstrained environments and dynamic scenes are still limited, the underlying concepts are tackling the same issue of inaccurate feature states.

This new framework adapts concepts from high-level scene understanding to low-level scene analysis. That is, instead of inferring high-level semantic knowledge from the scene and establishing relations to the objects, accurate knowledge of the low-level scene structure is to be related to the objects over time. This is motivated by the fact that the inaccuracy of the object models are caused by insufficient and inaccurate information, while the entropy in structure is generally higher due to less constraints in feature selection. The former set of indicator features, inherently determined by a pre-defined semantic, is augmented by accurate structural knowledge. Consequently, besides defining instruments for the extraction of accurate scene structure, new concepts for object modeling are needed which incorporate the structural knowledge in such a way that the accuracy of estimated states is increased. The focus is therefore on fusing knowledge from different domains.

The basic idea of this current research work is to additionally observe generic cues in a scene that are good to track and not a-priori related to any specific object. Such cues can be seen as the structure of the scene and further relations between said cues follow certain syntax, such like position and motion [19-21, 22]. The aim is to identify and observe those structure cues which are capable of increasing the accuracy of the inferred states of the object-related semantic models. Since current investigations are focusing on accuracy of 3D localization and orientation, the structure cues must cover high accurate spatial states, both in space and time. There are several ways of fusing information of structure cues and semantic features. High interest is attracted to the fact that it is possible to accurately estimate states of temporally occluded semantic features due to certain relations between semantic and structure. This is achieved by organizing structure cues in a pyramid giving different levels of abstraction and relations. The basic assumption is that there are weak relations between semantic and structure. This is in opposite to current scene analysis methods which always assume strong relations between semantic objects and indicator features. That is, the difference between scene structure and semantic features is basically not taken into account.

In this paper the previous concept, as was partly proposed in [4], will be implemented into a system (Figure 1) for the spatio-temporal analysis of articulated faces. Here, focus is put on the entire process for the analysis of the scene structure. That is, object modeling is not discussed. Further the fusion of structure and semantic knowledge will be published in a next paper. The current system is based on methods for stereo motion, graph theory, adaptive information fusion and multi-hypotheses-tracking (for discussion see section 2). The current system will be demonstrated (section 3) and examined (section 4). Future work will be discussed in section 5.

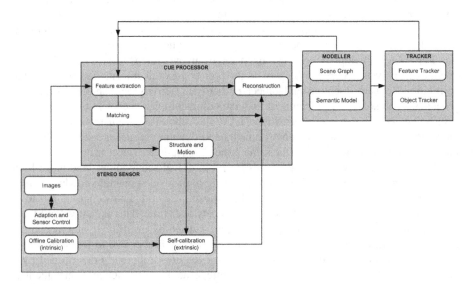

Fig. 1. Concept for Spatio-Temporal Scene Analysis comprising stereo sensor, cue processor, modeler and tracker

2 Previous Work

Methodically, the presented system is based on methods from structure and stereo motion like in [11, 12, 13], about spatio-temporal tracking of faces such as [14, 15], evolution of cues [16], cue fusion and tracking like in [17, 18], and graph-based modeling of partly-rigid objects such as [19, 20, 21, 22].

In all other studies no structure-based concept was developed like the one originally proposed in section 1. This report further contributes a detailed implementation of parts of the framework (section 3) and discusses experimental results (section 4).

3   System Design and Outline

3.1   Preliminaries

The system will incorporate a stereo sensor with verged cameras which are strongly calibrated as described in [23]. The imagers can be full-spectrum or infrared sensors. During operation, it is expected that only the relative camera motion becomes un-calibrated, that is, it is assumed that the sensors reside calibrated intrinsically.

The general framework as presented in Figure 1 will be implemented with one type of structure cue, a simple graph covering the spatial positions and dynamics (i.e. ve-locities). Tracking will be performed with a Kalman filter and a linear motion model, while re-calibration is performed via an overall skew measure of the reconstructed rays. The specific implementation is shown in Figure 2.

3.2   Feature Detection and Extraction

Detecting points of interest is one significant task in the framework. Of special inter-est in this work is the observation of human faces. Important structural characteristics of human faces show radial symmetric properties such as eye corners, nostrils, tip of the nose, mouth corners, and birth marks. The Fast Radial Symmetry Transform (FRST) is well suited for detecting such cues of interest.

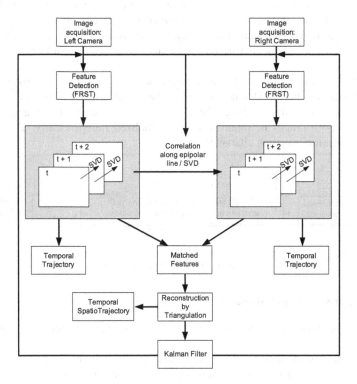

Fig. 2. Applied concept for tracking of faces

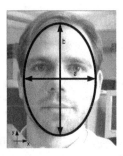

Fig. 3. Reducing the search space of a human face with an ellipse in one view

To reduce the search space in the images, a mask indicating the area of interest is evolved over the time. In this context, an ellipse (Figure 3) is suitable which was also demonstrated in other works [24]. Consequently, all subsequent steps are limited to this area and no further background model is needed currently.

The FRST, further developed by Gareth Loy and Alexander Zelinsky [5], determines radial symmetric elements in an image based on a corresponding gradient image. This algorithm is based on evaluating a gradient image and judging the symmetry contribution of each pixel to a certain centre of symmetry.

The transform can be split into three parts (Figure 4). From a given image the gradient image is produced (1). Based on this gradient image, a magnitude and orientation image is build for a radii subset of radii of interest (2). Based on the resultant orientation and magnitude image, a resultant image is assembled, which encodes the radial symmetric components (3). The mathematical relations would exceed the current scope.

The transform according to [5] was extended by applying a post-processing step which normalizes the transformed images. The transformation's output is a signed intensity image according to the gradient's direction. To be able to compare consecutive frames, both ranges of intensities are normalized independently. This measure yields illumination invariant characteristics.

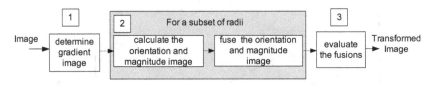

Fig. 4. Data flow of the Fast Radial Symmetry Transform (FRST)

3.3 Temporal and Spatial Matching

Two cases of matches are to be established: the temporal (inter-view) and stereo matches.

Due to the two cameras, two image sequences are available. Applying FRST on two consecutive images in the left view, as well as in the right view, gives a bunch of features through all images. Further, the tracking module gives information of former and new positions of known features.

The first task is to find repetitive appearing features in the left sequence. The same is true for the right stream. The second task is defined by establishing the correspondence between features from one left image in the right view.

Temporal matching is based on the Procrustes Analysis, which can be implemented via an adapted Singular Value Decomposition (SVD) of a proximity matrix $\mathbf{G}$ as shown in [7] and [6]. The basic idea is to find a rotational relation between two planar shapes in a least-squares sense. The pairing problem fulfills the classical principles of similarity, proximity, and exclusion. The similarity (proximity) $G_{i,j}$ between two features i and j is given by:

$$G_{i,j} = e^{-r_{i,j}^2/2\sigma^2} \quad (0 \le G_{i,j} \le 1) \tag{1}$$

where r is the distance between any two features in 2D and σ is a free parameter to be adapted. To account for the appearance, in [6] the areal normalized correlation index $C_{i,j}$ was introduced:

$$G_{i,j} = \left[e^{(-C_{i,j}-1)^2/2\gamma^2} \right] e^{-r_{i,j}^2/2\sigma^2} \tag{2}$$

The output of the algorithm is a feature pairing according to their locations in 2D between two shifted frames in time from one view. The similarity factor indicates the quality of fit between two features.

Spatial matching takes place with a correlation method combined with epipolar properties to accelerate the entire search process as a consequence of shrinking the search space to epipolar lines. Some authors like in [6] also apply SVD-based matching for the stereo correspondence, but this method only works well under strict setups, that are fronto-parallel retinas, so that both views show similar perspectives. Therefore, a rectification into the fronto-parallel setup is needed. But since no dense matching (dense disparity estimation) takes place [23], the correspondence search along epipolar lines is suitable.

The process of finding a corresponding feature in the other view is carried out in three steps: First a window around the feature is extracted giving the template. Usually, the template shape is chosen as a square. Good results for matching are gained here for edge length between 8 and 11 pixel. As a second part of the correlation approach, the template is searched for along the corresponding epipolar line of the feature in the other view (Figure 6). According to the cost function (correlation score) the matched feature is found, otherwise none is found due to occlusions.

Taking only features from FRST in the other view along the epipolar lines into account lead to less matches since due to perspectivity both views cover features which are not detected in the other view.

3.4 Reconstruction

The spatial reconstruction takes place via triangulation with the found correspondences in both views. In a fully calibrated system, the solution of finding the world coordinates of a point can be formulated as a least-square problem which can be solved via singular value decomposition (SVD). In Figure 10, the graph of a reconstructed pair of views is shown.

3.5 Tracking

This approach is characterized by a 3D feature position estimation, which is carried out by a Kalman filter currently. An introductory description of the filter can be found in [8]. A diagram of the currently used filter is shown in Figure 5.

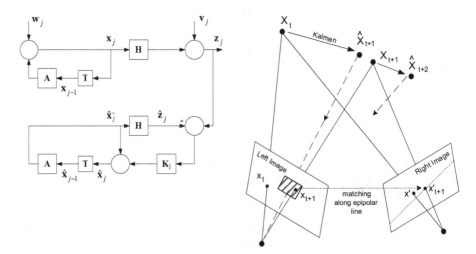

Fig. 5. Kalman Filter as block diagram [10]   **Fig. 6.** Spatio-Temporal Tracking using Kalman-Filter

A window around the back-projected estimated feature in 2D reduces the search space for the temporal as well as the spatial correspondence search in the successive images. Consequently, computational costs for detecting the corresponding features are limited. Figure 6 illustrates this shortly described approach.

Furthermore, features which are temporarily occluded can be tracked over time in case they can be classified as belonging to a group of rigidly connected features. The graph and the cue processor can estimate their states from the state of the clique to which the occluded feature belongs.

The Kalman filter comprises a reliable process model. Here, currently a linear model is chosen. Since the face moves in a three dimensional world, the state vector contains the current X-, Y- and Z-position. Furthermore, in order to give the possibility of modeling the current facial velocity, velocity components in all directions are also components of the state vector. Thus, the state is the following 6-vector:

$$\mathbf{x} = \left[X, Y, Z, V_X, V_Y, V_Z \right] \tag{3}$$

The process matrix $\mathbf{A}$ maps the previous position with the velocity multiplied by the time step to the new position:

$$\mathbf{P}_{t+1} = \mathbf{P}_t + \mathbf{V}_t \Delta t \tag{4}$$

The velocities are mapped one-to-one. The measurement matrix $\mathbf{H}$ maps the positions from $\mathbf{x}$ identically to the world coordinates in $\mathbf{z}$.

3.6 Re-calibration

Currently, the re-calibration of the relative camera motion (relative extrinsic parameters) is optimized in a least-squares sense in such a way that the distance of the reconstructed skew rays is minimized. The intrinsic parameters are held fixed.

4 Experimental Results

An image sequence of 40 frames is taken exemplarily here. The face as seen in the sequence moves from the left to the right and vice versa. The eyes are directed into the cameras, while in some frames the gaze is shifting away.

4.1 Feature Detection

The first part of the evaluation proves the announced property and should verify the ability of locating radial symmetric elements, especially in faces. In first evaluation sequences, the radius is the varying element by a fixed radial strictness parameter alpha. The algorithm yields the exemplarily transformed images seen in Figures 7 and 8.

Fig. 7. Performing FRST by varying the subset of radii and fixed strictness parameter (radius increases). Dark and bright pixels are features with a high radial symmetric property.

Fig. 8. Performing FRST by varying the strictness parameter alpha and fixed radius (alpha increases)

The images are processed by the earlier described FRST algorithm. The parameter for the FRST is a radii subset of one up to 15 pixels. The radial strictness parameter is 2.4. With exceeding a radius value of 15 pixels, the positions of the pupils is highlighted clearly. The same is true for the nostrils. By exceeding the radius of 6, the areas of the nostrils are affected accurately and therefore apparently emphasized.

The influence of the strictness parameter alpha is significant as the image sequence in Figure 8 reveals. The higher the strictness parameter, the more contour fading can be noticed. According to the task of finding features in human faces, the contours have no contribution. Thus, to mask this undesired effect for this test image, the strictness parameter was chosen around 3.

The transform was further examined under varying illumination and perspective. The internal parameters were optimized accordingly with different sets of face images. The results obtained are conforming to those in [5].

4.2 Matching

The temporal matching is performed as described with the above mentioned FRST parameters via the modified SVD method based on the Procrustes Analysis. Figure 10 shows one example of the spatial matching in this sequence, where 21 characteristic features were extracted.

4.3 Reconstruction

The following presents based on sample images the quality of the matching techniques and the resultant reconstructions. Therefore, on a face image, the FRST algorithm is applied. The matching process on the corresponding right image is performed by applying areal correlation along epipolar lines. More sophisticated approaches are presented in another paper [9]. Furthermore, as explained in the previous section, a reconstruction by triangulation is performed.

Figure 9 shows the left and right view, which is the basis for reconstruction. As one can see, due to applying the FRST algorithm, 21 enumerated features are detected. The reconstruction based on the corresponding right view is shown in Figure 10. As one can see, almost the entire bunch of features from the left view (Figure 9, top) is also assigned in the right image. Due to the different camera perspective, feature 1 and 21 are not covered on the right image and consequently not matched. Although the correlation assignment criteria is quite simple, namely the maximum correlation value along an epipolar line combined with a threshold value, this method yields a good matching success. All features, except the stumble of feature 18, are assigned correctly. Due to the wrong pairing, the relative relations between those two coordinates are not given by the epipolar geometry. Accordingly, a wrong triangulation and consequently a wrong reconstruction of feature 18 is the outcome as one can inspect in Figure 10.

4.4 Tracking

In this paragraph the tracking approach will be evaluated. The previous sequence of 40 frames was tracked.

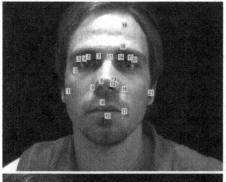

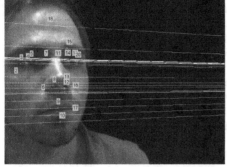

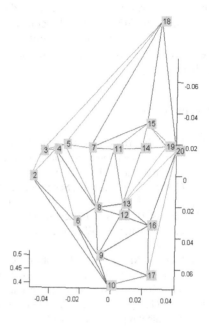

Fig. 9. Left Image with applied FRST, serves as basis for reconstruction (top); the corresponding right image (bottom)

Fig. 10. Reconstructed scene graph of world points from a pair of views selected for reconstruction (scene dynamics excluded for brevity). Best viewed in color.

The covariance matrices are currently deduced experimentally. This way the filter works stable over all frames. The prediction by the filter and the measurements have common trajectory. However, the chosen motion model is only suitable for relatively smooth motions.

The estimates of the filter were used to further reduce the search space of the features in the frame. The centroid of all features in 2D was used as an estimate of the center of the ellipse.

5 Future Work

At the moment there are different areas under research. Here, only some important should be named: robust dense stereo matching, cue processor incorporating fusion, scene graphs, model fusion, auto- and re-calibration.

6 Summary and Discussion

This report introduces current issues on driver assistance systems and presents a novel framework designed for this kind of application. Different aspects of a system for spatio-temporal tracking of faces are demonstrated. Methods for feature detection, for tracking in the 3D world, and reconstruction utilizing graphs were presented.

While all methods are at a simple level, the overall potentials of the approach could be demonstrated. All modules are incorporated into a working system. Future work is indicated.

References

[1] Im, S.-B., Hwang, K.-S., Cho, S.-B.: A Bayesian network framework for vision based semantic scene understanding. In: 16th IEEE Int. Conf. on Robot & Human Interactive Communication, Jeju, Korea, August 26-29, pp. 839–844 (2007)

[2] Cusano, C., Ciocca, G., Schettini, R., Brambilla, C.: Semantic Labeling of Digital Photos by Classification. In: Proc. of Internet Imaging IV, vol. SPIE 5018, pp. 296–303 (2003)

[3] Makris, D., Ellis, T.J.: Learning Semantic Scene Models from Observing Activity in Visual Surveillance. IEEE Transactions on Systems Man and Cybernetics - Part B 35(3), 397–408 (2005)

[4] Steffens, M., Krybus, W., Kohring, C.: Ein Ansatz zur visuellen Fahrerbeobachtung, Sensorik und Algorithmik zur Beobachtung von Autofahrern unter realen Bedingungen. In: VDI-Konferenz BV 2007, Regensburg, Deutschland (2007)

[5] Lay, G., Zelinsky, A.: A fast radial symmetry transform for detecting points of interest. Technical report, Australien National University, Canberra (2003)

[6] Pilu, M.: Uncalibrated stereo correspondence by singular valued decomposition. Technical report, HP Laboratories Bristol (1997)

[7] Scott, G., Longuet-Higgins, H.: An algorithm for associating the features of two patterns. In: Proceedings of the Royal Statistical Society of London, vol. B244, pp. 21–26 (1991)

[8] Welch, G., Bishop, G.: An introduction to the kalman filter (July 2006)

[9] Steffens, M.: Polar Rectification and Correspondence Analysis. Technical Report Laboratory for Image Processing Soest, South Westphalia University of Applied Sciences, Germany (2008)

[10] Cheever, E.: Kalman filter (2008)

[11] Torr, P.H.S.: A structure and motion toolkit in matlab. Technical report, Microsoft Research (2002)

[12] Oberle, W.F.: Stereo camera re-calibration and the impact of pixel location uncertainty. Technical Report ARL-TR-2979, U.S. Army Research Laboratory (2003)

[13] Pollefeys, M.: Visual 3D modeling from images. Technical report, University of North Carolina - Chapel Hill, USA (2002)

[14] Newman, R., Matsumoto, Y., Rougeaux, S., Zelinsky, A.: Real-Time Stereo Tracking for Head Pose and Gaze Estimation. In: FG 2000, pp. 122–128 (2000)

[15] Heinzmann, J., Zelinsky, A.: 3-D Facial Pose and Gaze Point Estimation using a Robust Real-Time Tracking Paradigm, Canberra, Australia (1997)

[16] Seeing Machines: WIPO Patent WO/2004/003849

[17] Loy, G., Fletcher, L., Apostoloff, N., Zelinsky, A.: An Adaptive Fusion Architecture for Target Tracking, Canberra, Australia (2002)

[18] Kähler, O., Denzler, J., Triesch, J.: Hierarchical Sensor Data Fusion by Probabilistic Cue Integration for Robust 3-D Object Tracking. Passau, Deutschland (2004)

[19] Mills, S., Novins, K.: Motion Segmentation in Long Image Sequences, Dunedin, New Zealand (2000)

[20] Mills, S., Novins, K.: Graph-Based Object Hypothesis, Dunedin, New Zealand (1998)

[21] Mills, S.: Stereo-Motion Analysis of Image Sequences, Dunedin, New Zealand (1997)

[22] Kropatsch, W.: Tracking with Structure in Computer Vision TWIST-CV. Project Proposal, Pattern Recognition and Image Processing Group, TU Vienna (2005)

[23] Steffens, M.: Close-Range Photogrammetry. Technical Report Laboratory for Image Processing Soest, South Westphalia University of Applied Sciences, Germany (2008)

[24] Kieneke, S., Steffens, M., Krybus, W.: Analysis and Implementation of Methods for Face Tracking. Technical Report Laboratory for Image Processing Soest, South Westphalia University of Applied Sciences, Germany (2007)

Spatio-Temporal Scene Analysis Based on Graph Algorithms to Determine Rigid and Articulated Objects

Stephan Kieneke[1,2], Markus Steffens[1,2], Dominik Aufderheide[1,2],
Werner Krybus[1], Christine Kohring[1], and Danny Morton[2]

[1] South Westphalia University of Applied Sciences, Luebecker Ring 2,
59494 Soest, Germany
{kieneke,steffens,aufderheide,krybus,kohring}@fh-swf.de
[2] University of Bolton, Deane Road, Bolton BL3 5AB UK
d.morton@bolton.ac.uk

Abstract. We propose a novel framework in the context of structure and motion for representing and analyzing three-dimensional motions particularly for human heads and faces. They are captured via a stereo camera system and a scene graph is constructed that contains low and high-level vision information. It represents and describes the observed scene of each frame. By creating graphs of successive frames it is possible to match, track and segment main important features and objects as a structure of each scene and reconstruct these features into the three dimensional space. The cue-processor extracts feature information like 2D- and 3D-position, velocity, age, neighborhood, condition, or relationship among features that are stored in the vertices and weights of the graph to improve the estimation and detection of the features and/or objects in the continuous frames. The structure and change of the graph leads to a robust determination and analysis of changes in the scene and to segment and determine these changes even for temporal and partial occluded objects over a long image sequence.

1 Introduction

Tracking and segmentation by using graph matching is well known and common in the field of computer vision and pattern recognition. Due to enormous requirements in performance by applying heuristic segmentation and tracking methods in the field of stereo vision, one new aspect is to apply graph algorithms to segment and track features and objects over a long image sequence.

This paper presents a new aspect to segment and track scene objects particularly of human heads and faces based on the approaches by [1, 2, 3, 6]. These approaches concentrate on using graph matching algorithms especially for stereo vision images. Former approaches usually applied graph-based methods only on monocular camera systems [13] whereat the proposed approach combines the advantages of the graph-based algorithms with the additional information that a stereo camera system is able to deduce.

This additional information is inferred by using a cue processor [8], which is the main source of information for the attributed relational graph (ARG) [5]. The cue

A. Gagalowicz and W. Philips (Eds.): MIRAGE 2009, LNCS 5496, pp. 254–264, 2009.

processor is used to transform relevant scene information into valid feature and object information that are the main data for the ARG.

The ARG is constructed to find and detect correspondences between objects of successive frames and ensures the tracking of human head and face segments over a long image sequence. Therefore, the ARG comprises vertices and edges whereby the vertices describe relevant objects that are delivered by the cue processor and the edges contain significant relationships between them.

Since the graphs are constructed from the actual and previous frames, the object recognition process is viewed as an inexact graph matching problem, which consists of finding correspondences between the set of vertices of the previous graph and the actual graph. This step is accomplished through an optimization algorithm that is based on the minimization of a cost function related to the weights of the edges.

The last step of the tracking process is based on the results of the segmentation, and a graph pyramid [6] is used to cluster nodes of the lower level segmentation graph, which belongs to similar rigid or articulated objects of the scene. Consequently, graphs from higher levels ensure the ability of tracking partial occluded objects since a motion model can be constructed of visible features that can be related to other partially occluded objects. Even objects that disappeared from the scene entirely can be tracked by using these higher level graphs as long as these objects reoccur again within a defined time frame.

2 Related Work

There is a variety of publications on segmentation and tracking by using graph-based algorithms, dating back over 30 years, with applications in many fields. In this section, we briefly consider some related work that is most relevant to the proposed approach.

Early graph-based methods were mainly focused on segmentation rather than tracking objects over a long image sequence because of the huge amount of computations involved.

Due to rapid progress in computer engineering and increasing computational power the graph-based methods become more popular and nowadays tracking via graph-based algorithms is well established [5, 6, 7].

Hence, the technique proposed herein was mainly motivated by the results and theoretical discussions presented in [1, 2, 3] and the idea of using a graph pyramid for tracking objects over long image sequences was proposed in [6].

The idea presented in [1, 2, 3] is to create so-called interval graphs that will contain nodes according to the 3D feature points extracted from stereo image sequences and weighted edges between them that represent the 3D distance between each node.

Whenever the distance between nodes changes over time, it leads to a change of weights for nodes that are not rigidly connected to each other and the edges between them will be dissolved. According to the relationship of nodes and feature points from the scene, non-broken edges can be seen as a rigid cluster of nodes that represent a rigid object in the real scene. Consequently, it is possible to segment and identify rigid objects in the scene and to track these objects over successive frames.

According to the NP-hard problem of using complete graphs for tracking feature points over continuous frames, this paper extends the method by integrating the idea from [6] to handle the correspondence problem in tracking by graphs. Thus, the present framework can be seen as a hybrid method that combines the benefits of both approaches.

The idea presented in [6] is to create a so called graph pyramid that is build up hierarchically from lower level graphs to higher level graphs, whereat higher level graphs contain and cluster main important information of lower level graphs. Therefore, it is possible to track rigid objects even if there is partial occlusion and to avoid the NP-hard problem because of the fewer nodes and complexity of higher level graphs.

Further tracking methods that use the Kalman or similar filtering algorithms are quite common but these kinds of methods are more feasible at dealing with single points rather than whole objects that describe entire segments in the image scene.

Therefore, graph-based approaches are more suitable for such tasks and are already adapted in [4, 5, 11, 12]. However, most of these methods are working with monocular images and were applied only to synthetic or laboratory scenes.

The primary goal of the present project is to apply graph-based tracking onto real-world scenes so to track human face and head movements in 3D space under varying lighting condition.

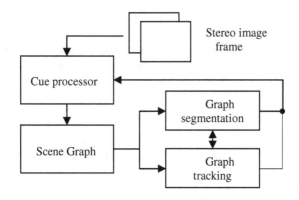

Fig. 1. Methodical overview of the graph-based tracking approach

This paper is organized as follows. An overview of the entire approach for recognizing and tracking objects over long image sequences is described in section 3. Section 4 concentrates on the cue processor [7] that will transform and provide all relevant scene information to be utilized within the graphs. The subsequent section gives a more detailed overview of the scene representation, which contains the structure and builds up the graph. Section 6 handles object segmentation, whereas object tracking over a long image sequence is explained in section 7. Section 8 describes the current status from this early stage of the project and finally section 9 and section 10 conclude the paper and present issues for future work.

3 Methodological Overview

The complete methodology can be seen in Figure 1, whereas it can be divided into four subtasks: (a) extract relevant features from the frame by using a cue processor, (b) fuse features to create an ARG to assign graph attributes between continuous frames, (c) to create a graph pyramid to cluster main objects for tracking, and (d) to estimate a motion model of rigid scene objects over a long image sequence.

This framework is based on the concept of spatio-temporal scene analysis, which is proposed in [7] for the identification, segmentation and tracking of features from successive frames in the field of stereo vision.

The ability to utilize the 3D scene analysis is a significant improvement over traditional feature tracking techniques. Monocular tracking analysis is restricted to the image plane, which leads to the loss of the depth information, whereas this approach consequently incorporates spatial information.

The introduced subtasks will be explained in more detail in the following sections.

4 Cue Processor

The cue processor, which was proposed first time in [8], extracts cues and depth information from the scene. Additionally this processor captures the feedback of the graph segmentation and tracking methods to improve the feature extraction procedures. It incorporates information like the age of the feature (e.g. number of frames since the feature occurred the first time), status of feature (see section 6.1), and an adaptive motion model.

In former papers only focus was put on feature-based or intensity-based methods for segmentation and tracking. Here, the cue processor handles both, intensity- and feature-based methods for extracting features from the scene. Hence, the present graph-based tracking and segmentation process is provided with holistic information which leads to higher accuracy.

5 Scene Representation

The scene is analyzed by several feature- and intensity-based methods within the cue processor. This cue processor was presented in the previews section. It delivers all important features that are necessary for the graph-based approach. In this section we describe how to incorporate the scene information into a graph that gives the ability to recognize the scene over continuous frames without loss of information.

5.1 Graph Representation

Graph-based feature tracking techniques generally represent the problem in terms of a graph. A graph $G(V, E)$, characterized by a set of vertices (nodes) V and edges E, whereby each node $v \in V$ corresponds to a feature in the scene, and the edges

$e \in E$ connect certain pairs of neighboring features or objects. Several details about the extracted feature are stored in the corresponding node.

The weights of the edges represent, based on the requirements of the graph-based approach, different relationship between their nodes. Currently edges are described by the related feature position in the 3D scene based on the idea in [1]. Additional feature information like age, neighborhood, similarity etc. influences the weights.

6 Stereo-Motion Segmentation

In this section we briefly explain the idea of this present work to recognize objects, extracted by using the cue processor, over successive frames by applying some kind of motion correspondence between our implemented graphs.

6.1 Graph-Based Matching

To recognize the graph between continuous frames, a well known method is used called graph matching. Graph matching compares two graphs by establishing a correspondence between their nodes and edges that reflect the structure. In the current work, the constructed graph from the previous frame is compared with the graph that is constructed from the actual frame and algorithms like the Hungarian method or Delaunay triangulation are applied to find the correspondence between their nodes and edges.

The proposed graph matching algorithm is called weighted graph matching (WGM) because of the weights of the edges and the goal to find the correspondence between sub-graphs with the smallest possible total weight. Therefore we describe the matching by using a bipartite graph, whereas this graph contains nodes from graph G_{t-1} and G_t (actual frame). Figure 2 outlines the graph matching process by using a bipartite graph.

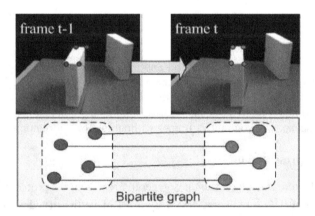

Fig. 2. Example of graph matching process by using a bipartite graph to find correspondences between successive frames

The connections between the nodes represent the smallest total weights that correspond with the correct assignment of nodes from G_{t-1} and G_t. Further information about this assignment method is described in section 6.3.

Often the number of nodes in the graph changes dynamically because of random scene changes. Thus it is crucial to maintain an invariant property to assure that there is an even number of nodes in the bipartite graph to handle the matching problem. The change of nodes can occur e.g. related to occlusion, failed feature detection and/or objects that disappear from the field of view and reappear after several frames. Hence, a method is used to recognize these occluded features in the case of reappearance. Graph-based methods are able to handle these dynamical changes by classifying nodes into visible and invisible. In case some nodes of the early frames do not correspond with the actual graph, these nodes are added to the actual graph, yet, with the status invisible. In case of reappearance the status will change to visible again, thus the tracking process is not interrupted. Former methods that incorporate similar ideas are using so called dummy nodes [4, 10].

6.2 Objects with Slightly Elastic Behaviors

Especially for the application of the present work, to segment and track human heads and faces in three dimensional spaces, the objects have slightly elastic behaviors. Hence objects based on our graph-based tracking approach need the ability to handle this kind of elastic deformations. Therefore we add some dissimilarity to the bipartite-graph by using e.g. interval arithmetic, which is also established in [2, 3].

The nodes of the graph represent the three dimensional location of feature points and the weights describe the distance between them. When some weights change over time it means that the represented features move away from each other and hence these features do not belong to the same rigid object. By using this simple idea it is possible to segment objects over a long period of time. Applied to the present work objects cover some elastic behavior, thus positions of objects in 3D space can be represented by interval measures. By using interval arithmetic features that belong to rigid objects are assigned correctly even if they slightly vary from the optimal position.

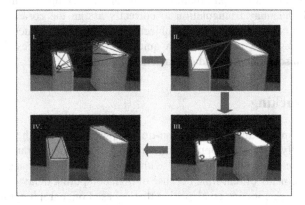

Fig. 3. Segmentation of rigid objects based on (I) Graph representation, (II) Delaunay triangulation, (III) Elimination of changing weights, (IV) final segmentation of two rigid objects

However, after each frame the graph information about the distance must be updated and weights that contain empty intervals are related to a different rigid object.

6.3 Assignment Problem

In general, nodes of a graph are unsorted, which means for finding correspondences between two different graphs a method has to be applied that maps nodes of two different graphs. This special assignment is known as the assignment problem and can be solved by several methods. In this present work, the Hungarian method is used to assign nodes of two different graphs. Therefore, the bipartite graph structure, which is described in the previouss chapter, is used as input for this method. More precisely the number of nodes and the weights are used to create a cost matrix M.

$$M = \sum_i \sum_j w\left(v_i, v_j\right) \tag{1}$$

Where $v_i \in V_{t-1}$ are nodes of G_{t-1} and $v_j \in V_t$ are nodes of G_t. Now the Hungarian method performs a minimization on the elements of the cost matrix to find the optimal correspondence. With the constraint that motion between two frames is small, weights are constructed by calculating the distance between V_{t-1} and V_t.

6.4 Delaunay Triangulation

To increase the speed of segmenting rigid objects, a Delaunay triangulation is applied based on the 3D model of the scene, which is reconstructed by the stereo data.

In the previews sections the method to segment rigid objects is explained but in general it needs several frames and not all edges can be eliminated between two different rigid objects. Hence a hypothesis is taken into account whereby the rigid object is described by planar areas and these areas can be constructed with finite number of triangles. This method is called triangulation and was also proposed in [2, 3].

The whole segmentation process is shown in Figure 3 and contains the transformed scene representation into graph notation and the applied Hungarian method, Interval arithmetic and Delaunay triangulation to correctly assign the corresponding nodes, and extraction of two independent rigid objects by deleting non-related weights.

Hence, the present work proposed a quick segmentation process by combining various common segmentation and optimization algorithm.

7 Object Tracking

Beside accuracy in tracking objects, a real-time requirement is another very important demand especially for online applications in tracking at video frame. Most of the former papers perform object tracking with pixel-by-pixel approaches but do not fulfill the requirements of real-time because of high computing delays.

The present work contains an algorithm that is based on a graph pyramidal decomposition of the scene according to the ideas in [6, 9].

The ability to compute 3D-motion is a significant improvement over traditional motion analysis techniques. Monocular motion analysis is restricted to the image plane that yields the loss of the depth information. The present work avoids this loss of information because of the stereo reconstruction, which is performed by the cue processor.

Therefore, motion of objects is more sophisticated and related to clusters of coherent features it is also possible to track partial occluded objects. In the following sections the process of the described method is explained in more detailed.

7.1 Graph Pyramid

To increase the performance of the tracking process it is quite useful to cluster similar objects and/or nodes that belong to the same rigid object into a single object. This process can be accomplished by using graph pyramids that is also used in frameworks such as [6, 9]. The graph representation of the scene is called a low level graph and a graph that clusters similar nodes is called high level graph because of the fewer nodes, less complexity, and higher level of abstraction. In Figure 4 a typical graph pyramid construction is shown.

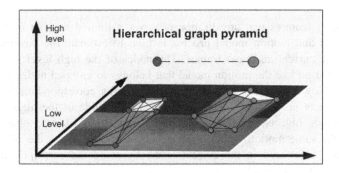

Fig. 4. Hierarchical graph pyramid representation

It is quite clear that by using graph pyramids, nodes that belong to the same objects are merged together and are represented by only one single node that contains all information that belongs to the rigid object like age, motion, etc.

7.2 Motion Model

To enhance the tracking of objects over long image sequences a motion model is incorporated based on the graph pyramid and the individual correspondences between nodes in the bipartite graph. After the graph-based optimization methods have assigned the corresponding nodes in the bipartite graph, each feature is reconstructed by using stereo triangulation to extract the actual position in 3D space. After that, the actual position is compared with the previous position and a 3D motion model is established for each feature. Equation 2 covers the 3D motion model.

$$\dot{R} = -t - \omega \times R \qquad (2)$$

Where $\dot{R} = (V_x, V_y, V_z)$ is the velocity of each feature, $R = (X, Y, Z)^T$ is the feature position in 3D space, $t = (t_x, t_y, t_z)^T$ is the translational velocity and $\omega = (\omega_x, \omega_y, \omega_z)^T$ is the rotational velocity.

With this additional information of each feature node, the cost matrix, which is used for the assignment problem, is adapted yielding better matching results.

7.3 Tracking of Partial Occluded Objects

Occlusion represents one of the most difficult problems in motion analysis. In this section the idea of [2, 3] and the proposed hybrid motion estimator for partial occluded objects will be explained in more detail.

Over a long image sequence it is quite usual that features disappear and reappear in the scene. This can happen when objects overlap or objects disappear and reappear from field of view. In this section we briefly describe how to handle these kinds of occlusions.

For tracking features that are disappeared over a limited number of frames, the graph pyramid and motion model process is used to estimate the position of these features in the current and next frames. The nodes of the high level graph contain shared information like the motion model that belongs to grouped nodes of the lower level graph. However, this low level graph represents corresponding features that belong to objects in the scene. That means that each node of the high level graph represents single objects. By the assumption that features of rigid objects are described by the same motion model, the current method estimates the 3D-postion of features that disappeared in the current frame by using the corresponding motion model of the high level graph.

Therefore, this process estimates the position of occluded or disappeared features as long as similar features reappear close to the estimated position. Otherwise after a limited number of frames the feature is finally discarded from the low level graph.

8 Experimental Evaluation

At time of writing, experiments with laboratory scenes were acquired only under stable lighting conditions and un-textured backgrounds. Partially occluded tracking was considered conceptually within the present framework rather than implemented in experiments.

We are currently investigating into adding other stable features to enhance the tracking performance and extending the algorithm for tracking objects in real-world environments rather than synthetic and/or laboratory scenes.

9 Summary

This paper describes the early stages of a research project in combining spatial and temporal analysis for a more accurate tracking and segmentation of partly rigid objects over a long image sequence.

The main important part and original work in this report is the additional usage of the 3D information that can be used to segment and track objects over a long image sequence and ensures high accuracy especially on partial occluded objects. Objects that move apart from each other can be recognized and identified as two separate objects by using stereo vision and graph-based segmentation. Traditional graph-based tracking approaches are monocular systems, which are not able to handle depth information. Using stereo vision higher computing performance is required. Certainly a further aim of the whole project is to implement the cue processor into dedicated hardware, such as field programmable hardware like FPGAs, that extract stereo features in real-time.

Work on graph-based matching moves into the focus because of increasing computing resources. Graph-based matching does not only compute and estimate positions of objects based on current measurements, further information of previous frames (such as age, state of occluded features etc.) are also integrated. This is done by calculating the weights of the graph and defining a proper cost function that describes the relationship of objects between consecutive frames. Hence graph-based optimization and assignment methods are the basis of future systems covering the ability of taking the scene structure into account.

Until now performance tests were not accomplished because of the early stage of this project and the primary relevant aim of accuracy and stability. At time of writing there are no experimental results for partial occluded objects and only experiments with laboratory scenes instead of real-world human head and face scenes were used.

10 Future Work

The next stage in this present framework is to improve the results in rigid object tracking with relation to robustness, accuracy and performance. The present project will concentrate on human face tracking under random scene changes. A further aspect of future work will be the implementation of the graph concepts in software.

References

[1] Mills, S.: Stereo-Motion Analysis of Image Sequences. Department of Computer Science, University of Otago, Dunedin, New Zealand (1997)
[2] Mills, S.: Graph Based Object Hypothesis. Department of Computer Science, University of Otago, New Zealand (1998)
[3] Mills, S., Novins, K.: Motion Segmentation in Long Image Sequences. Department of Computer Sciences, University of Otago, Dunedin, New Zealand (2000)
[4] Meyer, F., Gomila, C.: Graph-based object tracking. In: ICIP, vol. II, pp. 41–44 (2003)
[5] Graciano, A.B.V., Cesar Jr., R.M.: Graph-based Object Tracking Using Structural Pattern Recognition. Institute of Mathematics and Statistics – USP, Sao Paulo, Brazil (2007)

[6] Kropatsch, W.: Tracking with Structure in Computer Vision TWIST-CV. Project Proposal, Pattern Recognition and Image Processing Group, TU Vienna (2005)

[7] Steffens, M.: SCEAN – Spatio-Temporal Scene Analysis. Technical Report, Laboratory for Image Processing Soest, Institute for Computer Vision – Computational Intelligence, South Westphalia University of Applied Sciences, Germany (2008)

[8] Loy, G., Fletcher, L.: An Adaptive Fusion Architecture for Target Tracking. Department of Systems Engineering, Australian National University, Canberra (2002)

[9] Conte, D., Foggia, P.: A graph-based, multi-resolution algorithm for tracking objects in presence of occlusions. In: Brun, L., Vento, M. (eds.) GbRPR 2005. LNCS, vol. 3434, pp. 193–202. Springer, Heidelberg (2005)

[10] Christmas, W., et al.: Structural matching in computer vision using probabilistic relaxation. IEEE Trans. On Pattern Analysis and Machine Intelligence 5(3), 267–287 (1983)

[11] Tang, F., Tao, H.: Object Tracking with Dynamic Feature Graph. Department of Computer Engineering, University of California, Santa Cruz (2005)

[12] Paixao, T.M., et al.: A Backmapping Approach for Graph-based Object Tracking. Institue of Mathematics and Statistics, University of Sao Paulo (2008)

[13] Conte, D., et al.: Graph Matching Applications In Pattern Recognition and Image Processing. In: IEEE International conference on Image Processing, vol. 2, 3, pp. II-21–24 (2003)

Low-Cost Multi-image Based 3D Human Body Modeling

Zheng Wang[1,2], Andre Gagalowicz[2], and Meijun Sun[1]

[1] Tianjin University, School of Computer Science, China
{WZheng,SunMeijun}@tju.edu.cn
[2] INRIA-Rocquencourt, Domaine de Voluceau BP105 78153 Le Chesnay, France
Andre.Gagalowicz@inria.fr

Abstract. A method for 3D human body modeling from a set of 2D images is proposed. This method is based upon the deformation of a predefined generic polygonal human mesh towards a specific one which should be very similar with the subject when projected on the input images. Firstly the user defines several feature points on the 3D generic model. Then a rough specific model is obtained via matching the 3D feature points of the 3D model to the corresponding ones of the images and deforming the generic model. Secondly the reconstruction is improved by matching the silhouette of the deformed "d model to those of the images. Thirdly, the result is refined by adopting three filters. Finally texture mapping and skinning are implemented.

1 Introduction

Realistic human body models play a crucial role in many applications such as multimedia games, virtual reality, teleconferencing, digital art and towards the future for free viewpoint video, e.g. Carranza et al[1]. Precise and accurate 3D human body models are needed especially in many computer vision-based motion tracking system e.g.[3], [4], [5], [6].In the work by [7], [8], they also stressed the importance of the quality 3D model. However, obtain quality customized 3D model is a longstanding problem in animation and much work should still be realized before a near-realistic performance is achieved. This paper presents an algorithm for accurate 3D human body reconstruction from a small set of images. It is the first step for future precise 3D tracking. Although some interactive operations are involved for a trade-off between efficiency in use and feasibility in practice, experiment results indicate that our modeling system is much more efficient than using existing manual modeling tools. The obtained customized model is very realistic and accurate. It could also be further used, for example, for low cost model-based human motion tracking in video.

1.1 State of Art

In the literature, the existing vision-based reconstruction systems can be broadly divided into 2 categories:active and passive methods. When active methods use

A. Gagalowicz and W. Philips (Eds.): MIRAGE 2009, LNCS 5496, pp. 265–276, 2009.

an optical pattern projected into the scene, such as laser scanning system which is made of laser transmitter and receiver, passive techniques use the images of the patterns.

The 3D laser-scanner systems capture the entire surface of the human body in about 15 to 20 seconds with resolution of 1 to 2mm. However these devices are very expensive and require expert knowledge to interpret the data and build animated model [2]. Another drawback of this sort of system is it requires the subject to stay still and rigid for the whole duration of scanning (about 15 seconds for full body coverage) which is quite impossible in practice. It is well established that humans can stay still no more than 4–5 seconds.

Our approach is passive and it is a very low-cost reconstruction method due to its greater flexibility in scene capturing. Most of the existing methods e.g. by Hilton et al[14] or by Weik[15] make use of shape-from-silhouette related approaches, which require the subject to be segmented from the image background and the cameras to be calibrated beforehand by using a calibration tool of any kind. More recently Remondino[16] proposed a 3D calibration method from un-calibrated views, which uses feature correspondents, but requires the subject to remain still and rigid for about 40 seconds for capturing the whole body. In addition, the model reconstructed by such methods can contain non-manifold problem e.g. holes and open edges.

1.2   System Input

The goal of the proposed modeling system is to find a low-cost method to reconstruct accurately a 3D customized human body model, being given ONLY a small set of images taken from different viewpoints and also uses a 3D generic model with 24067 vertices and 48130 triangular faces. Figure 1 shows the input of this modeling system.

Fig. 1. System input: A generic model and six images taken from different views

The strategy of our framework is motivated by the method in [17], which was used for the construction of human faces. We brought many improvements described in this paper while keeping the same general idea.

2   Camera Calibration and Model Reconstruction

The principle of the first part of our modeling system is to deform the generic model according to several reconstructed characteristic points. The following subsections will describe all this procedure step-by-step.

2.1   Characteristic Points Selection

Figure 2 shows an example of the selected feature points on the 2D images corresponding to the 3D points. These correspondences can be established via an interactive point matching tool that we have developed. Although automatic body part recognition had been studied in e.g. Yaniz et al[19], however in our wide-baseline and cluttered environment, automatic feature detection becomes unstable. In our set-up, we utilize a set of 32 surface characteristic points. These characteristic points will provide an over-determined set of information and sufficient view coverage for camera calibration and reconstruction of points.

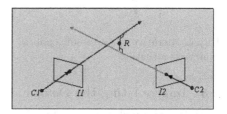

Fig. 2. Example of features points on 3D generic model corresponding on the 2D image

Fig. 3. Triangulation of projected rays,when the rays do not intersect images (R is the reconstructed point)

2.2   Interactive Calibration and Characteristic Points Reconstruction

The adopted technique for camera calibration is based upon POSIT by DeMenthon[9]. POSIT algorithm minimizes the difference between the projection position of 3D points and their 2D positions in the image until the error which is the maximum difference of pixel location is less than a threshold or when it reaches a maximum number of iterations. The POSIT method furnishes each camera's extrinsic parameters. Our way to determine the intrinsic parameters is to consider POSIT as a function of the intrinsic parameters, a function we minimize using the simplex algorithm. Obviously, one of the advantages of our camera calibration set-up is that the calibration tool is the subject itself. Thus, we do not need any special calibration tool.

The reconstruction of 3D characteristic points is performed using the triangulation of projected rays which start from the optical center of the camera and go through the characteristic points in the image. The algorithm takes into account

that the rays may not intersect exactly due to the fact that calibration is not perfect. To reconstruct the points, we minimize the sum of the squares of distances
to the rays from all possible views. It only can be done when the points are seen
in more than one image. Figure 3 shows an example of the reconstruction.

As Figure 4 shows when one pass 3D characteristic points reconstruction is
finished, we can apply POSIT to gain a more precise camera calibration data
and then perform 3D reconstruction. Repeat this loop. Figure 5 shows that the
iterative process converges after about 30 iterations. The re-projection error at
convergence is about 1.1 pixels with a standard deviation of 0.9 pixels.

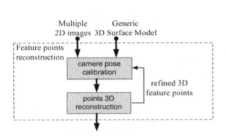

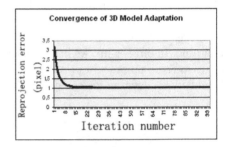

Fig. 4. Iterative process of calibra-
tion/reconstruction

Fig. 5. The convergence of the calibra-
tion/reconstruction loop

2.3 Interpolating the Deformation

Until this moment, we have obtained the position of reconstructed characteristic
points and their original position on the generic model. Now we need to deform
the rest of the vertices. The radial basis functions (RBF) are adopted to achieve
this goal. The distance between each vertex and each characteristic point is
chosen as a value of the function, because it provides a smooth interpolation.
Using RBF for data interpolation has been researched and used successfully
in e.g. Ruprecht el al[12]. More details about RBF technique can be found in
[10][11].

2.4 Silhouette Contour Adaption

After the deformation of the generic model based on characteristic points we
obtain only a geometry of body. However, some local parts such as the curves
on the arms and legs of the subject are not precisely reconstructed. In order to
correct this defect, the adaption by using silhouette contours of the 3D model and
the limbs of each image is performed. The following describes the approximate
process of the algorithm.

 - Automatic extracting of silhouette contours from the 3D mesh.
 - Interactively extracting the silhouette from 2D images by using Bezier curves
   drawing tools. Because we are working in a clutter environment, so the automatic

edge detection algorithm for image segmentation, e.g. Canny et al[18], does
not work perfectly here.

- Finding a good correspondent for each projected silhouette vertex with respect to the image curves by using a recursive subdivision matching algorithm.
- Use RBF to adapt the initial specific model.

More details about silhouette adaption can be found in [13]. The results of
reconstruction are shown in fig.6

Fig. 6. Reconstruction results of front and back-side view

3   Filters

The method above enables us to obtain a global surface geometry of the subject.
However, as figure 7 shows, there are some local distortions on the geometry
surface because of the non perfect camera calibration and small errors in the
silhouette matching.

Fig. 7. Local distortions on geometry

To resolve these problems, we analyzed the distortions and classified them
into three categories which are vertical bulges, large-scale and small-scale twists.
Three sorts of filters are designed to solve the corresponding distortions: smoothing filters, slice filters and neighbor triangle normal filters.

3.1   Smoothing Filter

Some noticeable vertical bulges can be spotted on the arm surface in the middle
of figure 7. We would like to eliminate the bulges and meanwhile preserve the

correct body features as much as possible. So we designed a smoothing filter to deal with this sort of problem. A vertex of the specific geometry is denoted as $P(x, y, z)$, $Q_i(x, y, z)$ is the neighbor vertices of P,set

$$P'(x, y, z) = \frac{1}{N} \sum_{i=1}^{N} Q_i(x, y, z)$$

Set D as the distance from P to P', if D is greater than a threshold ρ whose value is set to the mean value of the sum of D, then replace P with P'. The result of applying smooth filter to the whole body is shown in figure 8.

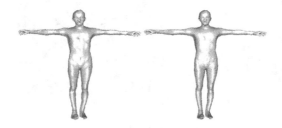

Fig. 8. Left:before smoothing; Right:after smoothing

3.2 Slice Filter

A very serious distortion can be noticed on the feet of the specific model. Because of the large distorted range, the smoothing filter does not work. The basic idea for detecting and fixing this problem is to slice the mesh of generic and specific model along the axis of every part of body. Then locate the distorted part of mesh by comparing each pair of slice. The following shows the detailed process of the slice filter.

- Segment whole body into several parts including head,neck,arms,legs and bust. The axis of every part is estimated by using principle component analysis(PCA) method.
- Each part is segmented in many slices orthogonal to its axis via calculating the intersection of the geometry and the cutting planes . To ensure the success of comparison, the number of slices on the generic model and specific model are equal. Figure 9 shows one typical sliced curve of each part.
- Overlap every pair of sliced curves and compare the variation of the distance between the two borders. A slice will be marked as "bad" one if the maximum distance between the two curves is greater than a threshold whose value is related to the mean distance of one pair slice. We can spot that the pair of slices on the feet is distorted seriously.
- After all the bad slices are detected, all vertices of each bad slice are replaced by a scaled copy of the slice of the generic model.

Figure 10 shows a result after using the slice filter for the feet.

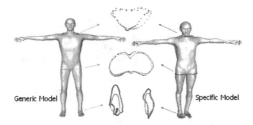

Fig. 9. Example of slice on every part. The distortion on the feet of specific model can be detected by comparing the shapes of their slices.

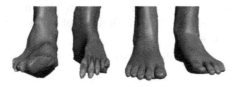

Fig. 10. Result using the slice filter on the feet

3.3   Neighbor Triangle Normal Filters

Some small area distortions are hard to be detected and corrected by using the two filters introduced above. For example a face with a twisted nose and a defective mouse even if the remaining is normal. To solve this problem a neighbor triangle normal filter denoted as NTNF is presented here. To use NTNF a precondition must be that on small matched areas the generic and the specific geometry are similar and fortunately it is the case for most human body parts. To each vertex P on the generic and its equivalent on the specific model, we compute the angle difference between the normals of the matched triangles attached to P and its equivalent. If the maximum angle is greater than a threshold, than P will be denoted a suspicious vertex. After checking all the vertices on the specific model, we obtain several suspicious vertex groups and their corresponding vertex groups of the generic model. According to a chosen bounding box of each pair of vertex groups, a scaled replacement is performed from the generic model to

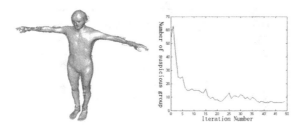

Fig. 11. The suspicious vertex groups(marked in blue) and their bounding boxes (marked in red). Left: Initial detected suspicious vertex groups. Right: Plot of the number of suspicious vertex group vs the number of iterations.

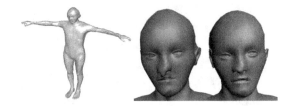

Fig. 12. Left: Finally suspicious vertex groups after using NTNF. Right: Green face(non-filtered) vs Red face(Filtered).

the specific one in order to suppress those suspicious vertex groups. After one pass of this replacement, we can reiterate the process on remaining suspicious vertex groups. Figure 11 shows that the number of suspicious vertex group stop decreasing after 45 iterations.

Figure 12 shows the effects on the body and face after having used NTNF.

4 Texture Mapping

In order to perform a model-based 3D tracking of a human body movement with feedback it is fundamental to dispose of a precise textured model of the person. Due to illumination variations, it is difficult to obtain a consistent one which is really necessary for a good tracker performance. We have been investigating two kinds of texture mapping methods.

4.1 View Point Independent Method

Because for almost each patch on the reconstructed model, its texture can be obtained from several views. The idea is to select the best view. We adopt the view point independent algorithm to solve this problem. The principle of this algorithm is the following: a view will be regarded as the best one if the angle between the camera viewing direction and the normal of the patch is maximum. Figure 13 shows the texture mapping results by using this method with 6 images.

Fig. 13. Texture mapping via view point independent method

4.2   Cylinder Unwrapping Method

From figure 13 we observed that the texture is not consistent because the illuminations in the 6 images are very different. To deal with this problem we are designing a cylindrical unwrapping method. It is implemented in the following way:

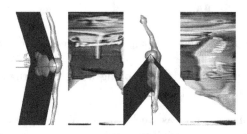

Fig. 14. Examples of generating texture by slicing the model. Left two images: The front view slicing and its generated texture. Right two images: The side view slicing and its generated texture. The definition of the texture is 512 width and 1024 height.

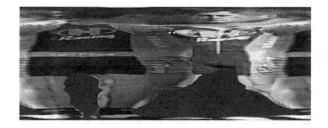

Fig. 15. Synthesized Texture

- Estimate first the 3D vertical axis of the specific model centered around the inertia center of the body volume
- construct the vertical cutting plane containing the camera focal center and this axis.
- Slice vertically the specific model with 2 vertical planes, one to the left of the cutting plane and the other one to the right side of this cutting plane. These planes are indexed by the angles -teta1 and +teta2 (they may be equal or not). This defines an angular portion of the 3D human body (see figure 14.
- We sample the interval [-teta1, +teta2] with N1 values, and the body total height by N2 values and build a N1xN2 image containing the color of the point corresponding to its height N2 and its angle N1 on the 3D body.
- we perform the same operation for the next view in order to cover the complete [0, 360] interval and chose each teta1 and Teta2 value such that the corresponding [height, teta] images overlap and finally perform a linear angular interpolation between neighboring images.Figure14 shows a 2D [height, teta] image corresponding to the left part of the figure. The full [0, 360] blending of the images is shown in in Figure 15.

Fig. 16. Texture mapping with Synthesized Texture

– Finally we perform the texture mapping using the former image. The final
  result is shown in Figure 16.

This work is still underway as we have to detect occlusions and adapt the
technique (using multi-layer visibility considerations) in order to use multiple
[height, teta] images for the same view. Occluded parts of one view will be textured by the other views and angular interpolation will be run on these multiple
images in parallel.

5 Skinning

For our future work on 3D human body tracking, the skinning is implemented
via adopting a vertex blending algorithm [20]. A specific skeleton is obtained
after deforming the generic skeleton with RBF. The weighting parameters used
for connecting the skin to the skeleton can be set in 3DS Max and exported into
a XML files. Figure 17 shows the result of skinning of one arm. However, this
skin deformation technique sometimes produces non-natural results. For future
work, dual quaternion algorithm [21] will be adopted to obtain better skinning
results.

Fig. 17. Skinning

6 Conclusion

In this paper a new technique for modeling a 3D human body of a particular
person based on a limited set of images acquired from different viewpoints with

wide baseline has been introduced. First, characteristic points clicked both on a generic model and on these images are required. Then we establish an initial model by using a camera calibration/feature-point reconstruction loop and interpolating the sparse reconstructed points using RBF. The initial model is refined by using a silhouette contour adaption which consists in matching the projection of silhouette edges of the initial model with the image limbs. To obtain a higher quality model, three filter are implemented to eliminate different defects distributed on the mesh. Finally the model is texture-mapped by using a novel method which synthesizes images used for reconstruction to yield an integrated texture with continuous illumination changes. Furthermore, a vertex blending skinning is implemented as the foundation for future motion tracking.

In the future, we are planing to improve the quality of texture mapping via automatically adapting slice angle according to silhouette of 3D model and incorporating occlusion considerations. Then we will use this customized model to track the target subject.

References

1. Carranza, J., Theobalt, C., Magnor, M.A., Seidel, H.: Free-Viewpoint Video of Human Actors. In: Proceedings of the SIGGRAPH2003 Conference, pp. 569–577 (2003)
2. Paquette, S.: 3d scanning in apparel design and human engineering. IEEE Computer Graphics and Applications 16(9), 11–15 (1996)
3. Garvrila, D.M., Davis, L.S.: 3-D model-based tracking of humans in action: a multi-view approach. In: CVPR, San Franscisco, USA, pp. 73–80 (1996)
4. Moeslund, T.B., Granum, E.: A survey of computer vision-based human motion capture. Computer Vision and Image Understanding 81, 231–268 (2001)
5. Sidenbladh, H., Black, M.J., Sigal, L.: Implicit probabilistic models of human motion for synthesis and tracking. In: Heyden, A., Sparr, G., Nielsen, M., Johansen, P. (eds.) ECCV 2002. LNCS, vol. 2350, pp. 784–800. Springer, Heidelberg (2002)
6. Wren, C., Azarbayejani, A., Darrell, T., Pentland, A.: Pfinder: real-time tracking of the human body. IEEE Trans. PAMI 19(7), 780–785 (1997)
7. Grard, P., Gagalowicz, A.: Human Body Tracking using a 3D Generic Model applied to Golf Swing Analysis. In: MIRAGE 2003 Conf., INRIA Rocquencourt, France (March 2003)
8. Kakadiaris, I.A., Metaxas, D.: Three-dimensional human body model acquisition from multiple views. International Journal of Computer Vision 30, 191–218 (1998)
9. Dementhon, D.F., Davis, L.S.: Model-based object pose in 25 lines of code. International Journal of Computer Vision 15, 123–141 (1995)
10. Roussel, R., Gagalowicz, A.: Morphological Adaptation of a 3D Model of Face from Images. In: MIRAGES 2004 (2004)
11. Buhmann, M.D.: Radial Basis Functions: Theory and Implementations. Justus-Liebig-Universitat Giessen, Germany
12. Ruprecht, D., Muller, H.: Free form deformation with scattering data interpolation methods. In: Farin, G., Hagen, H., Noltemeier, H. (eds.) Geometric Modeling (Computing Suppl. 8), pp. 267–281. Springer, Heidelberg (1993)
13. 3D Modeling of Humans with Skeletons from Uncalibrated Wide Baseline Views

14. Hilton, A., Beresford, D., Gentils, T., Smith, R., Sun, W., Illingworth, J.: Whole-body modelling of people from multi-view images to populate virtual worlds. The Visual Computer 16(7), 411–436 (2000)
15. Weik, S.: A passive full body scan using shape from silhouette. In: Proc. ICPR 2000, Barcelona, Spain, pp. 99–105 (2000)
16. Remondino, F.: 3-D reconstruction of static human body shape from an image sequence. Computer Vision and Image Understanding 93, 65–85 (2004)
17. Roussel, R., Gagalowicz, A.: Morphological adaptation of a 3D model of face from images. In: MIRAGE 2003 Conf., INRIA Rocquencort, France (March 2003)
18. Canny, J.: A Computational Approach to edge detection. IEEE Trans. PAMI 8(6), 679–698 (1986)
19. Yaniz, C., Rocha, J., Perales, F.: 3D Part Recognition Method for Human Motion Analysis. In: Proceedings of the International Workshop on Modelling and Motion Capture Techniques for Virtual Environments, pp. 41–55 (1998)
20. Lander, J.: Graphic content: skin them bones:game programming for the web generation (February 14, 2006),
http://www.darwin3d.com/gamedev/articles/col0598.pdf
21. Kavan, L., Collins, S., Zara, J., O'Sullivan, C.: Geometric Skinning with Approximate Dual Quaternion Blending ACM Transaction on Graphics 27(4) (2008)

Modified Histogram Based Fuzzy Filter

Ayyaz Hussain, M. Arfan Jaffar, Abdul Basit Siddiqui, Muhammad Nazir,
and Anwar M. Mirza

Department of Computer Science, FAST National University of Computer
and Emerging Sciences Islamabad, Pakistan
ayyaz.hussain@nu.edu.pk, arfan.jaffar@nu.edu.pk,
basit.siddiqui@nu.edu.pk, muhammad.nazir@nu.edu.pk,
anwar.m.mirza@nu.edu.pk
http://www.nu.edu.pk

Abstract. In this paper, a fuzzy based impulse noise removal technique has been proposed. The proposed filter is based on noise detection, fuzzy set construction, histogram estimation and fuzzy filtering process. Noise detection process is used to identify the set of noisy pixels which are used for estimating the histogram of the original image. Estimated histogram of the original image is used for fuzzy set construction using fuzzy number construction algorithm. Fuzzy filtering process is the main component of the proposed technique. It consists of fuzzification, defuzzification and predicted intensity processes to remove impulse noise. Sensitivity analysis of the proposed technique has been performed by varying the number of fuzzy sets. Experimental results demonstrate that the proposed technique achieves much better performance than state-of-the-art filters. The comparison of the results is based on global error measure as well as local error measures i.e. mean square error (MSE) and structural similarity index measure (SSIM).

Keywords: Image restoration, fuzzy logic, structural similarity index, impulse noise.

1 Introduction

Image restoration is an important branch of image processing, which deals with the reconstruction of images by removing noise and blurriness, and making them suitable for human perception. Images can become corrupted during any of the acquisition, pre-processing, compression, transmission, storage and/or reproduction phases of processing [1],[2]. Spatial image restoration technique can be divided into two broad categories namely conventional and blind image restoration techniques [3]. Information about the degradation process is generally known in case of conventional image restoration techniques. This known information can be used in developing a model which is further used to restore the corrupted image back to its original form. Conventional techniques are used to solve motion blur, system distortions, geometrical degradations and additive noise problems.

A. Gagalowicz and W. Philips (Eds.): MIRAGE 2009, LNCS 5496, pp. 277–284, 2009.
© Springer-Verlag Berlin Heidelberg 2009

Recently, more focus has been placed on the blind image restoration [3], where the image has to be restored directly from the degraded image without any prior knowledge about the degradation process. Main objectives in developing blind image restoration technique are to remove noise along with preserving the image details. Smoothing a region of the degraded image might destroy an edge and/or texture information while sharpening edges might lead to amplification of unnecessary noise. In the sequel, we present a spatial image restoration technique which is based on histogram statistics and fuzzy logic to remove impulse noise along with edge preservation.

A number of approaches have been developed for the impulse noise removal. Tukey [4], Astola et al. [5], and Pitas et al. [6] have used median filtering to remove impulse noise. It has been observed that the median based filter cannot give good results when noise rate is high. Furthermore, number of fuzzy based image restoration techniques has been developed for impulse noise removal. For instance, the histogram based fuzzy filter (HFF) [7], novel fuzzy filter (NFF) [8] and genetic based fuzzy image filter (GFIF) [9] are the examples of the most recent fuzzy filters. HFF is able to outperform the rank-order filter (such as median filter) for the whole range of the corruption rate ranging from 0.1 to 0.9 without any training. NFF gives superior performance than the median filter for highly corrupted images, however it does not preserve the image details well. NFF also uses histogram of the original image or image database to find the fuzzy parameters, which shows that it is not a pure blind technique. GFIF performs well for the whole range of corrupted images but the major drawback of GFIF is its extensive training as well as original image or image database is required to calculate the fuzzy sets.

In this paper, we propose a modified histogram based fuzzy filter (MHFF) to remove impulse noise from low as well as highly corrupted images. The proposed filter consist of noise detection, fuzzy set construction through fuzzy number construction algorithm, histogram estimation and fuzzy filtering process. Experimental results show that MHFF gives much better results than state-of-the-art fuzzy based filters as well as median filter for impulse noise removal. Main Contribution of the proposed technique includes:

- It is a pure blind image restoration technique which gives better results than state-of-the-art filters without any training.
- Proposed technique uses the fuzzy number construction algorithm[8] instead of principle of histogram potential[7] to construct fuzzy sets.
- Sensitivity analysis of the proposed technique is performed by varying the number of fuzzy sets.

The rest of the paper is organized as follows. In section 2, system architecture of the MHFF is presented. Section 3 presents the fuzzy filtering process. Fuzzy set construction has been discussed in section 4. Experimental results and sensitivity analysis of the proposed technique are described in section 5. Finally conclusion is drawn in section 6.

2   System Architecture of the MHFF

In this section, system architecture and its working are presented. Block diagram of the system is shown in figure 1. In the first step, set of noisy pixels N_{pixels} are detected using noise detection process. To determine N_{pixels} , noise detection process scan the image using a window of size 3x3 from left to right and top to bottom. The central pixel of the sub-image will belong to the set N_{pixels} if it is minimum, maximum, less than some threshold T or greater than 1-T. The set of noisy pixels and the corrupted image histograms are used to estimate the histogram of the original image using the following equation.

$$H_{est}(i) = \frac{H_{corr}(i) - H_{noisy}(i)}{\sum\limits_{g=0}^{255} (H_{corr}(g) - H_{noisy}(g))} \tag{1}$$

where H_{corr} and H_{noisy} are the histograms of the corrupted image and set of noisy pixels respectively. H_{est} represent the estimated histogram of the original image. We have considered 8-bit gray scale images so gray level ranges from 0 to 255.

Estimated histogram of the original image is used to construct fuzzy sets. MHFF is designed to create five fuzzy membership functions namely very dark (vdk), dark (dk), medium (md), bright (br) and very bright (vbr). Therefore, each intensity pixel under the considered window is treated as the fuzzy variable with membership degree in the fuzzy set vdk, dk, md, br and vbr. Membership functions identify the degree of brightness for each input pixel. Following equation shows the trapezoidal shaped membership function used in MHFF [11].

$$f_A(x) = \begin{cases} 0, & x < a_A \\ \frac{(x-a_A)}{(b_A-a_A)}, & a_A \leq x < b_A \\ 1, & b_A \leq x < c_A \\ \frac{(d_A-x)}{(d_A-c_A)}, & c_A \leq x < d_A \\ 0, & x \geq d_A \end{cases} \tag{2}$$

The trapezoidal function of fuzzy set A ε *vdk, dk, md, br, and vbr.* This fuzzy set is denoted by the parameters $\begin{bmatrix} a_A & b_A & c_A & d_A \end{bmatrix}$ Section 3 presents the details about the calculation of these parameters through fuzzy number construction algorithm [8].

3   Fuzzy Filtering Process

Fuzzy sets (Section 4) and estimated histogram are used in fuzzy filtering process. Fuzzy filtering process consists of fuzzification, de-fuzzification and predicted intensity calculation processes. These components of fuzzy filtering process are described one by one as follows:

Fuzzification. A window of size 3x3 is used to scan the image from left to right and top to bottom. In each window where the central pixel is detected noisy by the noise detection process is considered as the candidate for fuzzy filtering process. In the first step, each pixel under the candidate window is considered as a fuzzy variable and its degree of brightness is calculated using fuzzy membership functions. This process of calculating the degree of membership is known as fuzzification process.

Defuzzification. In this step, all the outputs from the previous step belonging to each membership function are separately used for defuzzification. Resultantly, the outputs of this step will be five crisp values, calculated using equation 3 [11].

$$
D_A = \begin{cases} \dfrac{\sum\limits_{i=1}^{9} f_A(x_i)*x_i}{\sum\limits_{i=1}^{9} f_A(x_i)} & if \ \sum\limits_{i=1}^{9} f_A(x_i) > 0 \\ 0 & otherwise \end{cases} \tag{3}
$$

where D_A represents output of the defuzzification process associated with fuzzy membership function having fuzzy set A ε vdk, dk, md, br, and vbr. This fuzzy set is denoted by the parameters $\begin{bmatrix} a_A & b_A & c_A & d_A \end{bmatrix}$ where x_i denotes the corresponding pixel value $i = 1, 2, \ldots, 9$ where $f_A(x_i)$ represent the membership degree of x_i in fuzzy set A.

Predicted Intensity Process. Finally in order to choose best estimate of the corrupted pixel under the considered window, predicted intensity is computed. It is calculated using the mean of the non-noisy pixels under the 3x3 pixel window as shown in figure 1.

4   Fuzzy Set Construction

The proposed technique uses the fuzzy number construction algorithm [8] instead of using the principle of histogram potential [7] to calculate the parameters of the trapezoidal fuzzy membership functions. Estimated histogram of the original image and the number of fuzzy sets to be constructed are given as input to the algorithm. This algorithm gives the parameters of the fuzzy sets as output.

In this paper, we use the luminance fuzzy variables with five linguistic terms. The fuzzy sets for an image include very dark (vdk), dark (dk), medium (md), bright (br) and very bright (vbr). These fuzzy sets can be represented by the following equation.

$$
\left.\begin{aligned} vdk &= \begin{bmatrix} a_{vdk} & b_{vdk} & c_{vdk} & d_{vdk} \end{bmatrix} \\ dk &= \begin{bmatrix} a_{dk} & b_{dk} & c_{dk} & d_{dk} \end{bmatrix} \\ md &= \begin{bmatrix} a_{md} & b_{md} & c_{md} & d_{md} \end{bmatrix} \\ br &= \begin{bmatrix} a_{br} & b_{br} & c_{br} & d_{br} \end{bmatrix} \\ vbr &= \begin{bmatrix} a_{vbr} & b_{vbr} & c_{vbr} & d_{vbr} \end{bmatrix} \end{aligned}\right\} \tag{4}
$$

The detailed algorithm for constructing the fuzzy sets can be found in [9].

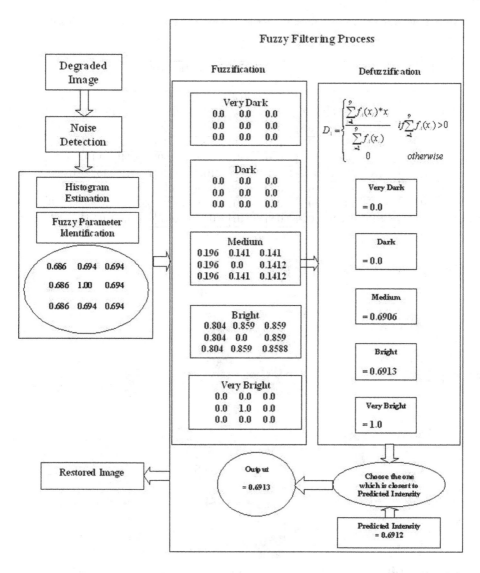

Fig. 1. Block Diagram of MHFF

5   Experimental Results and Sensitivity Analysis

In order to test the quality of the proposed technique, we compare MHFF with the other state-of-the-art filtering techniques such as HFF, NFF and median filter. The quantitative measures used for comparison are mean square error (MSE) and structural similarity index measure (SSIM) [10]. Representative results for a typical "Lena" image are shown in table 1, where the image is corrupted with impulse noise (salt and pepper) with noise level ranging from 0.1 to 0.9.

Table 1. Comparison of de-noising methods for Lena image degraded with salt and pepper noise having corruption rate varying from 0.1 to 0.9

Method	Quality measure	Noise Corruption Rate								
		0.1	0.2	0.3	0.4	0.5	0.6	0.7	0.8	0.9
NFF	MSE	40.6	59.6	82.4	125.17	222.04	561.7	1475.9	4964.3	13655
	SSIM	0.9	0.8	0.8	0.79	0.72	0.58	0.38	0.16	0.05
HFF	MSE	85.2	89.2	101.5	119	177.82	375.61	1095.1	3382.3	8899.2
	SSIM	0.8	0.8	0.8	0.84	0.8	0.69	0.44	0.19	0.06
MF	MSE	68.1	133.2	325.8	1028.4	2210.2	4326	7597.7	11509	16342
	SSIM	0.8	0.8	0.7	0.49	0.28	0.15	0.07	0.04	0.02
MHFF	MSE	31.1	39.1	52.5	74.91	135.99	333.28	999.51	3185	8379.3
	SSIM	0.9	0.9	0.9	0.91	0.86	0.75	0.48	0.21	0.06

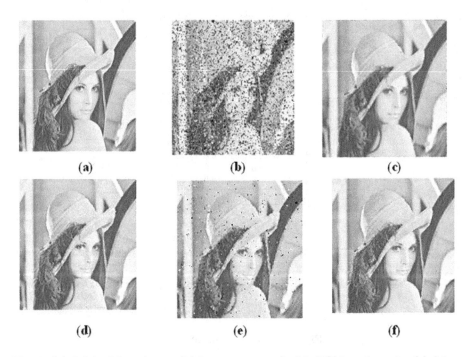

Fig. 2. (a) Original Lena image (b) Image corrupted with 30% impulse noise (c) After the HFF method (MSE=101.59)(d) After NFF method (MSE=82.48) (e) After MF method (MSE=325.82) (f) After the proposed MHFF method (MSE=52.52)

Results show that the MHFF gives much better results for low as well as highly corrupted images.

Visual performance of MHFF is shown in figure 2, which shows that the proposed technique (Fig. 4(f)) removes the impulse noise and preserves the image details better than the other competitive techniques. Sensitivity analysis of the

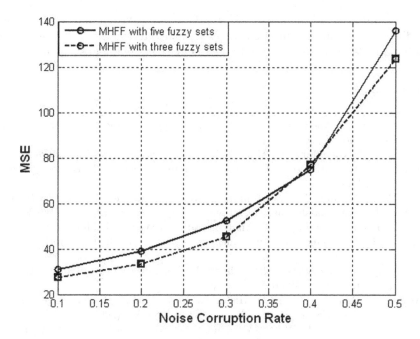

Fig. 3. Sensitivity Analysis of the proposed technique based on number of fuzzy sets

proposed technique is performed on the basis of number of fuzzy sets constructed through fuzzy number construction algorithm. Figure 3 compares the MHFF results with three and five fuzzy sets. It can be seen that MHFF with five fuzzy sets gives better performance but difference is quite less which shows that the proposed technique is robust against number of fuzzy sets used.

6   Conclusion

In this paper, a modified histogram based fuzzy filter (MHFF) is introduced. MHFF consists of noise detection, histogram estimation, fuzzy set construction and fuzzy filtering process. From the experimental results, we observed that the proposed filter gives much better results than the other filtering techniques. Sensitivity analysis of MHFF is also presented in this paper which is based on varying the number of fuzzy sets. In sensitivity analysis it has been proved that MHFF with five fuzzy sets gives better results than having three fuzzy sets but the difference is quite less which shows the robustness of the proposed technique against the number of fuzzy sets used. In future, we will extend this technique for uniform impulse noise removal.

Acknowledgment

The authors, Mr. Ayyaz Hussain, 042-310554-Eg2-360, Mr. M. Arfan Jaffar, 041-103525-Cu-031 and Mr. M. Nazir, 041-101648-Cu-013 would like to acknowledge the Higher Education Commission (HEC), Govt. of Pakistan and NU-FAST for providing funding and required resources to complete this work. It would have been impossible to complete this effort without their continuous support.

References

1. Gonzalez, R.C., Woods, R.E.: Digital Image Processing, 2nd edn. Pearson education Inc., London (2002)
2. Mirza, A.M., Chaudhry, A., Munir, B.: Spatially adaptive image restoration using fuzzy punctual kriging. Journal of Computer Science and Technology 22(4), 580–589 (2007)
3. Liu, P., Li, H.: Fuzzy Techniques in Image Restoration Research - a Survey (invited paper). International Journal of Computational Cognition 2(2), 131–149 (2004)
4. Tukey, J.W.: Exploratory Data Analysis. Addison-Wesley, Reading (1971)
5. Astola, J., Kuosmanen, P.: Fundamentals of Nonlinear Digital Filtering. CRC, Boca Raton (1997)
6. Pitas, I., Venetsanopoulos, A.: Nonlinear Digital Filters: Principles and Application. Kluwer, Norwell (1990)
7. Wang, J.H., Liu, W.J., Lin, L.D.: Histogram-Based Fuzzy Filter for Image Restoration. IEEE Trans. Syst., Man, Cybern. B 32(2), 230–238 (2002)
8. Lee, C.-S., Guo, S.-M., Hsu, C.-Y.: A novel fuzzy filter for impulse noise removal. In: Yin, F.-L., Wang, J., Guo, C. (eds.) ISNN 2004. LNCS, vol. 3174, pp. 375–380. Springer, Heidelberg (2004)
9. Lee, C.-S., Guo, S.-M., Hsu, C.-Y.: Genetic-Based Fuzzy Image Filter and its Applications to Image Processing. IEEE Trans. Syst., Man, Cybern. B 35(4), 694–711 (2005)
10. Wang, Z., Bovik, A.C., Sheikh, H.R., Simocelli, E.P.: Image Quality Assessment: from Error Visibility to Structural Similarity. IEEE Trans. on Image Processing 13(3), 1–14 (2004)
11. Hussain, A., Arfan Jaffar, M., Mirza, A.M., Chaudary, A.: Detail Preserving Fuzzy Filter (accepted)

Color Transfer in Images Based on Separation of Chromatic and Achromatic Colors

Jae Hyup Kim[1], Do Kyung Shin[2], and Young Shik Moon[2]

[1] Optronics System Group, R&D Center, Samsung Thales Co., Ltd,
Chang-ri 305, Namsa-Myun, Chuin-gu, Yongin-si, Gyunggi-do, Korea
[2] Dept. of Computer Science and Engineering, Hanyang University,
Sa-3-dong, Sangnog-gu, Ansan-si, Gyunggi-do, Korea
Jaehyup.kim@samsung.com, {dkshin,ysmoon}@cse.hanyang.ac.kr
http://www.visionlab.or.kr

Abstract. In this paper, we propose the method which transfers the color style of a source image into an arbitrary given reference image. Misidentification problem of color cause wrong indexing in low saturation. Therefore, the proposed method does indexing after image separating chromatic and achromatic color from saturation. The proposed method is composed of the following four steps. In the first step, using threshold, pixels in image are separated chromatic and achromatic color components from saturation. In the second step, separated pixels are indexed using cylindrical metric. In the third step, the number and positional dispersion of pixel decide the order of priority for each index color. And average and standard deviation of each index color be calculated. In the final step, color be transferred in Lab color space, and post processing to removal noise and pseudo-contour. Experimental results show that the proposed method is effective on indexing and color transfer.

1 Introduction

Recently as the multi-media technology has been developing rapidly, numerous pieces of information are made into digital signals and multimedia formats, digital composition tools such as digital cameras and scanners are getting better and better, and interests in the creation and modification of multimedia contents have been increasing steadily. However, it is not easy to express one's uniqueness freely by using the professional tools for multimedia such as Photoshop, Paintshop, Painter, and Image Ready, etc. Moreover, even if those composition tools for digital contents are widely developed for use, it is not easy to achieve an image with style that you really want unless you are an expert. Therefore, there have been on-going efforts for developments of a method that can transfer styles by using color, one of elements that exhibit style of an image. This can be achieved by transferring the colors of an original image in to colors of a reference image, and the style of the original image can be transferred very close to the reference image. In the previous studies on the color transfer of an image[1], a composer created by being influenced by the characteristics of the same composer, but a color transfer method by using a painting image that invokes different feeling has been investigated. However, if color is transferred by using the

A. Gagalowicz and W. Philips (Eds.): MIRAGE 2009, LNCS 5496, pp. 285–296, 2009.

existing methods for a nature image as an object instead of a painting image, many problems can occur. Recently, color transfer methods that use nature images have been frequently investigated, but in most cases, the number of colors used in nature images as application objects may be too small or change in color may not be drastic. Therefore, we need to develop a color transfer method for nature images in more various environments. This paper proposes a color transfer method that uses an image indexed by distinguishing chromatic and achromatic colors. There are several methods in the literature for color transfer; the color transfer method that uses the entire image as the object [2]; one that uses newly generated palette by extracting representative colors of an input image [3]; and, lastly, one that uses a cumulative histogram. The method that uses the entire image as an object for color transfer uses simple statistical tools to speed up the computational processing. It is effective to transfer color in an image with small number of colors, but, because it can not reflect the characteristics of a local color with increase in the number of colors, a mix up problem occurs with this method [2]. The method of color transfer that generates palette after extracting representative colors of an input image employs a painting image as an object. After being extracted from an input image, representative colors that dominate the overall style of an input image are selected, and from this color is transferred. However, after an image is segmented, representative colors are extracted by using down sampling method. Regions with the same colors are transferred in to different colors, and therefore, pseudo-contours that do not exist in the original image are generated. For this reason, one shortcoming of this method can be the unnatural resultant image [3]. In the method of cumulative histogram, histograms of original and reference images are matched for color transfer. Since this method uses the number of pixels, it does provide the information on space of an image, but due to lack of the information on space, a problem with this method is the generation of noise and pseudo-contour [4]. This paper proposes a method of color transfer that generates indexing by separating chromatic and achromatic colors of image in order to reflect the characteristics of a local color of an image. Also this method can be applied to not only nature images, painting images. By extracting the edge region from non-edge one, pseudo-contours and noise are removed by post-processing the non-edge region. This paper will explain the existing color transfer methods in the second chapter, and the method that we propose in the third chapter. The proposed method consists of four steps. A step that separates chromatic and achromatic colors by using threshold of saturation; a step that generates indexing of 11 index colors by using cylindrical metric for the separated pixels; a step that prioritizes each index colors and calculates the average and the standard deviation; and lastly a post-processing step that transfers colors by using channel a and channel b in the lab color space, and that eliminates the pseudo-contours and the noise of an image. The fourth chapter will analyze the proposed method through experimental results, and the fifth chapter presents the conclusion and the future projects.

2   Previous Works

In this chapter, we will survey the existing methods such as the method that transfers color by using the average and the standard deviation in the lab space, the method that uses palette, and the method that uses a cumulative histogram, as well as their characteristics and problems associated with each method.

Reinhard et al. proposed a transfer method of the original and reference images that uses the average of each channel in the Lab color space and the standard deviation of each pixel [1]. This method takes the entire image as an object, and uses the average and the standard deviation of sum of two images for color transfer. Therefore, the method is very efficient because it employs a simple computation that takes only 1 or 2 seconds. However, since it calculates the average and the standard deviation for the whole image, there are only one average and one standard deviation for the entire image. For this reason when we use an image with various colors, because the value of mixed colors are calculated for one average and one standard deviation, the original image will not be transferred in to the color that represents the reference image. Also by having one average and one standard deviation, there is a limitation for expressing the characteristics of local color of an image.

Greenfield et al. proposed a color transfer method that divides an image into pieces, extracts palette by using representative colors through down sampling, and use the result of extraction for color transfer [3]. This method used only oil painting images as its objects for a color transfer purpose. Therefore, they use only channel a that reflects the color information of red and green, and channel b that reflects the color information of blue and yellow, but do not use channel L that indicates the lighting out of all channels in the Lab space. Also because colors of palette from an input image are extracted, we are able to extract the colors that represent the image. However, among the colors of the original and reference images, by transferring colors that are close in space, the regions with similar colors are separated in to different regions and go through color transfer. Therefore, the resultant image becomes unnatural, and pseudo-contours that are not present in the original image can be created.

Neumann proposed a color transfer method by calculating cumulative histograms of hue, saturation, and intensity channels in the HSI color space and by using a function obtained from the probability density function and the cumulative density function [4]. In the method, we may notice that the cumulative histogram distributions in the resultant image and the reference image are the same if you transfer color by using a cumulative histogram in the first dimension. In the method that uses cumulative histograms for color transfer, there is no space information due to its use of the number of pixels for each channel, and because of this, a lot of pseudo-contours and noise are generated.

3 Proposed Method

The proposed method comprises of the following.

1. Separate chromatic and achromatic colors after measuring the saturation of an input image in order to prevent indexing error in the low saturation region.
2. In order to reflect the characteristics of local colors in an image, generate indexes into 11 index color by using the separated pixels. For the distance calculation, the cylindrical metric is used.
3. In order to transfer colors by grouping dominant colors in an image, each color index is prioritized by using the number of pixels and the degree of

dispersion. The average and the standard deviation for each color index are calculated.

4. Implement the color transfer by using channels a and b in the Lab color space. Also, in order to eliminate the noise and the pseudo-contours of an image, implement blurring after extracting the non-edge region.

3.1 Separating Chromatic and Achromatic Colors by Using Saturation

In this paper, in order to prevent indexing in to inappropriate colors under low saturation of each pixel of an image, the color index for each pixel is generated by separating chromatic and achromatic colors. For separation of chromatic and achromatic colors, threshold of saturation is used. Chromatic colors denote the colors that have saturation and have all three elements of color, which are hue, saturation, and intensity. In an achromatic color, there is only intensity but no hue and saturation. However, it is very rare that a common image is completely devoid of hue and saturation altogether, and, moreover, 'hue' one of elements that divide a color into chromatic and achromatic, is independent of the intensity component. Therefore, it has the most distinguishing power among three HSI channels of the color space (hue, intensity, and saturation). However, for the low saturation, the function that separates color gets very low, and the distinguishing power becomes unstable. Therefore, this paper presents a method that is able to separate chromatic and achromatic colors by using threshold of saturation [5], [6]. As saturation approaches to 0, it gets closer to intensity and it is hard to distinguish hue. As saturation approaches to 1, it gets closer hue and it is hard to distinguish intensity. Also, human eyes respond less sensitive to a change in saturation. Through these facts we can realize that hue and intensity are more important to the recognition of human eyes. Therefore, in order to limit changes in saturation, we need to separate an image into chromatic and achromatic colors for color transfer. The threshold value for saturation is set to 0.2 [7], [8]. Formula (1) shows the definition of the threshold for saturation.

$$\begin{cases} chromatic & \text{if } P_s \geq T_s \\ achromatic & \text{if } P_s < T_s \end{cases} \tag{1}$$

Here, T_s denotes the threshold value for saturation that separates chromatic and achromatic colors, and P_s pixel of an image. Using the threshold for saturation, colors are separated in to chromatic and achromatic colors, and indexes are generated with the separation by applying them to 11 index colors for an input image.

3.2 Indexing by Using Cylindrical Metric

Indexes are generated in order to reflect the characteristics of local colors of an image, and the cylindrical metric method is used to do this. The cylindrical metric is calculated in the HSI color space; that is, because this is a model that is more susceptible to the human sense by user-oriented expression, unlike the RGB color space that provides hardware-oriented expression such as CRT monitors, it enables generation of indexing in to appropriate colors that a user may be able to recognize easily [9], [10].

Index colors used to generate indexes employ the Basic Color Terms, first proposed by Berlin et al., who reported general Basic Color Terms by investigating similar types and the color range through 98 languages. Basic Color Terms are composed of red, green, yellow, blue, brown, purple, pink, orange, black, white, and gray [11], [12]. In order to measure the similarities between the pixels that are divided into chromatic and achromatic colors and 11 index colors, the cylindrical metric method is used to measure the distance between pixels in the HSI color space. The distance $D_{cyl}(x, y)$ between two pixels x and y can be calculated through formula (2), (3), (4), and (5).

$$D_{cyl}(x, y) = \sqrt{(d_I^2 + d_C^2)} \tag{2}$$

$$d_I = \left| I_x - I_y \right| \tag{3}$$

$$d_C = \sqrt{(S_x^2 + S_y^2 - 2S_x^2 S_y^2 \cos \theta)} \tag{4}$$

$$\theta = \begin{cases} \left| H_x - H_y \right| & \text{if } \left| H_x - H_y \right| < 180° \\ 360° - \left| H_x - H_y \right| & \text{otherwise} \end{cases} \tag{5}$$

Here, H denotes the hue value of a pixel, S the saturation value of a pixel, and I the intensity value of a pixel. By calculating similarities between each pixel and index colors, what color index each pixel belongs to is determined. For this, since generation of indexes excludes surrounding pixels when the distance between each pixel and index color is calculated, it is likely that a pixel is similar to the surrounding pixels, but has a different index because of a different index color value. In order to compensate this, a mask can be used to verify the index color of surrounding pixels and the central pixel. Through the verification procedure, the numbers of the central pixel and the surrounding pixels that have different index color are calculated. If there are more than 7 calculated values, it is considered a wrong index. Therefore, cases where surrounding pixels have more than 7 index colors should be sought for, and the index color at the time is modified. Of course, when the number of calculated surrounding pixels is than 7, index colors for the central pixel are used without modification. This is repeated until there is no index color that needs compensation.

3.3 Prioritizing Color Index

Priorities of index colors are determined in order to first transfer color for index color that dominates the overall style of an image, and in order to do this, the variation according to the number of pixels and configuration is used and calculated. When only the number of pixels pertaining to each index color is considered, and when the degree of scattering is big even though there are many pixels inside an image, the color that does not dominate the overall style of image may dominate, and a transfer color with an inappropriate order may occur. In order to solve this problem, priorities of variation calculated according to the locations between pixels that belong to each index color are determined. By variation according to the locations it means the calculation of distance between pixels in an image. For the purpose of color transfer, the average and standard deviation of the pixels that belong to each index color are calculated. Formula 6 is the average of pixels that belong to each index color, and formula 7 is the standard deviation of pixels that belong to each index color.

$$\mu^i = \frac{1}{n^i} \sum_{k=1}^{n^i} P_k^i \tag{6}$$

$$\sigma^i = \frac{1}{n^i} \sum_{k=1}^{n^i} (P_k^i - \mu^i)^2 \tag{7}$$

Here, i denotes the number according to each index color, μ^i the average of i-th index color, σ^i the standard deviation of i-th index color, and P the pixel of an image. When the number of index color in the original image is the same as that in the reference image, or when the number of index color in the reference image is more than that in the original image, there is no error that may be caused by the empty index color. However, if there is more index color in the original image, leaving empty the corresponding index color in the reference image, there will be a problem during the color transfer. In order to solve this problem, the average and the standard deviation are calculated by using index color existing in an image, and this kind of process will be implemented unless it damages the overall style of an image. Formula (8) calculates the average of empty index colors, and formula (9) calculates the standard deviation of empty index colors.

$$\mu^{bin\_i} = P_{BCTs}^i + \frac{1}{i} \sum_{k=1}^{i} (P_{BCTs}^k - \mu^k) \tag{8}$$

$$\sigma^{bin\_i} = \frac{1}{i} \sum_{k=1}^{i} \sigma^k \tag{9}$$

Here, P_{BCTs}^i denotes the value of the index color that corresponds to the i-th index color.

3.4   Color Transfer in Lab Color Space

There is also a color transfer method proposed by Reinhard et al. [2], who used the average and the standard deviation in the lab color space. They used the calculated values for color transfer, but this paper proposes a method of color transfer by using only channel a that represents red and green, and channel b that represents blue and yellow color information, but by omitting channel L that represents lighting in the Lab color space. Formula (10) is the color transfer formula for this method.

$$a' = \frac{\sigma^i_{reference}}{\sigma^j_{source}}(a - \mu^j_{source}) + \mu^i_{reference}$$

$$b' = \frac{\sigma^i_{reference}}{\sigma^j_{source}}(b - \mu^j_{source}) + \mu^i_{reference}$$

(10)

Because the color transfer method in this paper does not consider the relationship with the adjacent region, there will be the pseudo-contours and that noise that do not exist in the original image. The adaptive sobel filter that Chi Eun Mi et al. proposed can be used in order to solve this problem [13]. By locating edge using ASF, non-edge regions are smoothed by using the median filter. In general, a fixed threshold is used in order to determine the existence of edge. This kind of method has the drawback that it can not control the amount of edge depending on the change of intensity in an input image. In order to solve this, after a non-edge region is determined by using sobel edge filter, which determines a threshold value according to the local characteristics of the region that matches the mask and that applies to an image, the median filter is applied. Edges in the horizontal and vertical directions are extracted through the sobel mask, and are divided by the average intensity. If the divided value is greater than 1, it is an edge area; if it is smaller than 1, it is classified as non-edge. For the value that is classified as non-edge the median filter is applied. When the strength of vertical and horizontal edges obtained by the sobel mask is high, or when the average intensity is high, more edge is extracted. On the contrary when the strength of edge is low, or when the average intensity is low, less edge is extracted. That is, depending on the degree of change in light intensity within the mask region, each pixel is determined for whether it is edge or not. Therefore, we can obtain the edge extraction result that is more robust to change in light density, compared to the edge extraction obtained by using fixed threshold. The median filter is a method that takes the median value among those pixels by having surrounding pixels stand in a line, and that is very effective in eliminating impulse noises.

4 Experimental Result

In this paper, for the comparison purpose of the proposed method, tests for color transfer of an original image into a reference image by using a painting image of WebMuseum.Paris and Artframed, images of nature from Corel Database, and nature images collected from the Web [14], [15], [16].

4.1 Indexing Result

Figure 1 shows the result of index generation using such various methods. It can be seen that the color of the building in the original image is bearing light yellow. However, in the result from generation of index using Euclidean and cylindrical distances without applying chromatic separation, the building appears to bear yellow and white index colors. Also, if you look at the color of the sky, regardless of the fact that it went through chromatic separation or not, a color that is not blue is generated in the area where index is generated by using Euclidian distance. Also, in the resultant image that index color is generated by using cylindrical distance without undergoing chromatic separation, you can see that in the region of the sky the indexing is divided in to two different regions. Therefore, when indexing is generated without chromatic separation, two index colors can be generated in the image that gradation is applied to, or we can identify such generation that two improper index colors are generated for low saturation. In order to solve this, chromatic separation is applied, and in figure 4 we can identify that the index color for blue region and the color of the building's wall are effectively generated if you look at the result of indexing with chromatic separation.

Fig. 1. The indexing results (a) original image (b) proposed method(separation and cylindrical distance) (c) separation and Euclidean distance (d) no separation and cylindrical distance (e) no separation and Euclidean distance

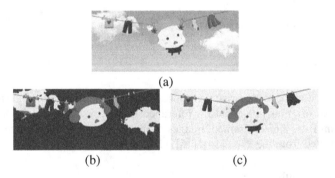

Fig. 2. The indexing result (a) original image (b) separation and cylindrical distance (c) no separation and cylindrical distance

As can be seen in figure 2, the colors of clothes, background, and blue jean and skirt on the laundry rope all have the same blue kind of color. These colors may look as if they are generated as blue index color in the proposed method. However, generation of index using the cylindrical distance without chromatic separation results in the white index color such as clouds, other than blue index color in the background with low saturation. Also, the color of the socks to the left of the character, results in the wrong index as the red index color due to low saturation.

4.2 Color Transfer Result

Figure 3 is the result of the method by Reinhard et al. that utilizes one average and one standard deviation in an original image and a painting image for color transfer. Dominant colors of the original and painting images are similar. However, because it uses one average and one standard deviation in the method that Reinhard et al. proposed, the color of the background and the color of the dinosaur in the reference image are mixed up resulting in the increase in the image intensity. Since the method that this paper proposes utilizes a method that separates saturation and generates index, it looks as though the color of the reference image is a lot better after color transfer.

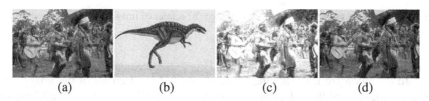

(a)   (b)   (c)   (d)

Fig. 3. The color transfer result (a) original image (b) reference image (c) Reinhard's method (d) proposed method

(a)   (b)   (c)   (d)

Fig. 4. The color transfer result (a) original image (b) reference image (c) Reinhard's method (d) proposed method

As can be seen in figure 4, the method that Reinhard et al. proposed mixes up blue, the color of rocks, and the color of stones in the reference image, and results in the transfer to the red color that does not exist in the reference image. In the method this paper proposed, the dominant green color in the original image and the dominant blue color in the reference image are properly matched during color transfer. Figure 5 and 6 shows the original and reference images and the result of the color transfer methods.

As can be seen in figure 5 (d), the background of the original image transferred into a similar color. However, you can see improper colors in the upper right and the lower left of the image. This kind of problem occurs when similar regions are divided into different regions and transferred into different colors, due to the region-based image segmentation, a kind of area-based segmentation. Also, if you look at figure 5 (e), the resultant image looks more natural because the kinds of color in the original

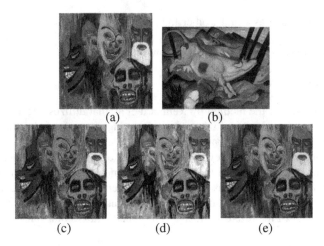

Fig. 5. The color transfer result (a) original image (b) reference image (c) proposed method (d) Greenfield's method (e) Reinhard's method

image and the reference image are similar. However, since only one average and one standard deviation are used and the characteristics of local colors of the image were not reflected, most of the colors in the resultant image exhibit a kind of yellow color. This problem can be solved by segmenting similar regions in to different colors, as shown in figure 5 (c) that used separation method of saturation-hue from indexing by using separation of saturation-hue by pixels. Also, because colors are indexed according to the index color item, you can identify that the color transfer was done effectively by applying the characteristics of local colors of an image.

Fig. 6. The color transfer result (a) original image (b) reference image (c) proposed method (d) Greenfield's method (e) Reinhard's method

The reference image in figure 6 (b) has mostly yellowish kind of a color with high intensity, but the background has the dark kind of a color. In the result that Greenfield et al. employed, saturation of an image is low due to the color of background, and a brown kind, instead of a yellow kind, are the majority of the color. Likewise, most of the color in figure 5 which is the result by using the method that Reinhard et al. proposed indicates the yellow kind, one of the representative colors of the reference image. Also, in the results of methods that Greenfield et al. and Reinhard et al. employed, changes in intensity occur many times, and because of this, you can notice that a lot of detailed description of length or bottom of tree disappears. In the result of the proposed method, by using indexing through saturation-hue separation, overall style of an image is maintained yellow, and the detailed description of an image is effectively shown by keeping the local characteristics of a color.

5 Conclusion

In this paper, a color transfer method of an image that generates indexing by using the threshold of saturation has been proposed. This proposed method can be applied not only to a painting image but also to a nature image, and it separates chromatic and achromatic colors by using the threshold of saturation. Therefore, indexing to a proper color is generated even in a low saturation environment that reduces the distinguishing power. Also, by using an adaptive edge mask and the median filter, the noises and pseudo-contours caused during the color transfer can be eliminated. However, in addition to hue there are many elements that affects the style of an image, and therefore, we need to investigate further color transfer methods on the style of an image by taking into consideration of other elements such as lighting other than color elements.

Acknowledgment

This research was supported by the MIC (Ministry of Information and Communication), Korea, under the ITRC (Information Technology Research Center) support program supervised by the IITA (Institute of Information Technology Assessment).

References

1. Gooch, A.A., Olsen, S.C., Tumblin, J., Gooch, B.: Color2Gray: Salience-Preserving Color Removal. ACM Trans. Graph. 24(3), 634–639 (2005)
2. Reinhard, E., Ashikhmin, M., Gooch, B., Shirley, P.: Color Transfer between Images. Int. J. of IEEE Computer Graphics and Applications 21, 34–41 (2001)
3. Greenfield, G., House, D.: Image Recoloring Induced by Palette Color Associations. J. of Winter School of Computer Graphics and CAD Systems 11, 189–196 (2003)
4. Neumann, L., Neumann, A.: Color Style Transfer Techniques Using Hue, Lightness and Saturation Histogram Matching. In: Int. Conf. on Computational Aesthetics in Graphics, Visualization and Imaging, pp. 111–122 (2005)
5. Skarbek, W., Koschan, A.: Colour Image Segmentation-A Survey. Inst. for Technical Informatics, Technical University of Berlin (1994)
6. Pujas, P., Aldon, M.: Robust Colour Image Segmentation. In: Int. Conf. On Advanced Robotics (1995)
7. Sural, S., Quin, G., Pramanic, S.: Segmentation and Histogram Generation Using the HSV Color Space for Image Retrieval. In: IEEE Int. Conf. on Image Processing, pp. 589–592 (2002)
8. Kim, T., Kim, S., Lee, K.: Contents based Image Retrieval using Adaptive Color Histogram and Directional Pattern Histogram. J. of IEEK 42, 119–126 (2005)

9. Nam, H., Kim, B., Kim, W.: Block based Color Image Segmentation using Cylindrical Metric. J. of IEEK 42, 285–292 (2005)
10. Shin, K.: Color Transfer of skin and illumination. Master's thesis, Seokang University, Korea (2003)
11. Berlin, B., Kay, P.: Basic Color Terms: Their Universality and Evolution. J. of Center for the Study of Language and Inf. (2001)
12. Chang, Y., Uchikawa, K., Saito, S.: Example-based color stylization based on categorical perception. In: ACM Int. Conf. Proc. Series, vol. 73, pp. 91–98 (2006)
13. Ji, E., Yoon, H., Lee, S.: Face Detection Method using Color and Edge Information. J. of KISS 29, 809–817 (2005)
14. http://www.ibiblio.org/wm/
15. http://www.artframed.com
16. http://wang.ist.psu.edu/docs/related/

Realistic Face Animation for Audiovisual Speech Applications: A Densification Approach Driven by Sparse Stereo Meshes

Marie-Odile Berger, Jonathan Ponroy, and Brigitte Wrobel-Dautcourt

LORIA/INRIA Nancy Grand Est
Marie-Odile.berger@loria.fr,Brigitte.Wrobel@loria.fr

Abstract. Being able to produce realistic facial animation is crucial for many speech applications in language learning technologies. Reaching realism needs to acquire and to animate dense 3D models of the face which are often acquired with 3D scanners. However, acquiring the dynamics of the speech from 3D scans is difficult as the acquisition time generally allows only sustained sounds to be recorded. On the contrary, acquiring the speech dynamics on a sparse set of points is easy using a stereovision recording a talker with markers painted on his/her face. In this paper, we propose an approach to animate a very realistic dense talking head which makes use of a reduced set of 3D dense meshes acquired for sustained sounds as well as the speech dynamics learned on a talker painted with white markers. The contributions of the paper are twofold: We first propose an appropriate principal component analysis (PCA) with missing data techniques in order to compute the basic modes of the speech dynamics despite possible unobservable points in the sparse meshes obtained by the stereovision system. We then propose a method for densifying the modes, that is a method for computing the dense modes for spatial animation from the sparse modes learned by the stereovision system. Examples prove the effectiveness of the approach and the high realism obtained with our method.

Keywords: Face animation, densification, PCA with missing data.

1   Introduction

There is a strong evidence that the view of speaker's face visual information noticeably improves the speech intelligibility. Hence, having a realistic talking head could help language learning technology in giving the student a feedback on how to change articulation in order to achieve a correct pronunciation. In [7], Munhall and Vatikiotis provide evidence that lip and jaw motions affect the entire facial structure below the eyes. High levels of details are thus required to obtain highly realistic and perceptibly correct facial animation of the complete face.

Though the utility of visual information to speech perception has been known for a long time, progresses to develop talking faces which both look real and

A. Gagalowicz and W. Philips (Eds.): MIRAGE 2009, LNCS 5496, pp. 297–307, 2009.

convey linguistic relevant information are more slower. What causes the intelligibility enhancement afforded by the visible component of speech is difficult to determine. For these reasons, many works argued the necessity for animating face directly from visible articulatory data either in 2D [3] or in 3D [2,4,5,6]. Most of these data are 3D meshes of the face which can be sparse -when they are acquired using markers or sensors glued on the face- or dense using laser scanner acquisition. It must be noted that do not operate with a sufficient speed for speech acquisition (an acquisition rate of 120 Hz is required to acquire fast articulatory gestures of consonants). As a result, high resolution scanners can be used for sustained sounds as vowels but it is not obvious that they can be used to acquire all the dynamics of speech production.

Most methods for animating face are based on the extraction of the basic modes of spatial deformations during speech using principal component analysis (PCA). They span a space which describes at best all the plausible face deformations corresponding to speech production. Constructing such a space requires to physically match the points of the meshes at each time instant. This task is easy when sparse meshes are considered but it becomes complex for dense meshes. In this latter case, a generic head template is generally used [5,6] to align all the scans using prominent features such eyes, nose. Obtaining rough alignment is largely automatic but accurate alignment requires a manual adaptation of the model to the speaker specificities. In this paper, we favor a fully automatic method and propose an automatic matching process guided by the sparse stereo meshes.

Though PCA can be used both on sparse or dense meshes, it is difficult to obtain the speech dynamics from dense meshes since the set of visemes that can be acquired is limited to sustained sounds due to technical limitations in scanner acquisition. It is thus not obvious to recover the complete speech dynamics from a reduced set of 3D scans. For these reasons, we propose in this paper an approach that makes use both of a small set of 3D scans acquired for sustained sounds and of the speech dynamics learned from a stereovision sequence of the face painted with markers. The main idea is to transfer the dynamics learned on the sparse meshes onto the 3D dense meshes in order to generate realistic dense animations of the face. This paper is an extension of our previous work [1]. In this past work, only one dense mesh was used which turned out to be not accurate enough for generating sounds which are too far from the reference dense mesh. We thus propose in this paper significant extensions of the work in order to transfer the dynamics learned from the sparse meshes onto the dense face.

Computing PCA modes for face animation from sparse meshes acquired by stereovision is difficult because all the markers may not be observable at every time instant. When the mouth is closed, points may become invisible, or at least not sufficiently visible, to be correctly detected and reconstructed. Though this difficulty is often ignored in the literature, this make the PCA more complex and needs to resort to PCA with missing data techniques [8]. A novel way of computing the face modes taking into account possible unobservable points is thus presented in section 3.

Our approach for computing the dense modes is presented in section 4. It borrows concepts from transfer techniques [10] used in computer graphics to map an object onto another. We specifically used this technique in section to physically and automatically match the dense meshes using the underlying sparse mesh to guide the transfer. This avoids doing manual adaptation of a generic mesh to our talker. Finally, we propose a densification method which allows the basic modes of the dense head to be computed from the sparse modes computed from the learning sequence. As a result, this allows us to animate a dense head using only a reduced set of 3D scans.

We demonstrate experimentally in section 5 that the proposed method is efficient and allows us to obtain very realistic dense faces.

2   System Overview

Our method requires the acquisition of a set of 3D dense meshes of the talker. In our study, these dense meshes were acquired with the Inspeck mega capturor (www.inspeck.com) for 15 sustained sounds (vowels, fricatives and lip closure). These dense meshes are also called visemes in the following.

In order to learn the face kinematics, a classical stereovision system with two cameras was used to record a corpus. The acquisition rate of the cameras is 120 images/frames which is sufficient to capture fast movements of the articulators (further details on this system can be found in [11]). Markers were painted on the talker's face in order to make the matching and the reconstruction stage automatic. With 45 points on the lips and a total of 209 points on the part of the face that is influenced by speech, the recovery of face kinematics is quite detailed. Our experiments proves that between 5 to 7 PCA modes are sufficient to describe the face kinematics.

The experimental set-up and the input data are shown in Fig.1. The first raw exhibits a stereo image pair of the talker. Note that the points located on the top of the head are used to compensate for head motions. Fig. 1.c is an example of a sparse mesh obtained with the stereo system. Finally, Fig.1.d is the dense mesh acquired for the /a/ sound.

3   PCA with Missing Data for Computing the Sparse Modes

The 3D coordinates of each marker can theoretically be computed at each time instant of the learning sequence using the stereovision system. In the following, the 3D sparse mesh of the face computed at time instant t ($t \in [1..T]$) are denoted $X_t = [X_{1,t}, ..., X_{N,t}]$, where N is the number of markers and $X_{i,t}$ are the 3D coordinates of the marker i at time instant t. Here, N=209 markers were painted on the face. The duration of the corpus recorded for learning kinematics was 6 minutes, giving rise to T=39000 stereo pairs.

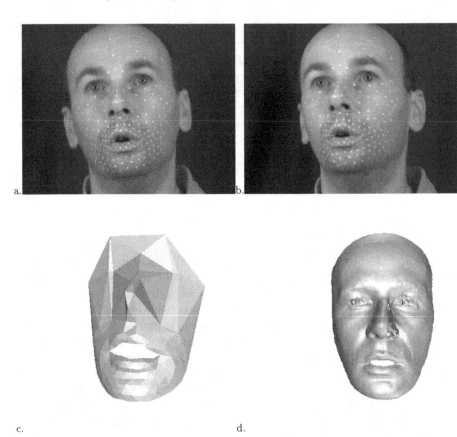

Fig. 1. Input data of the system: (a and b) a couple of stereovision images. White points/markers were painted on the face ; (c): reconstruction of the sparse mesh; (d): the 3D dense map obtained for the /a/ sound.

However, some markers may become unobservable during uttering when the lips are very close. Practically, these markers may not be reliably detected in the images by low level algorithms. It is especially the case for sounds like /u/, /o/, /i/. Markers may also become unobservable in one of the image pair due to slight head motions which make some points disappear from the field of view. Stereo reconstruction is thus not possible for these points. Practically speaking, 77% of the markers are always observable and thus reconstructed. 17% of the markers have a reconstruction rate in the range [70%, 95%] and 6% of the markers have a reconstruction rate less than 50% and are mainly located on lips (Fig .2). As a result, some data are may be missing in the X_t data.

Most of the algorithms for building principal subspaces are based on the decomposition of the covariance matrix of the input data and cannot be used when some data are missing, as in the present case. However, there exist probabilistic approaches where PCA is considered as a limiting case of a linear Gaussian model. Principal axis can then be computed using Expectation-Maximisation

(EM) algorithms [8]. Such algorithms can be extended to handle the problem of missing data [9,8].

In this paper, we adapt these ideas to our particular problem in order to compute the sparse modes. These ideas are also used in section 4 to compute the dense modes. We first compute the principal components for the markers $\mathcal{R}$ which are always reconstructed over the sequence. The components can be easily computed with classical methods since all the data are available for these markers. We then complete the entries of the principal components by introducing the markers which are not observable at every time instant using EM techniques.

Let $\{u_k^r\}_{k \leq q} \in \mathbb{R}^r$, be the q principal components computed from the set of markers which are always observable and let $\bar{X}^r$ be the mean of these meshes. Let $u \in \mathbb{R}^N$ be the extended basis we are looking for. Given a mesh X_t acquired at time instant t, let X_t^r be the reduced mesh, where only the markers always reconstructed are considered. X_t^r can be approximated on the q principal components as:

$$X_t^r \approx \bar{X}^r + \sum_{k=1}^{q} \alpha_{k,t} u_k^r \quad t = 1..T \tag{1}$$

The goal of the complete components is to approximate any mesh as a linear combination:

$$X_t \approx \bar{X} + \sum_{k=1}^{q} \beta_{k,t} u_k \quad t = 1..T \tag{2}$$

As X_t^r is a sub-vector of X_t - 77% of entries of X_t are the entries of X_t^r in our set-up- , it is likely that $\beta_k = \alpha_k$ is very close to the mean square solution of (2). In the same way, we can consider that the entries of $\bar{X}$ which correspond to always reconstructed markers are identical to the entries of $\bar{X}^r$.

We then build a linear system which incorporates all the observations on the non always observable markers. For each observation of a marker i at time instant t, equation (2) gives rise to the following linear equation :

$$X_{i,t} = \bar{X}_i + \sum_{k=1}^{q} \alpha_{k,t} u_{k,i}$$

Stacking all the equations for each visible marker at each time instant t leads to a linear system where the unknowns are the missing $u_{k,i}$ ($i \notin \mathcal{R}$) and the missing components of the mean $\bar{X}$. The $\alpha_{k,t}$ are computed by projecting the reduced meshes onto the reduced components $\{u_k^r\}$. Considering all the markers which can be reconstructed at some instant of the sequence, we obtained a set of linear equations which can be solved in the least square sense, giving rise to the missing entries of the principal components and of the mean.

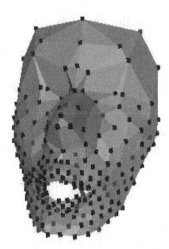

Fig. 2. The markers in black are the ones which are reconstructed in every frame and are taken into account in the reduced PCA. The markers in red are not always reconstructed in the sequence and are considered in the computation of the principal components using an iterative EM algorithm.

The principal components are then refined using the classical EM algorithm:

- Given the components u_k and $\bar{X}$, compute the coefficients $\alpha^{t,i}$ for all the meshes available in the sequence.
- given the $\alpha_{t,i}$, refine the $u_{k,i}$ and $\bar{X}$ by solving for each i the system of linear equations:

$$X_{i,t} = \bar{X}_i + \sum_{k=1}^{q} \alpha_{k,t} u_{k,i}, \quad t = 1..T$$

in the least square sense.

4   Generating Dense Modes from Sparse Modes

Dynamic realism is needed for linguistic expressions. Achieving realism needs to have a high resolution 3D model as well as knowledge about the facial kinematics. In our case, high resolution scanners of the face can only be obtained for sustained sounds as vowels. We thus propose in this paper to transfer the change in shape exhibited on the sparse meshes onto the dense mesh and to infer dense animation modes from the sparse ones.

Here gain, as in the computation of the sparse mode, we use a densification process to compute the dense modes from the sparse ones.

4.1   Overview

Let $X_1^{dense}, .., X_d^{dense}$ be the set of d dense visemes which were scanned. This set contains 15 visemes of the french vowels as well as a neutral expression and a face with closed mouth.

We first suppose that these meshes physically matched, which means that the vertex i in every mesh fits the same physical point in all the dense meshes. Additionally, we also suppose that the dense and the sparse meshes are registered. How to obtain such data will be considered in the next section.

The aim of PCA is to express each viseme as a linear combination of the dense modes:

$$X_i^{dense} \approx \bar{X}^{dense} + \sum_{k=1..q} \beta_k^i u_k^{dense}, i = 1..d \qquad (3)$$

The space spanned by the sparse modes is built so as to describe the plausible appearances of a face. There thus exists a sparse mesh X_i which best fits each dense mesh X_i^{dense}. Hence, we can compute coefficient $\alpha_{k,i}$ such that:

$$X_i \approx \bar{X} + \sum_{k=1}^{q} \alpha_{k,i} u_k \qquad (4)$$

where u_k is the set of sparse modes computed in section 3.

Unlike the preceding case, X is not dense in X^{dense}: the number of vertices of the sparse mesh is 209 whereas the dense mesh contains around 13000 vertices. However the vertices of the sparse mesh are distributed on the whole face and vertices are more present in mobile areas of the face during articulation. For these reasons, we also consider that $\beta_k = \alpha_k$ can be considered as a fair least square solution of equation (3).

These first estimates of β are then use to solve (3) in the least mean sense in the unknown modes u_k^{dense} and the mean $\bar{X}^{dense}$. Note that we estimate less than 7 modes. As we have 15 dense visemes, the system always has a solution.

4.2 Obtaining Physically Matched Dense Visemes

Dense visemes are recorded with a laser scanner and the recovered meshes do not physically match. Remeshing of the visemes must thus be performed.

However, physically matching points between deformable surfaces is known as a very difficult problem. We thus take advantage of the underlying sparse model to obtain a physically coherent parametrization.

The first step is to identify the sparse mesh which best fits each dense viseme. As the sparse shapes are described using a reduced number of sparse modes, this can be done by computing the rigid displacement which minimizes the distance between the dense mesh and the sparse mesh thank to an iterative closest point algorithm. We are thus looking for the displacement $\mathcal{T}$ and the coefficient α which minimize:

$$Min_{\mathcal{T},\alpha_1,..,\alpha_q} distance(\mathcal{T}^{-1}(\bar{X} + \sum \alpha_k u_k), Dense\ Mesh)$$

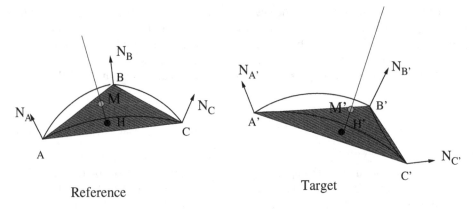

Fig. 3. How a physically registered dense mesh is obtained using the underlying sparse meshes

Doing this, we both identify the sparse mesh which best fits the dense viseme as well as the displacement between the two surfaces. For sake of simplicity, the registered dense visemes $T(V^{dense})$ are still denoted V^{dense} in the following.

It is important to note that the registered visemes contains all the vertices of the sparse model. This property is used to physically match the dense visemes under the control of the sparse meshes. We here consider the /a/ viseme as the reference mesh. Remeshing is achieved with respect to this reference viseme and is performed on the basis of an extended affine matching between each pair of corresponding facets for the sparse visemes.

Given a facet ABC of the sparse reference viseme and its 3D position $A'B'C'$ in the viseme to be matched (see figure 3), let M be a point of the reference dense visem. Each point of the dense mesh is associated to a sparse facet using the algorithm we described in [1]. The main idea is to define M as a function of the vertices of the facet and of the normals. The line HM, with H belonging to the facet, is affinely defined with respect to the facet: H is defined as the point with affine coordinates $(\alpha, \beta, 1 - \alpha - \beta)$ such that HM and $\alpha N_a + \beta N_b + (1 - \alpha - \beta)N_c$ are collinear. We then define $H'M'$ as the line with the same affine coordinates with respect to the facet $A'B'C'$ and its normals. M' is finally obtained as the intersection of this line with the dense mesh.

5   Results

Figure 4 shows the first dense mode computed from the sparse mode. Fig 4.a shows the mean shape of the sparse meshes and Fig. 4.b the computed dense mean shape. The second row exhibits the sparse first modes as well as the computed corresponding dense modes.

Figure 5 shows examples of dense faces computed from the corresponding sparse meshes. Given faces randomly taken within the sparse meshes in the database, the coefficients of the sparse mesh onto the sparse mode were computed. The dense

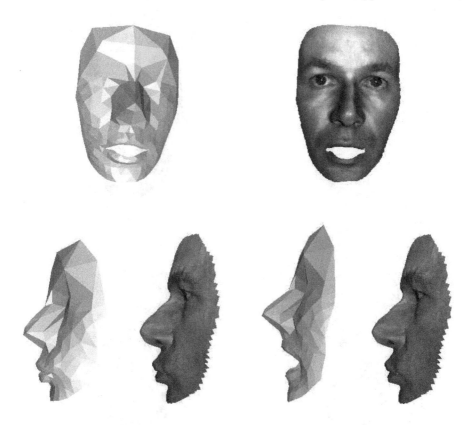

Fig. 4. First row: the mean shape of the sparse meshes and the computed dense mean mode. Second row: $mean - 3 \times u_1$ and $mean + 3 \times u_1$ are exhibited for the sparse modes and the computed dense modes.

face was then reconstructed from these coefficients using the computed dense modes. The first row 5 shows the sparse mesh and the corresponding dense face. Other reconstructed dense faces are shown in the second and third row. These examples prove the high realism of the obtained face with a global motion of the face under the eyes coherent with the speech gesture.

A full dense sequence is available on the webpage of the authors at http://www.loria.fr/~berger/teteParlante. Given a speech sequence of our corpus, the PCA coefficients were computed using the sparse modes and the dense head was generated for each image. The visual impression of the resulting sequence is very good and proved that our method is able to produce realistic facial animation.

It must be noted that the reconstruction is somehow incomplete in the inner region of the lips. This is due to the fact that the transfer procedure only allows the dense face in the neighborhood of the sparse mesh to be computed. We plan to investigate in future works methods to complete the lips from the acquired dense meshes. Lip modeling as realized in [5] is a possible research direction.

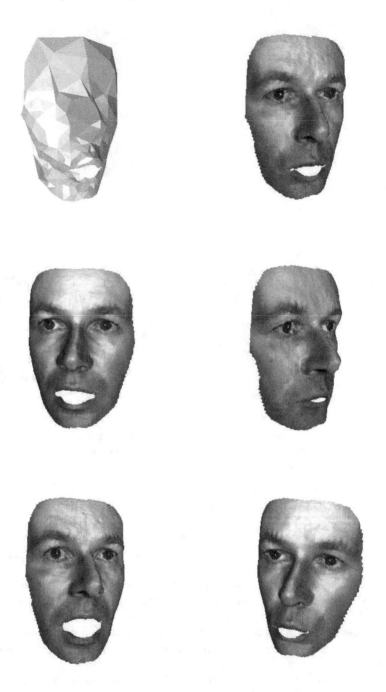

Fig. 5. Examples of dense faces generated using 7 dense modes. First row: a sparse mesh and the corresponding dense mesh. Second and third row: examples of dense faces computed from various sparse meshes.

6   Conclusion

We have proposed an approach that produces automatically highly realistic face animation. This method does not require expensive materials: two cameras are required for the stereovision system and only a reduced set of 3D scans of the talker are needed. The main strength of the approach is to transfer the speech dynamics which can be easily learned on sparse data onto the dense face. To this aim, we have proposed an original densification approach which allows to compute the dense modes from the sparse modes. This method produces highly realistic dense postures. In the near future, we plan to conduct perceptive evaluation of the head with the aim to prove that this very realistic face improves speech intelligibility.

References

1. Berger, M.O.: Realistic face animation from sparse stereo meshes. In: International Conference on Auditory-Visual Speech Processing 2007 (2007)
2. Blanz, V., Vetter, T.: A morphable model for the synthesis of 3d faces. In: SIG-GRAPH 1999, pp. 187–194 (1999)
3. Ezzat, T., Geiger, G., Poggio, T.: Trainable videorealistic speech animation. In: Proceedings of the 29th annual conference on Computer graphics and interactive techniques, pp. 388–398. ACM Press, New York (2002)
4. Kalberer, G., Van Gool, L.: Face animation based on observed 3d speech dynamics. In: Proceedings of Computer Animation 2001 Conference, pp. 20–27 (2001)
5. Kuratate, T.: Talking Head Animation System Driven by Facial Motion Mapping and a 3D face Database. PhD thesis, Nara Institute of Science and Technology (2004)
6. Muller, P., Kalberer, G., Proesmans, M., Van Gool, L.: Realistic speech animation based on observed 3d face dynamics. Vision Image and Signal Processing 152(4) (2005)
7. Munhall, K., Vatikiotis-Bateson, E.: The moving face during speech communication. In: Hearing by Eyes, vol. 2, ch. 6, pp. 123–139. Psychology Press, San Diego (1998)
8. Roweis, S.: Em algorithms for pca and spca. In: Advances in Neural Information Processing Systems, pp. 626–632. IEEE and ACM (1998)
9. Skočaj, D., Bischof, H., Leonardis, A.: A robust PCA algorithm for building representations from panoramic images. In: Heyden, A., Sparr, G., Nielsen, M., Johansen, P. (eds.) ECCV 2002. LNCS, vol. 2353, pp. 761–775. Springer, Heidelberg (2002)
10. Summer, R., Popovic, J.: Deformable transfer for triangle meshes. In: Procceding of SIGGRAPH (2004)
11. Wrobel-Dautcourt, B., Berger, M.-O., Potard, B., Laprie, Y., Ouni, S.: A low-cost stereovision based system for acquisition of visible articulatory data. In: Proceedings of the 5th Conference on Auditory-Visual Speech Processing (AVSP), Vancouver Island, BC, Canada, July 2005, pp. 65–70 (2005)

Meshless Virtual Cloth

Weiran Yuan, Yujun Chen, and André Gagalowicz

INRIA Paris-Rocquencourt, France
Yuan.Weiran@inria.fr,
Chen.Yujun@inria.fr,
Andre.Gagaowicz@inria.fr

Abstract. A systematic description of a novel physically-based virtual cloth simulation method using meshless models is carried out in this paper. This method is based upon continuum mechanics and discretized without explicit connections between nodes. The mechanical behavior of this cloth model is consistent and is independent of the resolution. Kirchhoff-Love (KL) thin shell theory is used as the basis of the cloth model. Approaches to the parametrization and boundary sewing problems are presented to suit with meshless models. Furthermore, a corotational method is proposed in order to take care of large deformation problems. As for the collision solution, a new shape-function-based collision detection method is developed for meshless parameterized surfaces. The experimental results show that our cloth simulation model based upon meshless methods can produce natural and realistic results.

Keywords: Cloth simulation, meshless methods, collision handling.

1 Introduction

Cloth simulation has become a very popular research topic in recent years in computer animation and textile industry domain.

Till now, mass-spring methods are the traditional models for cloth simulation. However this new meshless method has advantages that the mass-spring methods do not have. The mass-spring approach has the advantage of easy implementation and low computation. But it has the natural drawbacks related to non-continuum configuration. For instance, the material cannot be simulated consistently and the results depend on the mesh resolution; the spring parameters do not reflect well the physical behavior of textile. In the textile industry realistic cloth behavior is required. A possibility relies on the use of continuum mechanics such as finite element methods [7] or meshless methods [4], to solve the problem. By continuum methods, material behavior can be reproduced accurately, independently of discretization.

Meshless methods have been introduced in computer graphics in recent years and gained increasing attentions as alternative computational methods [8,12] to the traditional mesh-based methods, such as FEM . Meshless methods have

A. Gagalowicz and W. Philips (Eds.): MIRAGE 2009, LNCS 5496, pp. 308–320, 2009.

attracted more and more attention due to their flexibility in solving engineering problems. Among these methods, Meshless Local Petrov-Galerkin method (MLPG) [2] has been considered as a general framework or a general basis for the other meshless methods [1].

In this paper, we present a new meshless cloth simulator with Kirchhoff-Love (KL) thin shell theory. The special features of cloth require the use of a thin shell approach which bring several problems to traditional meshless methods. For instance, cloth has different stiffness as in-plane and bending directions. In most of cloth simulation approaches, the treatments of bending models is done by an angular expression which is not accurate [13]. This means that realistic material parameters and resolution independence cannot be expected. However, our method can provide both accuracy and continuum representation. The discretization is based upon a meshless method, which means that the discretization is independent of the geometric subdivision into finite elements. The requirements of consistency are met by the use of a polynomial basis of quadratic or higher order.

Another advantage of meshless models with KL theory is that the parametric coordinates are fixed during the dynamic simulation, due to the fact that the relative position of neighbor nodes on the background of the cloth surface will not change unless the cloth topology is changed (e.g. tearing cloth). As general meshless methods without KL model (i.e. without parametric space) use the global Euclidian space as background coordinate system, so they don't have fixed background coordinates when displacement exists. The fixed parametric coordinates of KL model speeds up the search of local neighbors of nodes and simplifies the 3 dimensional problem to be solved to a 2 dimension one.

When large deformations are involved, the nonlinear equations make the simulations costly. The finite strain, known as geometrical nonlinearity, is closely linked to the invariance of the measure under rotations. We use co-rotational formulation to attach the parameterized local coordinate system of nodes. In addition, we compute the rotation field by an efficient iteration scheme. This allows us to use stable co-rotated strains.

The collision problem is a difficult problem for the meshless method, since the model has not the explicit connections and triangles. It makes the traditional collision detection invalid. We propose a detection method based upon the moment matrix from shape functions. The shape functions construct the meshless approximation and provide a natural indicator to track the surface. The detection method presented in this paper can detect a contact region using a simple criterion.

This meshless cloth simulation approach can both profit by the mechanics foundation and produce realistic simulation results.

2 Meshless Models for Cloth Simulation

Kirchhoff-Love (KL) theory is used to define the cloth surface and a meshless discretization and interpolation of KL surface is also presented to simulate physical behavior of cloth.

2.1   Thin-Shell Model

The Kirchhoff-Love (KL) theory assumes the shell to be thin which suits for cloth. The shell in the 3D space is described in a global cartesian coordinate system $\mathbf{E}^I$. The idea of the thin shell is mapping the 3D cloth space to a 2D thin shell space since cloth is thin. Figure 1 illustrates the model of KL thin shell theory. The pair $(\varphi, \mathbf{a}_3)$ defines the position of an arbitrary point of the shell, φ gives the position of a point on the shell mid-surface, and $\mathbf{a}_3$ is a unit vector (normal to the shell surface). The configuration Ω can be put down as

$$\Omega = \left\{ \mathbf{x} \in R^3 \middle| \mathbf{x} = \varphi(\xi^1, \xi^2) + \xi^3 \mathbf{a}_3(\xi^1, \xi^2) \right\}, \tag{1}$$

with $\xi^1, \xi^2 \in \Lambda$ and $\xi^3 \in\, <h^-, h^+>$. Here Λ denotes the parametric surface, $<h^-, h^+>$ are the distances of the "lower" and "upper" surfaces of the shell from the reference surface. ξ^1, ξ^2 can be used to describe the cloth surface.

In the parametric surface Λ, we define

$$\mathbf{a}_\alpha = \frac{\partial \varphi(\xi^1, \xi^2)}{\partial \xi^\alpha}, \ \mathbf{a}_3 = \frac{\mathbf{a}_1 \times \mathbf{a}_2}{|\mathbf{a}_1 \times \mathbf{a}_2|}, \tag{2}$$

where Greek indices take the values 1 and 2. The details and deductions can be found in [15].

2.2   Meshless Interpolation

For a so-called meshless implementation, a meshless interpolation scheme is required, in order to approximate the trial function over the solution domain.

Figure 1 illustrates the KL thin shell representation and the meshless sub-domains. Consider a sub-domain Ω_s, the neighborhood of a point $\mathbf{x}$, which is local in the solution domain. To approximate the distribution of function u in Ω_s, i.e. position, displacement or body force, over a number of scattered nodes

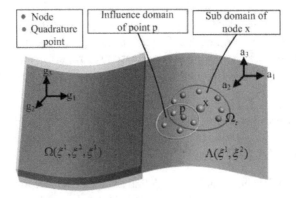

Fig. 1. Thin shell model and discretization

$\{\mathbf{x}_i\}$, $(i = 1, 2, ..., n)$ (with ξ_i^1, ξ_i^2 as the parametric coordinates), the local interpolation u of $\forall \mathbf{x} \in \Omega_s$ (with ξ^1, ξ^2 as the parametric coordinates), by augmented radial basis function (RBF) can be expressed in a standard form. With $\boldsymbol{\Phi}(\xi^1, \xi^2)$ being the shape functions:

$$\boldsymbol{\Phi}^T(\xi^1, \xi^2) = \left[\mathbf{R}^T(\xi^1, \xi^2), \mathbf{P}^T(\xi^1, \xi^2)\right]_{(n+m)} \mathbf{G}_{(n+m)*(n+m)}, \tag{3}$$

where the square brackets denote to composite a new matrix or vector by a set of elements. A subscript with round brackets denotes the dimension of the matrix or vector. $\mathbf{R}^T(\xi^1, \xi^2) = \left[R_1(\xi^1, \xi^2), R_2(\xi^1, \xi^2), ..., R_n(\xi^1, \xi^2)\right]_{(n)}$ is a set of RBFs centered around the n scattered nodes, $\mathbf{P}^T(\xi^1, \xi^2) = \left[p_1(\xi^1, \xi^2), p_2(\xi^1, \xi^2), ..., p_m(\xi^1, \xi^2)\right]_{(m)}$ is a monomial basis of order m. The monomial basis is used to overcome the lack of completeness. $\mathbf{G}$ is a matrix composited by R and p at the scattered nodes:

$$\mathbf{G} = \begin{bmatrix} \mathbf{A} & \mathbf{B} \\ \mathbf{B}^T & 0 \end{bmatrix}_{(n+m)*(n+m)}^{-1},$$

the elements of matrix $\mathbf{A}$ and $\mathbf{B}$ are defined as:

$$A_{i,j} = R_j(\xi_i^1, \xi_i^2), \; B_{i,j} = p_j(\xi_i^1, \xi_i^2), \; i,j = 1, 2, ..., n. \tag{4}$$

The interpolation of u can be expressed as:

$$u(\xi^1, \xi^2) = \sum_{i=1}^{n} \Phi_i(\xi^1, \xi^2) u_i, \tag{5}$$

where Φ_i is the ith element of the shape functions $\boldsymbol{\Phi}$ and u_i is the given value of function u at node i. Note that to evaluate the derivatives of the function $u(\xi^1, \xi^2)$ one needs only to differentiate the shape functions $\boldsymbol{\Phi}(\xi^1, \xi^2)$. This means that the smoothness of the displacement approximation depends on the smoothness of the shape function. To compute the strains one needs only second order derivatives of displacement with respect to the parameters ξ^1, ξ^2.

In comparison, the finite element method approximates the field variables within an element by interpolating their values only at the nodes on the element. For non-continuum methods such as mass-spring methods, variables of a node are computed explicitly and do not have interpolation schemes.

2.3   Acquisition and Parametrization of Models

The model data of the cloth model could be obtained from a CAD system. To form a garment in a CAD system, all the pieces of cloth 2D patterns are defined and sewing information is added. 2D patterns are simple shapes. In the parametric surface, a 2D pattern should have:

- An exterior boundary described by a single, non-self-intersecting curve of arbitrary shape.
- Any number of interior boundaries (holes) of arbitrary shape without self-intersection.

So that the intrinsic coordinates of each patch are single-valued.

In order to be able to use the meshless surface approximation, the surface must be parameterized, i.e. the parametric surface Λ of equation (1) must be defined to build the shape function in ξ^1, ξ^2.

A good choice seems to be the intrinsic coordinates defined for simple surfaces (e.g. quadrics), but they are difficult to define for general surfaces (e.g. the cylinder shape of a sleeve). The approach used in the present work is based upon the fact that a polygonalization of the cloth surface at hand is usually readily available in form of the 2D patterns that buildup arbitrary garments, with any number of holes and both closed and open surfaces. Therefore, the direct and simple way to parameterize the cloth surface is to use the warp and weft directions on every single patch. Therefore, the parametrization is easy to implement. Another advantage of using the warp and weft directions is that from these two directions, it is easy to build an anisotropic cloth.

2.4   Border Sewing

A garment can be obtained by combination of one or several 2D patterns. The first step to construct the garment is to identify the position of the pieces in the 3D model and to sew them together. Then we need to connect the sewing line in the cloth in order to have a continuum movement. Lagrange multipliers can be used to constrain the sewing points to have same displacement. As discussed before, although cloth has been sewed together, each patch needs to be parameterized separately, i.e., for a garment composited by several pieces of cloth, the shape function needs to be computed in each single piece, and not computed across the neighboring pieces. The global stiffness matrix can be combined with all pieces of the cloth. Because the Lagrange Multipliers enlarge the global stiffness matrix, we only use lagrange multipliers method to constrain the sewing borders. Since the constraint condition is an implicit constraint, we can get the explicit condition after the equation is solved. As for another type of constraint which is used for solving the collision response, we use the method presented by Baraff[3]. This method applies the filtering process to the stiffness matrix and velocity matrix, which is an explicit constraint and will not enlarge the stiffness matrix.

Compared to the method based upon mesh, such as mass-spring method or finite element method, the meshless method has great advantages in dealing with the variational topology, since it does not need to destroy and rebuild the meshes when topology changes. While for the mass-spring method and the finite element method, the re-meshing processing will cause big problems to the computation.

3   Collision Detection of Meshless Cloth Models

Collision is an essential aspect in cloth simulation. As the meshless methods do not maintain the mesh of the surface, the only way to use the conventional mesh-based collision detection methods is to define segments between nodes in neighborhood and use the segment-to-segment detection in self-collision of cloth or segment-to-face detection of cloth-body collision (suppose that only cloth is simulated in meshless, i.e., the body is based upon polygonal meshes). Unfortunately, in self-collision of cloth, segment-to-segment detection alone is not sufficient. The problem will be resolved if we have a segment-to-meshless-surface collision detection method.

Therefore, we proposed a new segment-to-meshless-surface collision detection method based upon shape functions for meshless surfaces. The segment in this method can be defined both from two nodes in a local domain of cloth and from a segment of mesh-based body model. This method should be applied in one frame, that only the current configuration is needed for collision detection.

At first we assume a general parametric surface $\overset{\frown}{\Lambda}$ which is defined by all the discrete nodes via the approximation function (RBF in our implementation) over the local domain of cloth. We get $\Lambda \subset \overset{\frown}{\Lambda}$ and $\forall M \in \overset{\frown}{\Lambda}$ is defined with the discrete point in Λ.

Take a segment PQ as an example, as shown in figure 2. If PQ penetrates $\overset{\frown}{\Lambda}$, we assume the intersection point is M. Then the problem can be separated into two steps. At first we determine whether the intersection point M exists and compute the parametric coordinates of M; Secondly, on $\overset{\frown}{\Lambda}$, determine whether M is inside Λ.

At first we search a node X as a minimum of $|1 - \frac{XP \cdot \mathbf{a}_3^X}{|XP|}|$ and $X \neq P$, which implies that PX almost perpendicular to the sub-surface near X. Define P' as the real projection of P on $\overset{\frown}{\Lambda}$, i.e. $P = P'(\xi^1, \xi^2) + \xi \mathbf{a}_3^{P'}(\xi^1, \xi^2)$. Therefore P' is near to X and the parametric coordinates of P' can be written as $(\xi_X^1 + XP \cdot \mathbf{a}_1^X, \xi_X^2 + XP \cdot \mathbf{a}_2^X)$. Then we can easily draw the conclusion that if $XP \cdot \mathbf{a}_3^X$ have the same sign with $QP \cdot \mathbf{a}_3^X$ and $|XP \cdot \mathbf{a}_3^X| \geq |QP \cdot \mathbf{a}_3^X|$ (note that $|XP \cdot \mathbf{a}_3^X| = |MP \cdot \mathbf{a}_3^X|$), then PQ will have an intersecting point M with $\overset{\frown}{\Lambda}$, and the position of M will be $M = P + PQ\frac{|PM|}{|PQ|} = P + PQ\frac{XP \cdot \mathbf{a}_3^X}{PQ \cdot \mathbf{a}_3^X}$. So the parametric coordinates of M can be expressed as $(\xi_X^1 + XM \cdot \mathbf{a}_1^X, \xi_X^2 + XM \cdot \mathbf{a}_2^X)$. Figure 2 illustrates the algorithm of finding M as described above.

So soon as we get the parametric coordinates of M, the second step is simplified to a collision problem of meshless domain in a two dimensional space.

In our meshless interpolation methods, the construction of shape function of RBF method requires the computation of matrix

$$\mathbf{N} = \mathbf{G}^{-1} = \begin{bmatrix} \mathbf{R}_0 & \mathbf{P}_0 \\ \mathbf{P}_0^T & 0 \end{bmatrix} \qquad (6)$$

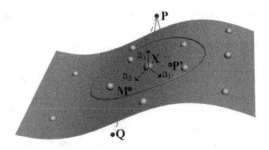

Fig. 2. Collision detection of meshless surface

It has been found that this matrix provides a natural indicator to track the surface of the continuum object. Based upon the matrix $\mathbf{N}$ from shape function, the collision detection in the two dimensional parameterized space can be done simply and accurately. [11] proved that the internal position of a continuum domain and the external can be distinguished by checking the determinant of the matrix $\mathbf{N}$. Inside the continuum domain of any shape, $det\,\{\mathbf{N}(x)\}$ has a positive value, and outside the domain $det\,\{\mathbf{N}(x)\} \to 0$. The usefulness of above property is that one can accurately track the position of any continuum without knowing the exact shape of its boundary, which is almost impossible to know in general meshless models. Therefore we can track if the intersection point M is on the internal cloth domain or on the outside (or holes on cloth).

The collision response is available through different approaches. Physically, it can be done by adding a penalty force or constraints between the contacting nodes. The $\mathbf{a}_3$ of the intersection point M (choose between $\mathbf{a}_3$ and $-\mathbf{a}_3$ according to the position of P) can be used as an indication of the direction to eliminate collisions and $\tau = \frac{|PQ|}{|PM|}$, the ratio of crossed part of PQ, as an indication of the amount of penalty force or position of constraints.

4   System Solving

As deduced in [15], the Green-Lagrange strain tensor of the shell is found to be of the form:

$$E_{IJ} = M_{IJ} - \xi B_{IJ}, \tag{7}$$

the non-zero components of the tensors M_{IJ} and B_{IJ} are in turn related to the deformation of the shell, where in this section uppercase Latin indices take the values 1, 2 and 3. The following contents will use upper-script to denote variables in the reference configuration. For instance $\bar{\varphi}$ is a point on the reference surface; and in the current configuration, variables are without upper-scripts.

we can write the non-zero part of in-plane and bending strains as:

$$M_{\alpha\beta} = \frac{1}{2}(\bar{\mathbf{a}}_\alpha \cdot \mathbf{u}_{,\beta} + \mathbf{u}_{,\alpha} \cdot \bar{\mathbf{a}}_\beta), \tag{8}$$

$$\begin{aligned} B_{\alpha\beta} = &-\mathbf{u}_{,\alpha\beta} \cdot \bar{\mathbf{a}}_3 \\ &+ \frac{1}{|\bar{\mathbf{a}}_1 \times \bar{\mathbf{a}}_2|}[\mathbf{u}_{,1} \cdot (\bar{\mathbf{a}}_{\alpha,\beta} \times \bar{\mathbf{a}}_2) + \mathbf{u}_{,2} \cdot (\bar{\mathbf{a}}_1 \times \bar{\mathbf{a}}_{\alpha,\beta})] \\ &+ \frac{\bar{\mathbf{a}}_3 \cdot \bar{\mathbf{a}}_{\alpha,\beta}}{|\bar{\mathbf{a}}_1 \times \bar{\mathbf{a}}_2|}[\mathbf{u}_{,1} \cdot (\bar{\mathbf{a}}_2 \times \bar{\mathbf{a}}_3) + \mathbf{u}_{,2} \cdot (\bar{\mathbf{a}}_3 \times \bar{\mathbf{a}}_1)], \end{aligned} \tag{9}$$

where α and β take the values 1 and 2, and a comma followed by a subscript indicates partial differentiation with respect to the corresponding coordinate. The only variable of equation (8) and (9) is the displacement field $\mathbf{u}$ of the middle surface, therefor $\mathbf{u}$ furnishes a complete description of the deformation of the shell. So we regard $\mathbf{u}$ as the primary unknowns of the analysis. It follows from these relations that, by virtue of the assumed Kirchhoff-Love kinematics, all the strain measures of interest may be deduced from the deformation of the middle surface of the shell.

Owning to the linear relationship between M, B and $\mathbf{u}$ from equation (8) to (9), and the approximation of displacement $\mathbf{u}$ by the nodal displacement $\mathbf{u}_I$ in equation (5), we obtain:

$$M(\xi^1, \xi^2) = \sum_{i=1}^{n} \mathbf{R}_m^i(\xi^1, \xi^2)\mathbf{u}_i, \tag{10}$$

$$B(\xi^1, \xi^2) = \sum_{i=1}^{n} \mathbf{R}_b^i(\xi^1, \xi^2)\mathbf{u}_i, \tag{11}$$

where $\mathbf{R}_m^i$ and $\mathbf{R}_b^i$ are matrices relating nodal displacement to in-plane and bending strain.

The dynamics equation in form of the second order ordinary differential equation in time is:

$$\mathbf{M}\ddot{\mathbf{u}} + \mathbf{D}\dot{\mathbf{u}} + \mathbf{K}\mathbf{u} = \mathbf{f}, \tag{12}$$

where $\mathbf{M}$ is the diagonal nodal mass matrix, $\mathbf{D}$ is the viscosity matrix and $\mathbf{K}$ is the stiffness matrix, with the nodal displacement vector $\mathbf{u}$ and forces vector $\mathbf{f}$.

In MLPG approaches [2], one may write a weak form over a local sub-domain Ω_s of a point $\mathbf{x}_k$, which may have an arbitrary shape. As deduced in [15], after definition of the in-plane and bending strains and stress, the elastic strain energy can be rewritten, and a generalized local weak form corresponding to the stiffness matrix and nodal force vector of equation (12) will be:

$$\mathbf{K}_k^{IJ} = \int_{\Omega_s} \left[(\mathbf{R}_m^I)^T \mathbf{H}_m \mathbf{R}_m^J + (\mathbf{R}_b^I)^T \mathbf{H}_b \mathbf{R}_b^J \right] d\Omega, \tag{13}$$

$$\mathbf{f}_k^I = \int_{\Omega_s} \Phi_I \mathbf{q} d\Omega. \tag{14}$$

where $\mathbf{H}_m$ and $\mathbf{H}_b$ are matrices corresponding to the in-plane and bending part of the material law, which are the same as finite element method [6]. $\mathbf{q}$ denotes the applied force. The stiffness matrix is evaluated by numerical integration in the local domain Ω_s. Usually, Gaussian integration will be applied. Due to the arbitrary shape of the local domain, we do not need any mesh or background mesh for the integration of the weak forms.

4.1   Co-rotation Deformation

When considering a finite strain, the strain is nonlinear which leads to a nonlinear system. The solving of a nonlinear system is very costly. The co-rotational formulation aims at the elimination of the geometrical nonlinearity. The idea is to keep track of a rotated local coordinate system of every nodes of the body.

We linearized the strain so the strain in the former kinematic description is linear in displacement but not rotationally invariant anymore. However, if the rotation field R is known, the co-rotational strain formulation can be used and we obtain the rotated linear strain tensor on the rotated current configuration.

Via polar decomposition the deformation gradient tensor F can be split into a rotational tensor R and a pure deformation U as $F = RU$.

[9] proposed an efficient, quadratically convergent iteration scheme to extract rotation field from deformation gradient

$$R^0 := F \tag{15}$$

$$R^{n+1} := \frac{1}{2}\left(R^n + (R^n)^{-T}\right). \tag{16}$$

This allows a very fast and accuracy controlled method of computing R. The iteration is defined for square, nonsingular matrices only, but [9] also presents a preliminary QR decomposition enables the treatment of singular matrix.

With the rotated field $\mathbf{R}$, the stiffness matrix $\mathbf{Ku}$ becomes $\mathbf{RKu}^R$, where in the nodal view of node I, $u_J^R = R_I^T(\bar{x}_J + u_J - x_I) + (x_I - \bar{x}_J)$ is the displacement influence nodes J. Therefore, the dynamics system equation can be rewritten as

$$\mathbf{M\ddot{u}} + \mathbf{D\dot{u}} + \mathbf{RKR}^T\mathbf{u} = \mathbf{f} + \mathbf{RK}(\mathbf{R}^T - \mathbf{I})(\mathbf{x}_{node} - \mathbf{x}). \tag{17}$$

Now the system to be solved in each time step remains linear. Compared to classical linear meshless methods, the linear system change over time, when the reference configurations are updated.

Compared to co-rational approach of finite element [10] which remove rotations for each triangle separately, we treat rotations for each node on the local sub-domain and produced a continuous co-rotated deformation field instead of the element separated non-continuous field in finite element method. The computing of polar decomposition in this paper is carried out by a fast iteration method, while [10] compute the eigenvalues and eigenvectors of U^2.

4.2   Efficient Solving

For numerical solutions, the dynamics equations are translated into first order ODEs and any time integration scheme can be applied. The coefficient matrix of the arising linear system of equations with nodal velocities as primary unknowns is sparse and does have an almost symmetric distribution of non-zeros. We call this kind of matrix a quasi-unsymmetric sparse matrix (QUSM), which has a restriction of symmetric nonzero locations compared to general unsymmetric matrices. For simplicity only the system stiffness matrix which has positive principal minors is considered. In fact QUSM provides most of the advantages of symmetric matrix in the solution process.

General sparse solvers may be mainly divided into two categories: symmetric solvers and unsymmetric solvers. The symmetric solvers do not suit for meshless methods. Although a QUSM can be treated as general unsymmetric matrix and solved by general unsymmetric solvers, the solver proposed in this paper takes advantage of the properties of QUSM arising in meshless methods and delivers significantly higher efficiency.

In order to solve the system, we developed a direct solution method for the QUSM arising in the meshless methods. The new solver provides higher efficiency for LDU factorization on benchmark tests, so it speeds up the solution processes for linear systems of equations. This solver is accelerated by two-level unrolling techniques that employ the concept of master equations and searches for appropriate depths of unrolling during factorization (see details in [14]). The new solver for the QUSM can increase the efficiency of the solving of meshless methods.

5   Experimental Results

In this section we present our experimental results on our meshless cloth simulator. In our algorithm, since the shape function is computed in the parameterized surface Λ, $\mathbf{R}(\xi^1, \xi^2)$ and $\mathbf{P}(\xi^1, \xi^2)$ and $\mathbf{G}$ for a point is constant. So we store them in order to accelerate the whole computation process. When computing the neighbors id of x in the sub-domain Ω_s, the searching computation can be processed only once in the initialization. These conditions are due to the fact that the parameterized space and surface Λ is constant.

As for the selection of material parameter, the elasticity coefficient (Young's modulus) $E = 10000\text{N/m}$ and Poisson's ratio $\nu = 0.3$. The real parameters in continuum mechanics present the behavior of material constantly and directly, but they are hard to be obtained though experiments for cloth. Cloth has nonlinear curves of the parameters. Solutions by Kawabata evaluation system (KES), which contains a group of standard equipments which measure the physical parameters of the textile, have already developed for mass-spring methods [5]. KES includes 16 parameters, containing stretching, shearing, bending, compression and friction aspects of the textile. These parameters reflect the physical

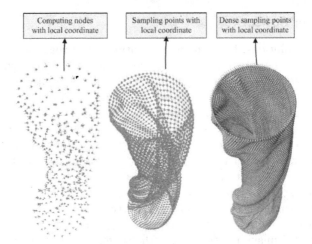

| Computing nodes with local coordinate | Sampling points with local coordinate | Dense sampling points with local coordinate |

(a) Computing nodes and sampling points

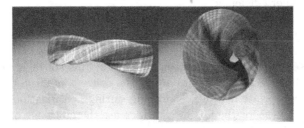

(b) Results of cylindrical sleeves with twisting forces

(c) Results of a cloth dropping on a pole

(d) Results of a cylindrical cloth draping on its middle line

Fig. 3. Results of meshless clothes

character of the specific textile. Therefore, with these KES parameters we can get deep understanding of the textile behavior and make the animation more vivid. KES parameters are considered to adopt in our meshless cloth simulation system in the future work, in order to make the animation meets more physical behavior. Nevertheless, the Poisson's ratio cannot be obtained from KES. The relationship between Poisson's ratio and KES parameters should be constructed in order to make the animation have more physical fundamental.

In the first example, we simulate a cylindrical sleeve constrained by a twisting force. Cylindrical sleeves are often used in cloth simulation research and can well judge the validity for buckling and folds for the simulator. Cylindrical sleeves are also the basic elements of virtual garments and available results can prove potential capabilities for virtual try-on. Figure 3(a) shows the original computational nodes with the local coordinates drawn in each node, and the thicker sampling points for rendering. In figure 3(b), the folds are clearly reproduced and the process of twisting produces realistic behaviors as observed with the real fabrics. The realistic shapes of meshless cloth simulation results are obtained because we use continuum physics and consistent way to model cloth. Sampling and rending of meshless surfaces can take advantage of the research achievements in point based graphics.

Another experiment is the cloth dropping on a pole. Figure 3(c) shows the dynamic behavior of the results. Figure 3(d) shows a cylindrical cloth without sewing which is hanging on its middle line. The examples below show small buckles and wrinkles produced by the simulator.

6 Discussion

This paper developed a meshless cloth simulation method combining the KL thin shell representation. The system is solved with collision handling. Meshless methods are usually computationally costly but as a continuum method, it has a more smooth interpolation field and a natural mechanical behavior. Nevertheless, the computation cost was reduced by the fact that KL model reduces a 3 dimensional cloth to a 2 dimensional parametric space as shown in Fig. 1.

We have to mention that although meshless methods present many aspects positive compared to mass-spring methods on cloth simulation, a major difference with the mass-spring method is that it needs to overcome the large deformation problem using co-rotational formula. While non-continuum methods as mass-spring methods have a natural property to treat large deformation because they measure strains by the state of springs instead of the difference of current configuration and reference configuration in continuum mechanics.

Overall, the proposed meshless method for virtual cloth simulation has theoretical fundamentals with the thin shell theory and meshless solving, at the same time the experimental results proves that this method can make the natural cloth simulation.

References

1. Atluri, S.N., Shen, S.: The basis of meshless domain discretization: the meshless local Petrov-Galerkin (MLPG) method. Advances in Computational Mathematics 23, 79–93 (2005)
2. Atluri, S.N., Zhu, T.: A new meshless local Petrov-Galerkin (MLPG) approach in computational mechanics. Comput. Mech. J. 22, 117–127 (1998)
3. Baraff, D., Witkin, A.: Large steps in cloth simulation. In: SIGGRAPH 1998: Proceedings of the 25th annual conference on Computer graphics and interactive techniques, pp. 43–54. ACM Press, New York (1998)
4. Chang, J., Zhang, J.: Mesh-free deformations. Computer Animation and Virtual Worlds 15(3-4), 211–218 (2004)
5. Charfi, H., Gagalowicz, A., Brun, R.: Viscosity damping parameters of fabric related to a non-linear textile model. Textile Research Journal 76, 787–798 (2006)
6. Cirak, F., Ortiz, M., Schroder, P.: Subdivision surfaces: a new paradigm for thin-shell finite-element analysis. International Journal for Numerical Methods in Engineering 47(12), 2039–2072 (2000)
7. Etzmuß, O., Keckeisen, M., Straßer, W.: A fast finite element solution for cloth modelling. In: PG 2003: Proceedings of the 11th Pacific Conference on Computer Graphics and Applications, Washington, DC, USA, p. 244. IEEE Computer Society Press, Los Alamitos (2003)
8. Guo, X., Li, X., Bao, Y., Gu, X., Qin, H.: Meshless thin-shell simulation based on global conformal parameterization. IEEE Trans. Vis. Comput. Graph 12(3), 375–385 (2006)
9. Higham, N.J., Schreiber, R.S.: Fast polar decomposition of an arbitrary matrix. SIAM Journal on Scientific and Statistical Computing 11(4), 648–655 (1990)
10. Keckeisen, M.: Physical Cloth Simulation and Applications for the Visualization, Virtual Try-On, and Interactive Design of Garments. PhD thesis, Eberhard-Karls-Universitat Tubingen (2005)
11. Li, S., Qian, D., Liu, W.K., Belytschko, T.: A meshfree contact-detection algorithm. Computer Methods in Applied Mechanics and Engineering 190, 3271–3292 (2001)
12. Pauly, M., Keiser, R., Adams, B., Dutré, P., Gross, M., Guibas, L.J.: Meshless animation of fracturing solids. ACM Trans. Graph. 24(3), 957–964 (2005)
13. Thomaszewski, B., Wacker, M., Straßer, W.: A consistent bending model for cloth simulation with corotational subdivision finite elements. In: SCA 2006: Proceedings of the 2006 ACM SIGGRAPH/Eurographics symposium on Computer animation, Aire-la-Ville, Switzerland, pp. 107–116. Eurographics Association (2006)
14. Yuan, W., Chen, P., Liu, K.: A new quasi-unsymmetric sparse linear systems solver for meshless local Petrov-Galerkin method (MLPG). CMES: Computer Modeling in Engineering & Sciences (Journal) 17(2), 115–134 (2007)
15. Yuan, W., Chen, Y., Liu, K., Gagalowicz, A.: Application of Meshless Local Petrov-Galerkin (MLPG) method in cloth simulation. CMES: Computer Modeling in Engineering & Sciences (Journal) 35(2), 133–156 (2008)

New Human Face Expression Tracking

Daria Kalinkina and André Gagalowicz

INRIA-Rocquencourt, Domaine de Voluceau BP105 78153 Le Chesnay, France
{darya.kalinkina,andre.gagalowicz}@inria.fr

Abstract. In this paper we propose a new method for precise face expression tracking in a video sequence which uses a hierarchical animation system built over a morphable polygonal 3D face model. Its low-level animation mechanism is based upon MPEG-4 specification which is implemented via local point-driven mesh deformations adaptive to the face geometry. The set of MPEG-4 animation parameters is in its turn controlled by a higher-level system based upon facial muscles structure. That allows us to perform precise tracking of complicated facial expressions as well as to produce face-to-face retargeting by transmitting the expression parameters to the different faces.

1 Introduction

Tracking of human face is an essential task for a large number of applications in computer vision, such as teleconferencing, post-production for the film industry, video games, human-computer interaction and others. Although this research domain has received much attention over the last twenty years, reliable human face tracking in a complex environment still remains a challenge due to the great variations of the appearance of the face within a video sequence which can be caused by changes in facial expression, head movements, variations of lighting properties, etc.

There have been many publications dedicated to recovering facial motion from video – from simple coarse face localization to precise detection of the head pose with or without facial expressions.

Methods of coarse localization are generally based on probabilistic frameworks and use skin color distribution [10] or more sophisticated region-based level-set segmentation approaches [11]. Being very fast they can provide only planar head displacements and easily lose their target in case of occlusions since they don't possess any information about the properties of the 3D shape of the object they track.

There are a big number of techniques that use vision-based approaches for tracking also in 3D, and some of them are in addition capable to recognize seven basic facial expressions [20]: happiness, sadness, surprise, fear, anger, disgust, and the neutral one. Methods of this kind can for example be based on feature or edge detection [12, 13] (here a simplified 3D model is used to determine 2D-3D correspondences needed for camera calibration and thus to estimate the spatial position of the object in each frame), deformable patches [17], active contours [14, 15] or active shape models [16].

Usage of the 3D face model may give more flexibility and robustness to the tracker. In [18] authors employ a technique similar to ours by using a texture-based criteria for pose detection but their model is simplified to an ellipse roughly adapted

A. Gagalowicz and W. Philips (Eds.): MIRAGE 2009, LNCS 5496, pp. 321–331, 2009.

to the face dimensions during initialization. The same basic idea but with a more sophisticated parameterized 3D model is proposed in [19] which allows detecting not only the head pose, but also the speech-related facial mimics. In [23] optical flow is used as a constraint on the motion of a deformable 3D model. Technique based upon the active appearance models [21, 22] works in real-time but is sensitive to lighting changes and occlusions and requires a preliminary learning process for every new face to track. In respect to parameterization of facial expressions a large group of methods can be distinguished that represent non-rigid deformation of the face as a linear combination of several basic shapes defined directly on the mesh [25, 27], or captured from special training video sequences [26]. Providing this way a compact representation for non-rigid tracking parameters, this approach however doesn't cover the whole range of possible facial expressions.

Many existing algorithms give good tracking results but still lack in precision that is necessary for post-production purposes, especially in capturing complex facial expressions. Development of an algorithm that would allow a completely automatic and robust realistic face-to-face retargeting from any kinds of video (without any constraints on the head movements or environment) still remains an open problem.

Our work focuses in particular on precise head pose and facial expression detection aiming at further use of tracking results in rotoscoping. In order to be able to detect complex expressions we've adopted a highly parameterized MPEG-4 face animation mechanism as a basis of our animation system. In addition our implementation is completely compliant with the virtual conversational agent "Greta" developed by the team of prof. Catherine Pelachaud [4] but unlike Greta's one it is parameterized with respect to the face geometry which makes it applicable to any kinds of faces produced on basis of our generic face model. This gives us two advantages: firstly it allows to refine our tracking algorithm by validating it on synthesized Greta-animated video sequences and secondly – to do the retargeting of any facial expression from the rich hierarchical semantic library of facial expressions that is used for Greta's animation to other faces.

The remainder of this paper is organized as follows. In the next section we will review the work already done in this field. Then we'll shortly describe the general idea of the tracking algorithm. In section 3 we'll speak about our face animation system and its employment in tracking. Section 4 is dedicated to GPU-acceleration of the algorithm, and, finally, the last section presents some results and future research directions.

2 Face Tracking Framework

The goal of the proposed algorithm is to perform automatic tracking of the human face in a video sequence providing as an output the head position and the facial expression in each image of the sequence. The general scheme of the algorithm framework is presented in Fig. 1. It is based upon an analysis/synthesis collaboration approach using a textured 3D model as a tool to detect face position and expression in each frame of the sequence by minimizing the mean-square error between the generated synthetic image of the face and the real one.

The performance of tracking (precision and stability of detection) relies heavily on the precision of the 3D model used. Such a 3D specific model is obtained through the deformation of a generic one and is positioned interactively on the first image of the sequence.

The following subsections will describe all these procedures step by step.

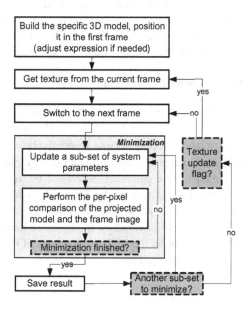

Fig. 1. Global tracking framework

2.1 System Initialization

The very first step of the algorithm is interactive and consists of building the specific 3D model of the person to be tracked. In order to perform full tracking of the head pose including strong rotations we need to construct a precise model and so some additional images of the person's head taken from different views are required. However if the face in the video always remains in the frontal position we can directly use only one of the sequence images.

The reconstruction of the specific face is performed in a semi-automatic way and consists of two basic steps. First, we adapt the generic model at the feature point level by iteratively performing 3D stereo reconstruction of the 2D characteristic points defined manually on the images followed by recalibration of the cameras related to each image. After that the adaptation is carried out at the silhouette level by matching the projection of the 3D contours of the model with the manually drawn 2D contours of the face on the image. At each step we compute the deformation vector for only a subset of vertices, interpolating the deformation over the rest of the mesh by the means of radial basis functions. The detailed description of the 3D reconstruction process which can be found in [1, 2].

As soon as the specific model has been obtained it acquires texture from the first frame of the sequence. Thus the algorithm adapts to the video capture lighting conditions. To do the texturing, the model should be accurately positioned on the sequence image so that the texture coordinates for all the visible vertices could be computed by projecting them onto the image plane. Being interactive, the positioning is however facilitated by using the automatic calibration system POSIT based upon 2D-3D point correspondences [6]. Due to occasional changes in lighting or possible head rotations that make appear previously hidden and therefore untextured facets, this procedure is repeated every 100 frames which is normally equal to 4 sec. of video stream.

2.2 Iterative Pose and Expression Detection

For each frame of the video sequence our algorithm searches for an optimal set of the system parameters $\sigma = (t, r, \lambda)^T$, where $t = (t_x, t_y, t_z)^T$ and $r = (r_x, r_y, r_z)^T$ are the global head translation and rotation vectors respectively and λ is the vector of parameters that control facial expression, by using an iterative minimization tool. The error function is represented by the per-pixel difference between the current video frame image and the textured projection of the 3D model which has been updated with respect to the new values of the parameter vector σ. Per-pixel difference is computed only for those pixels that are covered by the textured model projection and is represented by the Euclidean norm on the RGB components of the image. Mathematically the error function can be expressed as:

$$F_e = \frac{1}{N} \sum_{k=1..N} e_k(\sigma) \text{, where} \tag{1}$$

$$e_k(\sigma) = \|I(\sigma, k) - \hat{I}(\sigma, k)\|, \tag{2}$$

$I(\sigma, k)$ and $\hat{I}(\sigma, k)$ being the values of the kth pixel in the generated and the analyzed images respectively.

In order to avoid getting trapped in local minimum we use the simulated annealing minimization algorithm [8].

Due to the large total number of parameters to optimize we split them into several independent groups and perform minimization over each group separately. These groups are:

(1) rigid tracking (translation and rotation), which is performed in the first instance;
(2) expression tracking – mouth part (lips, jaw);
(3) expression tracking – eyes part (eyelids, eyeballs, eye squint);
(4) expression tracking – eyebrows part;
(5) expression tracking – nose, tongue (optional).

For groups (2)-(3) the error is computed only from the sub-image that represents the part of the face in question.

3 Facial Animation Parameterization

Our approach to the parameterization of the facial expression is based upon a hierarchical combination of two different animation systems: one is based upon the facial
animation parameters specified by MPEG-4 standard and the other – upon the facial
muscular structure. First we'll talk about how MPEG-4 standard defines the specification for the animation of a face model and our implementation of this specification,
and then we'll describe our combined animation system in detail.

3.1 Facial Animation System Based upon MPEG-4 Standard

In MPEG-4 standard facial animation is defined by two types of parameters [5]: a set
of Facial Definition Parameters (FDPs) that reflects the geometry of the 3D model,
and 68 Facial Animation Parameters (FAPs) that specify the animation part and are
closely related to muscle action. At the same time MPEG-4 only defines the action
related to each FAP leaving to the users its interpretation in terms of actual deformation of the 3D mesh.

As a basis of our implementation of the MPEG-4 facial animation system we've
chosen that one that is used for animating the virtual conversational agent "Greta"
created by the team of prof. Catherine Pelachaud [4]. This allowed us not only to
obtain parameters for expressive tracking but also to greatly refine tracking precision
in comparison with our previously used muscle-based parameterization [3] by learning from Greta's rich facial expression library.

In respect to the Greta's MPEG-4 implementation [9] we've constrained the action
of each FAP by splitting the surface of our generic model into 68 semantic zones,
each FAP affecting only a specified sub-set of them. The region of influence of each
FAP within this sub-set is in its turn defined as an ellipsoid centered at the control
point (FDP) corresponding to this FAP. The deformation magnitude d of each vertex
$v = \{x, y, z\}$ inside this ellipsoid can be expressed by the following formula:

$$d = \begin{cases} \frac{1}{1-cos(k)}(cos(k\rho) - cos(k)) & \text{if } \rho < 1 \\ 0 & \text{otherwise} \end{cases}, \text{ where} \qquad (3)$$

$$\rho = \sqrt{\left(\frac{x - x_0}{r_x}\right)^2 + \left(\frac{y - y_0}{r_y}\right)^2 + \left(\frac{z - z_0}{r_z}\right)^2} \qquad (4)$$

is the normalized distance to the control point $v_0 = \{x_0, y_0, z_0\}$; r_x, r_y and r_z are the
radii of the ellipsoids of influence and k is the parameter that defines the shape of the
deformation function. Being set as absolute values in Greta, r_x, r_y and r_z didn't allow
any transfer of Greta's animation mechanism to the facial models of other dimen-
sions/geometry because the visual effect of the animation of a FAP greatly depends
on the dimensions of its region of influence with respect to the 3D model (see Fig. 2
(a, b) for example). Since our goal was to build an animation system that would work
for any kind of faces produced from our generic model, we've created geometrical
dependencies for all the ellipsoids radii, so that the system could adapt automatically
to any face given as an input. At the same time our system should provide the same

effect on the Greta model reconstructed from our generic 3D face, as it does on real Greta. Video that demonstrates the validation of our adaptation of the Greta's MPEG-4 facial animation system on our generic model will be shown at the conference (it can be as well downloaded from ftp://ftp.inria.fr/incoming/mirages / daria/ comp_ v4. avi).

3.2 Hierarchical Animation System

Although MPEG-4 parameterization is potentially capable of reproducing a rich palette of realistic expressions, it has several significant drawbacks which makes it difficult to use FAPs directly as parameters for tracking. Firstly it gives too many parameters to minimize – especially in the mouth region where animation is controlled by a total of 28 FAPs! – that would lead to huge computing time and would make minimization more sensitive to local minimum traps. Secondly it doesn't provide any global constraints for the FDP's displacements: all of them cause only local deformation and are completely independent which means that a minimization tool will waste a lot of time on searching the optimum in the aggregate of non-existent facial expressions that cannot be reproduced by a real person. At last, the deformation function (3) being generalized doesn't take into account particular properties of the mesh geometry and thus is capable of producing unrealistic expressions as shown, for example, on Fig. 2 (c).

We've solved the problems described above by applying the specific spatial constraints onto the mesh deformation.

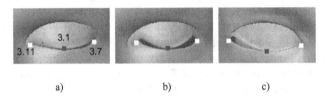

a)   b)   c)

Fig. 2. a) Correct animation of the eyelid (FDP 3.1 being the center of the deformation); b) Region of influence is not adapted to the mesh dimensions (here the ellipsoid of influence has a smaller X-dimension than it should; as a consequence the deformation becomes zero before reaching the borders of the eye – FDPs 3.11 and 3.7 – causing gaps between the under-eye part and the eyelid); c) Deformation function is not adapted to the mesh geometry (here the ellipsoid has the correct size but due to the specificity of the eyelid geometry the Y-deformation on the left should be stronger than on the right, so being uniform in all directions it leads to the gap visible on the left)

Mouth. In case of the mouth region our goal is to reduce the dimension of the parameter space. What we propose is to limit its degrees of freedom by constraining the displacements of the FAPs by the curves that are built on the 3D mesh surface with respect to the facial muscles structure in a way that all together they contain all the FDPs related to the mouth region (Fig. 3). Each curve is represented by a Bezier patch attached to the mesh vertices and therefore its shape can be controlled by four parameters: two control points and two tangent vectors. Being defined only once on our generic 3D model, these curves can be easily transferred to the specific model (which is actually used for tracking) by applying the same RBF interpolation matrix that was

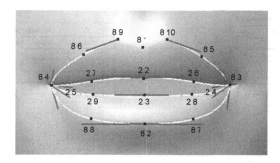

Fig. 3. Deformable curves in the mouth region; tangent vectors are shown in black color (numbered points refer to FDPs)

used at the modeling phase to its control points and tangents (see [2] for details) and thus adapt the curves to the new geometry.

Above the deformable curve structure we've built a higher-level animation mechanism – a set of expression units. Each expression unit controls simultaneously several curves providing their coherent deformations in respect to different facial expressions (Fig. 4). Currently we use four expression units: "mouth width" (wide/narrow), "mouth corners up/down", "lower lip up/down" and "upper lip up/down" which can be extended to six if we want to track asymmetric lip movements.

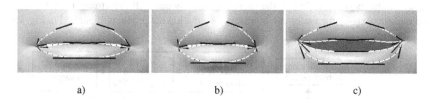

a)   b)   c)

Fig. 4. Examples of expression units and corresponding curve deformations (curves are shown in white with black tangents): a) neutral position; b) "mouth width" expression unit activated; c) all the expression units activated to a certain degree to emulate a smiling expression

These expression units are the parameters that replace FAPs in our mouth tracker. At each iteration of the minimization tool we compute "virtual" displacements of the FDPs from the deformed curves shape, than derive the corresponding FAPs values and apply them. So we still use MPEG-4 animation but indirectly. In that way we reduce the number of system parameters from 28 to 4 (6) but of course loose in precision. Since we aim at the precise detection of lips movements, we use the curve parameterization only as an approximation (which is sufficient in most cases) leaving the possibility to refine the result afterwards using FAPs directly on a more local level. The scheme of the mouth tracking algorithm is presented in the Fig. 5.

Eyes. We apply the same muscle-based approach to eyes animation as well but for a different reason. Here there's no need to reduce parameter space since we have only

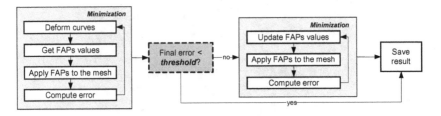

Fig. 5. Mouth tracking framework

two FAPs per eye (one for the eyelid and one for the under-eye region), so the purpose of using curves as constraints serves in this case to ensure the realism of the eye movement regardless of mesh geometry by exact matching of the upper and the lower eye boundaries. Eye-curves and their effect on mesh deformation are illustrated on Fig. 6.

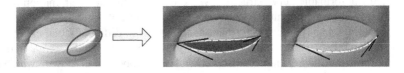

Fig. 6. Correction of the eyelid movement using deformable curves. Now the deformation of the eye boundaries is controlled by the curves and not by the FAP deformation function, so the eyelid matches exactly the under-eye boundary eliminating the problem illustrated in Fig. 2 c)

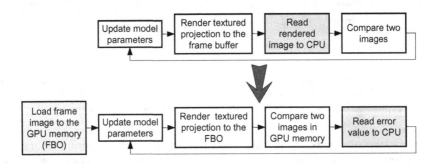

Fig. 7. Operations performed at each iteration and the GPU acceleration scheme

4   GPU Acceleration

Being very precise and stable our algorithm is however far from real-time requiring computational time of around 1,8 minutes per frame due to multiple cycles of minimization for a single image each containing 300-400 iterations. This time however can be reduced to 1 minute by transmitting a part of the computation to the GPU. Indeed if we look closely at the operations performed at each iteration (Fig. 7) we'll see that the most "expensive" one from the computational point of view is reading back the image with the textured projection from the GPU memory for comparison with the frame image. So the idea was to perform this operation directly on the GPU by using GPGPU reduction algorithm [24] and to transfer only the error value. Masking of the neglible pixels is done through the alpha-component and all the GPU-computation is implemented using pixel shaders.

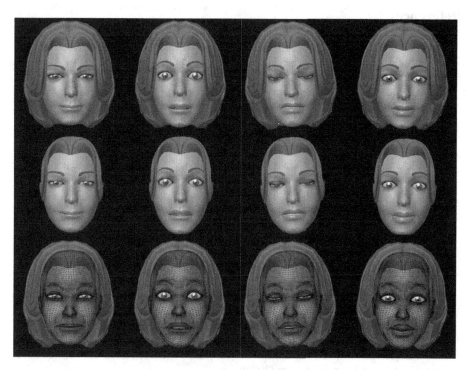

Fig. 8. Tracking result on the generated video sequence of Greta. The first row represents the frames from the original sequence; the second row – the reconstructed 3D model of Greta animated with the parameters detected by the tracker; and the last row – the same model without texture superpositioned over the frame images.

5 Conclusion

In this paper we proposed an algorithm for facial tracking in video based upon analysis/synthesis collaboration approach providing a new animation parameterization for capturing facial expression based upon MPEG-4 specification. Our algorithm was tested on a synthesized video sequence containing Greta's face animated with various types of facial expressions taken from the library that was developed by the team of Catherine Pelachaud and showed very good results (some screenshots from the sequence are presented in Fig. 8; the complete sequence can be downloaded from ftp://ftp.inria.fr/incoming/mirages/daria/tr_greta_res.avi). Our current research is dedicated to the texture update issue in order to make it more flexible (adaptive to changes in lighting conditions/head pose rather that fixed to the frames counter) and more robust to occasional tracking errors, because once texture is updated with wrong pixels, as it's our unique criterion, it will propagate the error over the whole sequence. Apart from that we're investigating other possibilities to accelerate our algorithm, for instance using pose prediction from statistical image analysis.

References

1. Kalinkina, D., Gagalowicz, A., Roussel, R.: 3D reconstruction of a human face from images using morphological adaptation. In: Gagalowicz, A., Philips, W. (eds.) MIRAGE 2007. LNCS, vol. 4418, pp. 212–224. Springer, Heidelberg (2007)
2. Roussel, R., Gagalowicz, A.: Morphological Adaptation of a 3D Model of Face from Images. In: MIRAGE (2003)
3. Roussel, R., Gagalowicz, A.: A hierarchical face behavior model for a 3D face tracking without markers. In: Gagalowicz, A., Philips, W. (eds.) CAIP 2005. LNCS, vol. 3691, pp. 854–861. Springer, Heidelberg (2005)
4. GRETA: Embodied Conversational Agent, http://linc.iut.univ-paris8.fr/greta/
5. Pandzic, I.S., Forchheimer, R.: MPEG-4 facial animation
6. Dementhon, D., Davis, L.S.: Model-based Object Pose in 25 Lines of Code. International Journal of Computer Vision 15, 123–141 (1995)
7. Buhmann, M.D.: Radial Basis Functions: Theory and Implementations. Justus-Liebig-Universität Giessen, Germany
8. Press, W.H., Teukolsky, S.A., Vetterling, W.T., Flannery, B.P.: Numerical Recipes in C
9. Pasquariello, S., Pelachaud, C.: Greta: A Simple Facial Animation Engine. In: Proc. of the 6th Online World Conference on Soft Computing in Industrial Applications (2001)
10. Schwerdt, K., Crowley, J.L.: Robust Face Tracking Using Color. In: 4th IEEE International Conference on Automatic Face and Gesture Recognition, pp. 90–95 (2000)
11. Bibby, C., Reid, I.: Robust Real-Time Visual Tracking Using Pixel-Wise Posteriors. In: Forsyth, D., Torr, P., Zisserman, A. (eds.) ECCV 2008, Part I. LNCS, vol. 5302, pp. 831–844. Springer, Heidelberg (2008)
12. Cipolla, R., Gee, A.: Fast Visual Tracking by Temporal Consensus. Image and Vision Computing, 105–114 (1996)
13. Drummond, T., Cipolla, R.: Real-time Tracking of Complex Structures with On-line Camera Calibration. In: British Machine Vision Conference, pp. 574–583 (1999)
14. Terzopoulos, D., Waters, K.: Physically-Based Facial Modeling, Analysis, and Animation. The Journal of Visualization, Analysis, and Animation 1,2, 73–80 (1990)
15. Terzopoulos, D., Waters, K.: Analysis and Synthesis of Facial Image Sequence Using Physical and Anatomical Models. IEEE Trans. on Pattern Analysis and Machine Intelligence, 569–579 (1993)
16. Lanitis, A., Taylor, C.J., Cootes, T.F.: Automatic Interpretation and Coding of Face Images Using Flexible Models. IEEE Trans. on Pattern Analysis and Machine Intelligence 17, 743–756 (1997)
17. Black, M.J., Yacoob, Y.: Tracking and Recognizing Rigid and Non-Rigid Facial Motions Using Local Parametric Models of Image Motion. In: International Conference on Computer Vision, pp. 374–381 (1995)
18. Malciu, M., Prêteux, F., Buzuloiu, V.: 3D Global Head Pose Estimation: A Robust Approach. In: Proceedings of International Workshop on Synthetic-Natural Hybrid Coding and Three Dimensional Imaging (1999)

19. Reveret, L., Essa, I.: Visual Coding and Tracking of Speech Related Facial Motions. In: CVPR 2001 Workshop on Cues in Communication (2001)
20. Ekman, P., Friesen, W.: Facial Action Coding System: A Technique for the Measurement of Facial Movement. Consulting Psychologists Press, Palo Alto (1978)
21. Ahlberg, J.: Using the Active Appearance Algorithm for Face and Facial Feature Tracking. In: ICCV 2001 Workshop on Recognition, Analysis and Tracking of Faces and Gestures in Real-Time Systems, pp. 68–72 (2001)
22. Fanelli, G.: Facial Features Tracking Using Active Appearance Models. University essay from Linköpings universitet (2006)
23. Douglas DeCarlo, D.: Optical Flow Constraints on Deformable Models with Applications to Face Tracking. International Journal of Computer Vision 38, 99–127 (2000)
24. GPGPU Reduction Tutorial,
    http://www.mathematik.uni-dortmund.de/~goeddeke/gpgpu/
25. Roivainen, P., Li, H., Forchheimer, R.: 3-D Motion Estimation in Model-Based Facial Image Coding. IEEE Transactions on Pattern Analysis and Machine Intelligence 15, 545–555 (1993)
26. Gokturk, S.B., Bouguet, J.-Y., Tomasi, C., Girod, B.: Model-Based Face Tracking for View-Independent Facial Expression Recognition. In: FGR (2002)
27. Pighin, F., Szeliski, R., Salesin, D.H.: Resynthesizing Facial Animation through 3D Model-Based Tracking. In: Proceedings of the 7th IEEE International Conference on Computer Vision, vol. 1, pp. 143–150 (1999)

A Model-Based Approach for Human Body Reconstruction from 3D Scanned Data

Thibault Luginbühl[1], Philippe Guerlain[2], and André Gagalowicz[1]

[1] INRIA Rocquencourt, MIRAGES,
Domaine de Voluceau,
78150 Rocquencourt, France
[2] UFR PEPS, Université de Haute Alsace
32 rue du Grillenbreit, BP 50568,
68008 Colmar Cedex, France
thibault.luginbuhl@inria.fr, philippe.guerlain@uha.fr,
andre.gagalowicz@inria.fr

Abstract. Human body scanners can quickly provide clouds of more than 200 000 points representing the human body's surface. Many new applications can be derived from the ability to build a 3D model of a real person, especially in the textile industry allowing virtual try-on approachs. However, getting a regular model, suitable for these applications from scanned data is not a straightforward task. In this paper, we propose a model-based approach to model a specific person. We use a generic model whitch is segmented and points are organized in slices. We adapt the sizes of each body limb and then fit each slice on the data, limb by limber.

Keywords: Human body modeling, model-based modeling.

1 Introduction and Related Works

Today, several scanning techniques can digitalize 3D human bodies. Yet, data provided by the devices are not immediately suitable for higher-level applications. Most of the time data are noisy, contain holes in the parts that the beam of the scanner cannot reach and, in our case, some parts of the body are missing completely (feet) or partially (see Fig 1 top of the head).

Several works have tackled the problem of reconstructing and repairing the scanned surface. An exhaustive survey of techniques concerning modeling and segmentation of human body from scan data can be found in [1]. So we will focus on some techniques directly related to our problem. We assume that people are asked to stay in a specific position during the scan ; therefore all our models will be static in a reference position (roughly the same that we chose for the generic model). Our aim is to fit the model to the available data and fill the missing part with it, by deforming the generic model so that the deformed model matches the available data, but keeps the topology of the generic model.

In order to fill the missing parts in scanned data, [2] proposed to define implicitly the surface defined by the point cloud, they fit a RBF function so that

A. Gagalowicz and W. Philips (Eds.): MIRAGE 2009, LNCS 5496, pp. 332–343, 2009.

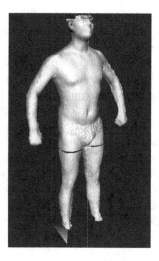

Fig. 1. A sample of reconstruction provided by SYMCAD [11]. Noise appears where the feet should be and on the head and also on many other places. This is why we need a generic model to be able to fill the missing parts.

the value of the function is 0 on the point cloud and then reconstruct the zero set using a marching tetrahedra variant. This method has shown precise results and can be adapted to take into account the noise in the point cloud. The limit of this approach in our case is that data are filled smoothly, this might not always be be the best and the most realistic solution, for example if we want to repair the sole of the feet. Furthermore, since the feet are completely missing in our data, we need more than a smooth filling of existing points.

Using a generic model to fill the parts where data are lacking has been done in [3] for faces. In this method, the target shape is tagged interactively with a set of landmarks corresponding to a similar set in the template model. A deformation is computed to fit this set, then the set is automatically refined to reach a higher level of precision. But for our application, we need to build a completely automatic system, therefore the manual (or even semi-automatic) selection of landmarks should be avoided.

Reconstruction of the complete human body using a generic model from uncalibrated wide baseline views was done by [4]. A set of 32 characteristic points is used. Camera calibration and 3D location of characteristic points is determined through an analysis-synthesis loop. The 3D location is calculated with the calibration estimation at each step, each view where a characteristic point is visible defines a ray in 3D space, the intersection of the rays is the location of the point. To deal with imperfect calibration a least square solution minimizing the distance to the projected rays is adopted. This leads to a set of deformation vectors for each characteristic point. Then, complete deformation is obtained by interpolating these deformation with RBF. After that, surface is refined using silhouette information: silhouettes' curves are extracted from each view, and a curve matching technique is employed to find deformation vectors. These deformation are also interpolated with RBF to get the refined surface. This approach has also been used with success for faces in [5].

A complete fitting technique of a template model on scanned data using automatically detected landmarks can be found in [6]. Here, the CAESAR database has been used, 74 landmarks have been put on people before they were scanned. An energy term to be minimized is defined as a weighted sum of errors (data, smoothness and markers error) and an optimization is performed at low and high resolution to fit the surfaces. PCA analysis was also proposed for markerless matching in the same article and also in [7] where the scanned bodies are aligned inside a fixed volume. A voxelisation of the volume is then computed where each voxel has the value of the signed distance to the surface, PCA is computed on this voxel grid to get the variability of shapes. Yet, this method requires processing to eliminate holes, gaps and noise before PCA. Both methods also rely on the database to get the eigenvectors to reconstruct specific bodies after.

When no markers are available on the scanned data, it is possible to detect landmarks automatically. [8], [9] propose a method based on fuzzy logic to locate feature points defined as sharp angles on 2D contours given by intersecting the data with a plane. [10] tried to find mathematical definitions of the human body features defined in ASTM and ISO. Then these features are located using image processing techniques. To be able to use image processing techniques, the point cloud is encoded as a 2D depth map. 21 features points an 35 lines have been extracted with this approach.

2 Our Approach

This paper is organized as follows. Section 2.1 describes our approach to fit a generic model to scanned data, section 2.2 explains how we adjust the number of slices in the limbs, section 2.3 describes how we fit each curve in simple cases, in section 2.4 we propose a solution for some problematic slices and in section 2.5, we discuss the current results and future work.

2.1 Fitting Technique

Our data come from a 3D-scanner using structured light called SYMCAD [11] and produced by a french company: Telmat [12]. The technique developped here is designed specifically for this scanner. Telmat is our partner in SimulVET, an ANR funded project for virtual try-on applications. Different types of output are available from the scanner: the point clouds (one for the front view and one for the back view), segmented surfaces (one triangulated surface for each member in each view) and a segmented model where in each member the points are organized in slices, one slice per centimeter. Segmentation was computed on 2D images, using a matching procedure between 2D pixels and 3D points. The segmentation distinguishes 5 parts: left and right leg, left and right arm and

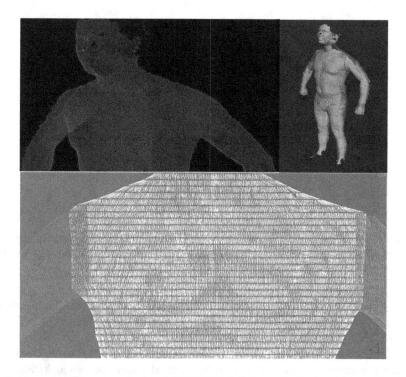

Fig. 2. Here are different outputs of SYMCAD [11] for a single acquisition. Top-left, a zoom on a part of the point cloud, top-right, the segmented surfaces with a specific color for each member and, at the bottom, the sliced model with one slice per centimeter. This last one is a zoom of the model in Fig 1.

torso. The sliced model is composed of 5 independent closed meshes, curves have been fitted to each segmented surface to link points from the two views and to form a single mesh of each member (see Fig 2).

We developed a fitting technique on this segmented model. Our textile application application needs a single regular mesh without holes to work correctly. So we built a generic model (see Fig 3) with the following properties:

1. It is a single mesh of the complete human body.
2. It has the right topology for our application.
3. It is segmented (left and right feet, legs, hands and arms, torso and head), a specific color was manually assigned to every point of each member.
4. Arms, legs and torso are organized in slices like the data coming from the scanner.

2.2 Adjusting the Number of Slices

A first step is to establish a correspondence between the slices of the generic model and the slices of the data. According to the size and the proportions

Fig. 3. Our generic model. We chose a specific color for each limb and we organized the points in slices as shown at the bottom of Fig 2. This mesh is regular with a correct topology that we intend to use for our virtual try-on application.

of a specific person we may need to add or remove some slices. Therefore two functions were created: *addSlice* and *removeSlice*. Both of them rely on a third one: *buildFacesBetweenSlices*.

buildFacesBetweenSlices builds the faces between two consecutive curves. We consider 2 consecutive curves of a limb, we read them couter-clockwise. Fig 4 shows two examples of the principle in 2D. In each cell we represent points from the lower and upper curve. A first triangle is build by taking a point on the lower curve, its closest point on the upper and the next point on the lower (the triangle is in white on the picture). Last points used in each curve are shown in yellow. We build new triangles while moving around the lower curve and finding the closest points on the upper. The first two cells show the case where the closest upper point of the next point of the lower curve is the current upper yellow point. In this case, we build a single triangle as shown in the second cell. The third and fourth cell explains what we do when the closest upper point is not the current yellow point. In this case, we consider all the intermediate points of the upper curve and link them to their closest neighbor on the lower curve. We finally build the corresponding triangles as shown in the last cell.

addSlice enables to add a slice between two others in a limb. It first deletes the faces that were existing between the two slices, then makes a copy of the lower slice and builds new faces between the new consecutive slices. The new points are placed in a middle position between their upper and lower neighbors so that the surface remains smooth.

removeSlice deletes the faces between the previous and the next slice, builds new faces between these two slices and delete the points of the slice to be removed.

After adding or removing a slice, points are translated to keep the regularity of the spacing between slices. When several slices need to be added or removed,

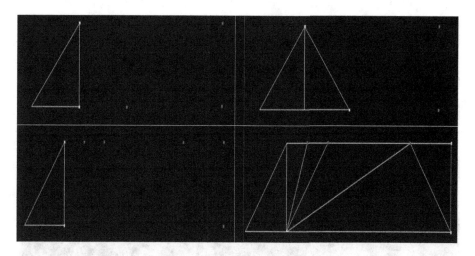

Fig. 4. Two examples of triangulation between two slices. In yellow we show the last current points on each curve used to build a triangle. In the first case we read the next point on the lower curve, its closest point on the upper curve is the current yellow point so we build a single triangle and go on reading the lower curve. In the second case, there are several points to insert on the upper curve, each point is linked to its closest point and the triangles are built as shown in the picture.

we add/delete them at different heights in the member and not where there is a joint such as the knee, the elbow or the shoulder not to deform too much a specific part of the member.

The number of slices to add or remove is determined automatically for each member. The scanner provides the position of the crotch ; with it we can know the height of the legs and then the number of slices needed since the spacing is a constant. For the arms and torso we just have to put the same number of slices as in the data.

2.3 Fitting Each Slice

After the previous adjustments we have almost everywhere a 1 by 1 correspondence between the slices on the model and the slices of the data. The only slices that have still and important difference are the ones linking the shoulders to the torso, those slices will be treated differently than with the following method.

For all the other slices we start by aligning the center of the curve of the generic model to the center of the curve of the data. Then, for each point of the slice of the generic model, we consider the ray starting from the center and going through the point. We move the point to the intersection of the ray and the curve on the data (see Fig 5).

However, this approach can't be used for the slices where the arms and the torso are involved. Fig 6 shows how these slices look like on the generic model and the data. In the generic model, we have one single curve going through the three limbs. On the other hand, as we have one mesh per limb in the data, there

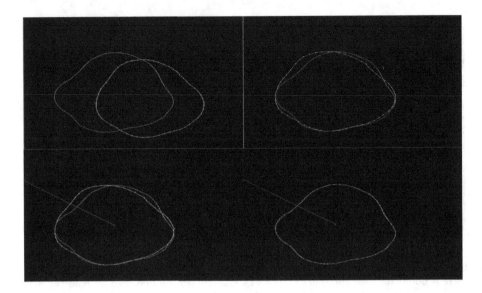

Fig. 5. This picture shows the different steps of the fitting. The points of the data are in red and those of the model are in green. First we have the curves in their initial position, we fit the bounding box of the generic model to the data, then for each point we draw a ray (here in blue) from the center to a point of the generic model, we move each point to its intersection with the data.

are three closed curves for each slice and those curves have overlapping parts. Furthermore, we see in Fig 6 that the segmentation is not precise enough because a large part of the torso is in the arm member. This, added to the fact that such slices do not correspond to a cylindrical shape and therefore don't have a usable center for our approach, makes it clear that we need to find another way.

To deal with these slices, we built an envelope of the points at the height of the curve we want to fit. Then we moved each point of the generic model according to the normal direction at the point, and found the closest intersection with the envelope. We define a threshold to avoid moving points which are too far from their original position. However, results were not good enough for this part. That is why we propose another technique for these slices using B-splines.

2.4 Fitting Problematic Slices Using B-splines

In order to find a solution for the slices where arms and torso are joining, we propose the use of B-splines. Our idea is to fit a B-spline functions to the 2D points, and use the parametric equation to move the points of the generic model.

First we need to fit a B-spline to the data points. Lower and upper envelopes of the slice are computed using the CGAL library [13]. The two envelopes are linked to get a complete envelope of the slice with points organized counterclockwise. Let m be the number of points of this envelope. We set the closest

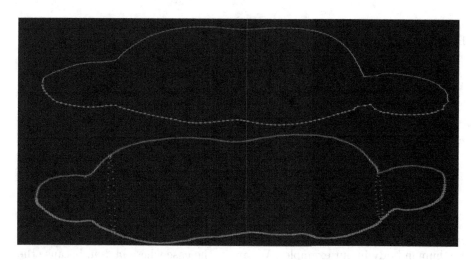

Fig. 6. Here are two curves where arms and torso are meeting. The first one corresponds to the generic model and the second one to the data. The first one is a single closed curve, we just added color to highlight front and back parts. The second is composed of 3 closed curves (left and right arm, torso) the red points inside enable to guess how each curve is closed. We can also see that the segmentation provided by the scanner is not precise enough in that area: red points, that show the limits of the segmented meshes, are far from the points with sharp angles which correspond to the real boundary.

point to the axis $x = 0$ with highest y coordinate as the origin of the curve. For each point $A_i, 0 \le i \le m-1$ on the envelope, we compute its chordal distance d_i defined by : $d_0 = 0$ and for $1 \le i \le m - 1, d_i = d_{i-1} + ||A_i - A_{i-1}||$ (it is a first order approximation of the arc length). We define $d_m = d_{m-1} + ||A_{m-1} - A_0||$ and we finally divide all the values by d_m to have a parametric representation in $[0, 1]$.

Then, we compute a least-square approximation B-spline function. We choose a degree d and a number of control points n. We build a periodic knot vector because we want to represent a closed curve :

$$t_k = \frac{k - d}{n - d}, 0 \le k \le n + d \tag{1}$$

The B-spline curve will be

$$C(t) = \sum_{k=0}^{n-1} B_{k,d}(t) P_k \tag{2}$$

where $P_k, 0 \le k \le n - 1$ are the control points to be found and $B_{k,d}$ is the basis B-spline function defined for $d = 0$ by

$$B_{k,0}(t) = \chi_{[t_k, t_{k+1}[} \tag{3}$$

with χ_I the characteristic function of the interval I and otherwise by the Cox-de Boor formula

$$B_{k,d}(t) = \frac{t - t_k}{t_{k+d} - t_k} B_{k,d-1}(t) + \frac{t_{k+d+1} - t}{t_{k+d+1} - t_{k+1}} B_{k+1,d-1}(t) \qquad (4)$$

Using all the points of the envelope, we get a linear system,

$$\begin{pmatrix} B_{0,d}(d_0) & B_{1,d}(d_0) & \cdots & B_{n-1,d}(d_0) \\ \vdots & \vdots & \ddots & \vdots \\ B_{0,d}(d_{m-1}) & B_{1,d}(d_{m-1}) & \cdots & B_{n-1,d}(d_{m-1}) \end{pmatrix} \begin{pmatrix} P_0 \\ \vdots \\ P_{n-1} \end{pmatrix} = \begin{pmatrix} A_0 \\ \vdots \\ A_{m-1} \end{pmatrix}$$

We use B-splines of the third degree. For this degree we found that choosing n between 30 and 40 is enough to get a good approximation of all curves of the human body in our examples. We are in the case where $m > n$, because the envelopes contain at least more than 100 points. So we choose the pseudo-inverse solution of the system.

With it we get the control points of the fitted B-spline. Now we have to move the points of the generic model. We consider for the curve of the generic model the same kind of parametrization that we use for the envelope (between 0 and 1 and with the origin at $x = 0$ with maximum z value). We build it the same way as we explained. We evaluate the fitted B-spline to all parameters of the points of the generic model and move them to the found value. Fig 7 shows the fitted B-spline to a curve from data, we see how it smooths some sharp angles.

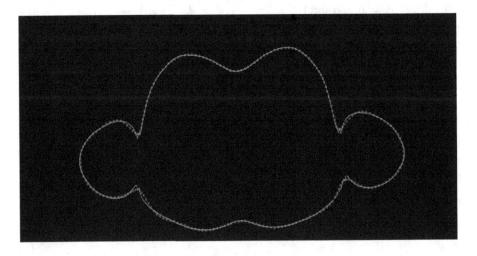

Fig. 7. An example of a fitted B-spline with $n = 35$ $d = 3$. In green the data curve, in red the fitted B-spline.

2.5 Results and Future Work

Once the slices are fitted, some parts of the generic model are translated in order to be aligned with the new positions (feet and hands). We took the hands of the generic model because data is too imprecise to make a correct fitting of this part. Fig 8 shows the results on a scanned person without using B-splines. First the data are shown then curves are fitted and finally a laplacian smoothing is used. Since the last curves of each limb in the data are often used to close the meshes, they contain some non-reliable points as we see in the third model in Fig 8, at the link between the legs and the torso. To solve this problem we put all the points of these curves at the middle position between their upper and lower neighbors.

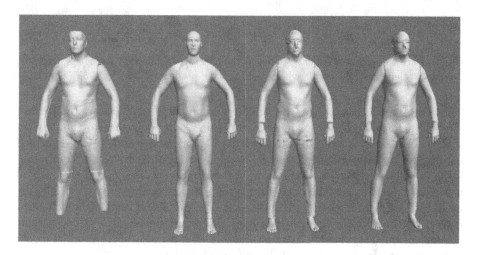

Fig. 8. From left to right: the data from the scanner, the generic model, the model after the fitting, the model after smoothing. Those results are without the use of B-splines.

Fig 9 shows the result around the shoulders of the 3D model with the use of B-splines, the approach gives promising results, there is no more self intersection as with previous method. However we still have to find a solution when there is an asymmetry (for example when an arm is higher than the other we may have curve where right arm and torso meet but not left arm and torso or vice versa).

The main problems we have to deal with now are the head and the links with parts taken from the generic model (links between legs and feet or arms and hands for example). Work is still in progress. For these parts, we will use more 3D information instead of 2D curves only, relations between a set of consecutive curves can be explored to improve regularity. Comparison with the original point cloud can also be used to refine the result after this step.

Several filtering techniques have also been developed in our team. These techniques may be useful to treat difficult parts like the linking regions between parts fitted to the data and parts taken completely from the generic model.

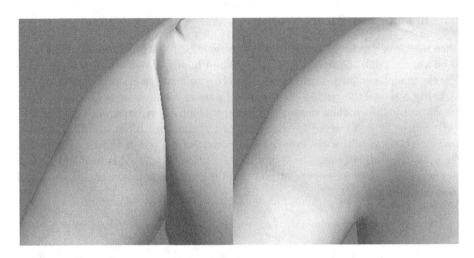

Fig. 9. Here we see the difference between the case where we don't use B-splines to fit the curves (left) and the result when B-splines are used with $d = 3$ $n = 35$ (right)

Those filtering techniques use information from the original generic model before the deformation. For example, a slice filter computes slices orthogonal to the principal axis of each member in the original model and in the deformed model, the variation of the distance between correspondent slices is computed, if the maximum variation is more than a threshold, a scaled copy of the original model is made for this slice. Another filter for smaller distortions can also be used, for each vertex on the original and its correspondent one on the deformed model, we consider the incident triangles and compute the angle between their normals. If this angle is more than a threshold a scaled copy of the original model is used to replace the vertices of the deformed model.

Finally, after the full geometry is recovered, the next step is to get texture information from the pictures taken in the scanner.

References

1. Werghi, N.: Segmentation and Modeling of Full Human Body Shape From 3-D Scan Data: A Survey. IEEE Transactions on Systems, Man, and Cybernetics, Part C 37(6), 1122–1136 (2007)
2. Carr, J.C., Beatson, R.K., Cherrie, J.B., Mitchell, T.J., Fright, W.R., McCallum, B.C., Evans, T.R.: Reconstruction and Representation of 3D Objects With Radial Basis Functions. In: SIGGRAPH 2001, Computer Graphics Proceedings, pp. 67–76 (2001)
3. Kähler, K., Haber, J., Yamauchi, H., Seidel, H.-P.: Head shop: Generating animated head models with anatomical structure. In: Proceedings ACM SIGGRAPH Symposium on Computer Animation (SCA) 2002, July 21-22, pp. 55–64 (2002)
4. Quah, C.K., Gagalowicz, A., Roussel, R., Seah, H.S.: 3D Modeling of Humans with Skeletons from Uncalibrated Wide Baseline Views. In: Gagalowicz, A., Philips, W. (eds.) CAIP 2005. LNCS, vol. 3691, pp. 379–389. Springer, Heidelberg (2005)

5. Kalinkina, D., Gagalowicz, A., Roussel, R.: 3D Reconstruction of a Human Face from Images Using Morphological Adaptation. In: Gagalowicz, A., Philips, W. (eds.) MIRAGE 2007. LNCS, vol. 4418, pp. 212–224. Springer, Heidelberg (2007)
6. Allen, B., Curless, B., Popovic, Z.: The space of human body shapes: reconstruction and parameterization from range scans. ACM Trans. Graph. 22, 587–594 (2003)
7. Ben Azouz, Z., Rioux, M., Shu, C., Lepage, R.: Analysis of Human Shape Variation using Volumetric Techniques. The Visual Computer International Journal of Computer Graphics 22(5), 302–314 (2005)
8. Wang, C.C.L., Cheng, T.K.K., Yuen, M.M.F.: From laser-scanned data to feature human model: a system based on fuzzy logic concept. Computer-Aided Design 35(3), 241–253 (2003)
9. Wang, C.C.L.: Parameterization and parametric design of mannequins. Computer-Aided Design 37(1), 83–98 (2005)
10. Leong, I.F., Fang, J.J., Tsai, M.J.: Automatic body feature extraction from a marker-less scanned human body. Computer-Aided Desing 39(7), 568–582 (2007)
11. SYMCAD website: http://www.symcad.com
12. Telmat website: http://www.telmat.com
13. Manual of the CGAL library, envelopes of curves in 2D:
   http://www.cgal.org/Manual/3.3/doc_html/cgal_manual/
   Envelope_2/Chapter_main.html

Region-Based *vs.* Edge-Based Registration for 3D Motion Capture by Real Time Monoscopic Vision

David Antonio Gómez Jáuregui and Patrick Horain

INRIA / Projet MIRAGES
B.P. 105, 78153 Le Chesnay Cedex, France
Institut TELECOM ; TELECOM & Management SudParis
9 rue Charles Fourier, 91011 Evry Cedex, France
{David.Gomez,Patrick.Horain}@IT-SudParis.eu

Abstract. 3D human motion capture by real-time monocular vision without using markers can be achieved by registering a 3D articulated model on a video. Registration consists in iteratively optimizing the match between primitives extracted from the model and the images with respect to the model position and joint angles. We extend a previous color-based registration algorithm with a more precise edge-based registration step. We present an experimental analysis of the residual error *vs.* the computation time and we discuss the balance between both approaches.

Keywords: 3D motion capture, monocular vision, 3D / 2D registration, region matching, edges matching.

1 Introduction

Research in motion capture by computer vision has been motivated by many target applications: human-computer interfaces, animation, interaction with virtual environments, video surveillance, games, etc. We focus on 3D human motion capture in real-time without markers [8]. This is a difficult problem because of the ambiguities resulting of the lack of depth information, partial occlusion of human body parts, high number of degrees of freedom, variations in the proportions of the human body and different clothing of each person [14].

In this work, we extend a previous work for 3D human motion capture by registering a 3D articulated human body model on video sequences using a color-based step followed by an edge-based step [7]. In this work we shall experimentally characterize the contribution of color and edge information to model matching. We present a detailed analysis of the precision and processing time achieved by the color-based registration and the edge-based registration steps.

This paper is organized as follows. First, in section 2, we describe previous works related to 3D human motion capture by computer vision. In section 3, we introduce our 2 steps approach based on matching color regions and the edges. Then, our performance characterization experiments and the results obtained are presented in section 4. Finally, in section 5, we conclude and discuss how a balance can be found between both steps while facing limited computation resource.

A. Gagalowicz and W. Philips (Eds.): MIRAGE 2009, LNCS 5496, pp. 344–355, 2009.
© Springer-Verlag Berlin Heidelberg 2009

2   Previous Work for 3D Human Motion Capture

Previous works rely on various appearance features such as color [10], [6], [7], edge [6], [7], [15], shape [1], [11], and motion [15], [9]. They can be divided into two main approaches: model-based and model-free approaches [14].

The model-based approaches use a 3D model of a human body and a matching cost function to find the 3D pose that best matches input images. Estimating the 3D pose from monocular images, is achieved by searching for the pose that minimizes some matching cost function [8]; some other works use human body part detectors to assemble the 3D pose using physical and proximity constraints [3]. Temporal coherence can be enforced with particle filters that allow multiple hypotheses matching [6], [15]. Some works use motion priors [16] to guide tracking within a motion model previously learned [16], or to learn a mapping between the pose space and a low-dimensional latent space in which the tracking occurs [13].

Model-free approaches do not use any 3D explicit human body model. Instead, they try to infer directly 3D poses from images. The learning-based approaches rely on training data to learn a function that maps the image observation to the 3D pose space [1]. Example-based approaches avoid this learning by saving in a database a collection of examples of 3D poses with their corresponding image descriptors and by searching this database and interpolating candidate poses for a 3D pose similar to the input image [11].

3   Our Approach for 3D Human Motion Capture

Our method consists in registering a 3D articulated model of upper human body on video sequences [8], [10]. Our 3D human model (figure 2a) has 3 global position parameters and 20 joint angles of the upper-body part (chest, arms, forearms, hands, neck and head). A 3D human pose is represented by a vector of parameters of the joint angles.

For each captured image frame, we extract color regions and edges. A colored silhouette and occluding edges are also computed for each candidate pose of the model. Our registration process consists in searching for the best matching correspondence between these primitives. We iteratively optimize registration using a color region-based criterion and then an edge-based registration (fig. 1) [7].

3.1   3D Human Model Calibration and Pose Initialization

To registering the 3D upper-body human model on a video sequence requires calibrating the body model to make it similar to the actor captured in the video. This is done by adjusting manually the 20 joint angles of the 3D model taking as reference the pose of the human in the first image and the dimensions (length, width and height) of each body part of the 3D model by watching the overlapping between each body part of the projected model and the human in the captured image (figure 2b).

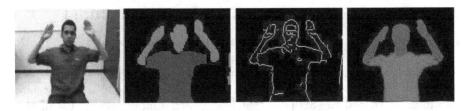

Fig. 1. Our prototype for 3D human motion capture. The images are respectively: the captured image, the segmented image, the edges in the foreground and finally the projection of the registered 3D human body model.

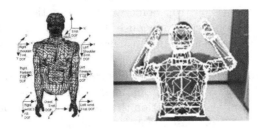

Fig. 2. 3D human model calibration step: a general design of our 3D human upper-body model with the 20 degrees (a), the 3D model projection superposed with the human in order to adjust the parameters of the model in the calibration and pose initialization steps (b)

3.2   Region-Based Registration

The human silhouette is detected by comparing the captured image with a reference image of the background. This silhouette (foreground) is segmented in two color classes (skin and clothes color). Color samples are extracted automatically from the first captured image. A skin color sample is taken in the face region found with Adaboost face detector [17]. A clothes sample is taken under the face. We model each sample using a simple gaussian model in a HSV color space. For each image, we project the 3D model [18] (using OpenGL API) according to the pose described in the vector of parameters. The 3D model is projected by rendering the skin and clothes colors. The matching between the 3D model projection and the segmented image is evaluated using a non-overlapping ratio:

$$F(q) = \prod_{c=1}^{m} \left(\frac{|A_c \cup B_c(q)| - |A_c \cap B_c(q)|}{|A_c \cup B_c(q)|} \right)^{\frac{1}{m}} . \tag{1}$$

where q is the vector of parameters describing a candidate 3D pose, m is the number of color classes, A_c is the set of pixels with the c color class in the segmented image, $B_c(q)$ is the set of pixels with the c color in the projection of the 3D model and $|X|$ represent the number of pixels in X.

This cost function is minimized using a downhill simplex algorithm [12] under biomechanical constraints. Further details can be found in [10].

Fig. 3. Limited precision of the region-based registration. The images are respectively: the captured image, the segmented image, the projection of the registered 3D human body model and finally the projection of the 3D model superposed with the segmented image. The pose of the 3D model differs from the pose of the actor because the region-based registration is not precise.

It is important to note that for convergence towards a position approximately correct, this method initially requires only a partial overlapping between colored regions. However, it is not precise because the number of pixels in the border regions is few compared to the number of pixels inside the region (fig. 3).

3.3   Edge-Based Registration

We propose a further edge-based registration step to improve the precision. It works by matching edges of the captured image and occluding edges of the 3D model [9], [15]. The initial 3D pose of this step is the final state output by the region-based registration. Edges in the input video image are extracted with a Deriche filter [4] in the foreground region of the image. Then, a chamfer algorithm [2] allows computing a map of the distance between each pixel of the image and the nearest edge.

The occluding edges of the 3D model are the lines of the surface where the observation direction is tangent to the surface [5]. These occluding edges can easily and efficiently be extracted with the OpenGL API by rendering the surface mesh with culling based on the normal orientation. First, the back facing triangles and their edges are rendered with some foreground color while the front facing triangles are eliminated, then the inside of front facing triangles is rendered with background color while the back facing triangles are eliminated, so only the occluding edges remain highlighted in the image. Then the mean distance between the projected occluding edges of the model and the edges in the input video image is computed by masking the previous distance map with the projected binary image of the 3D model occluding edges:

$$D_c = \frac{1}{N_p} \sum_i I_{DT}(p_i) \,. \tag{2}$$

where D_c is the mean edge distance, I_{DT} is the distance transform image, p_i are the pixels in the projected occluding edges of the 3D model. We minimize this function with the downhill simplex algorithm [12] as previously in the region-based registration step.

Our registration process basically consists in using downhill simplex optimization algorithm [12] to minimize the non-overlapping ratio and the mean edge distance. The downhill simplex method has the advantage that it requires computing only values of the function to be optimized rather than its derivates. We use a measure of the size of the simplex (defined by the ratio between the highest and lowest value of the simplex) as a convergence criterion.

4   The Optimization Process

Iterative optimization in a high dimensional space usually requires a large variable number of iterations to converge. Because we are interested in real-time motion capture, we have to limit the computation time and thus, the number of iterations per image. Unfortunately, limiting the number of iterations decreases the precision of the registration process (fig. 4). For this reason, we experimentally analyzed the performance (precision and process time) of our registration process by varying the number of iterations of both registration step (region-based and edge-based), searching for an optimal balance between the precision and processing time in both registration steps (region-based and edge-based).

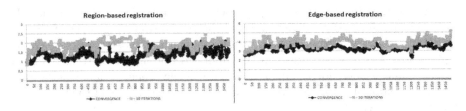

Fig. 4. Effect of a limited number of iterations on the residual error (ordinates). The abscissa is the image number in the video sequence. The left chart corresponds to the non-overlapping ratio minimized by region-based registration. The right chart corresponds to the mean edge distance minimized by edge-based registration. Black lines are the residual error at convergence while the gray line is limited to 30 iterations. A limited number of iterations decrease both the precision and robustness of registration.

We aim at real time tracking, so the available computation time for each captured image must be shared between the two steps of the registration process.

We used 6 video sequences showing various gestures with occlusions (e.g. arms crossed), fast movements, including in the depth direction (fig.5) and a person possibly not exactly facing the camera. These 160 x 120 pixels video sequences were captured using a Logitech QuickCam Pro 5000 webcam. The computation time varies with the number of iterations, the central processor (CPU) and the graphics card (GPU). Table 1 shows the computation time on two platforms[1] with varying number of iterations shared in our two-steps registration process.

---

[1] Experiments were run on a CPU Intel Pentium 4 3.6 GHz and a GPU NVIDIA Quadro FX 1400 (platform 1) and a CPU Intel Pentium M 1.4 GHz and a GPU NVIDIA GeForce 4200 Go (platform 2).

Fig. 5. The video sequences used in our experiments. The video sequence 1 (top left), the video sequence 2 (top center), the video sequence 3 (top right), the video sequence 4 (bottom left), the video sequence 5 (bottom center) and the video sequence 6 (bottom right) contains respectively 290 frames, 1497 frames, 1433 frames, 887 frames, 1032 frames and 551 frames. The first three sequences include various types of gesture. The sequence 4 includes principally gestures when arms are crossing each other. In the sequence 5, the person is not facing directly the camera. The sequence 6 includes movements in which the person is turning around himself.

Table 1. Computation time in milliseconds (average and standard deviation) with respect to the number of iterations shared in our two-step registration process on two platforms. In these experiments, 50% of the total number of iterations is given to each step.

Number of iterations	Avg. Time Platform 1 (ms)	Std. Dev. Time Platform 2 (ms)	Avg. Time Platform 2 (ms)	Std. Dev. Time Platform 2 (ms)
40	22.34	±6.58	88.18	±9.16
100	35.61	±6.68	137.24	±16.07
200	57.99	±6.84	223.30	±27.51
300	78.69	±6.90	300.13	±32.74
400	97.25	±7.12	370.06	±35.81
500	100.59	±9.80	436.21	±48.72

Table 2 shows the computation time of our 3D motion capture prototype[2] for images with higher definition. From these experiments, we found that the performance of the registration process with respect to the non overlapping ratio and the mean edge distance is similar for images with larger number of pixels (higher resolutions) since our approach does not require much accuracy in the segmentation and edge extraction.

4.1   Edge-Based Registration Precision Experiments

We recall that the region-based registration is used to initialize the registration process because is more robust, then, the edge-based registration is used to

---

[2] Computation times on a CPU Intel Pentium 4 3.0 GHz and a GPU NVIDIA GeForce 9600 GT.

Table 2. Average computation time in milliseconds for images with higher resolution. The image processing part includes background subtraction algorithm, color segmentation, edges extraction and distance transform computation. We show also the computation time of each matching cost function (non overlapping ratio and mean edge distance).

Image Resolution	Avg. Time (ms) Image processing	Avg. Time (ms) Non Overlapping Ratio	Avg. Time (ms) Mean Edge Distance
160 x 120	92.45	1.05	0.97
256 x 256	323.53	1.89	1.86
326 x 240	381.19	2.08	2.09
480 x 480	1187.92	3.09	3.05
512 x 512	1353.84	3.74	3.82
640 x 480	1581.86	5.29	5.58

improve the precision. In order to verify the increase of the precision given by the edge-based registration step with respect to the region-based registration, we experimentally compared the residual edge distance achieved by each step of the registration process (region-based and edge-based). Here, we iterated until convergence. Our results (fig. 6) show that the edge-based registration step allows correcting, for some images, incorrect region-based registrations that appear as peaks in the residual distance between edges. The figure 7 illustrates an example of such a correction.

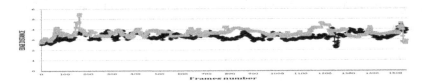

Fig. 6. Residual distance between edges achieved by the region-based registration (gray line) and the edge-based registration (black line)

4.2   Difficulties Encountered in Edge-Based Registration

After limiting the maximum number of iterations in the edge-based registration step, we encountered some difficulties related to the high instability and imprecision in the results. This is because the edges extracted from the image are not necessarily matched with the correct model edges by registration process, so the registration process is trapped in some local optimum. This issue can be limited by initializing the edge-based registration step using the usually smaller simplex after final iteration of the region-based registration step. Thus, the edge-based registration will start to search in a reduced search-space starting from the solution achieved by the previous step (region-based) avoiding being trapped in some local incorrect minimum. We can see the experimental results of the solution proposed in the figure 8.

Fig. 7. Incorrect region-based registration corrected by the edge-based registration. The images are respectively: the captured image, the projection of the 3D model superposed with the segmented image that shows an incorrect region-based registration, the 3D model occluding edges that shows the correction by the edge-based registration and finally the 3D model projection showing the corrected 3D pose.

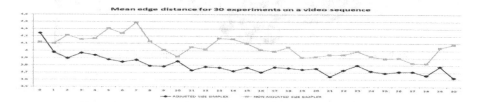

Fig. 8. Residual distance between edges on a video sequence. The abscissas is the number of iterations of the edge-based registration step (other iterations are devoted to the region-based registration step). The black line is the residual error with final simplex at step 1 (region) used as initial simplex at step 2 (edges). The gray line is the residual error using a large initial simplex. The residual distance between edges by reducing the size of the simplex (black line) is smaller because it can avoid more local wrong minimums.

4.3 Performance Experiments in Our Registration Process

In order to analyze the performance of our approach, we considered, for each experiment, the residual values of each evaluation function and also the number of failures (mistrackings) with varying numbers of iterations in the registration process.

We were interested in analyzing the performance from 1 to 500 iterations because the computation time is below 100 milliseconds (see table 1), thus allowing tracking at 10 Hz or more. In each experiment, we sampled the residual value of the non-overlapping ratio and the residual value of the mean edge distance. We considered only the mean residual for all the images in a video sequence. A way of measuring the robustness in each registration step is to count the number of failures for each experiment. We consider as failures or mistrackings the residual values above a defined threshold (a "peak") for each evaluation function. In this way, if the residual value is relatively large, we consider that the solution output by the optimization algorithm is a "bad" registration. We show the experiments results for the video sequence 2 in the next figures (fig. 9, 10, 11 and 12).

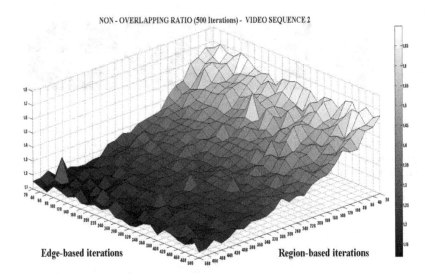

Fig. 9. Mean residual of the non-overlapping ratio (z-axis) with respect to the number of iterations of the region-based registration (x-axis) and the number of iterations of the edge-based registration (y-axis) achieved on video sequence 2. Experiments on video sequences 1, 3, 4, 5 and 6 showed similar results.

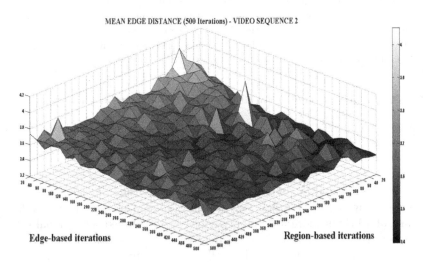

Fig. 10. Mean residual of the mean edge distance (z-axis) with respect to the number of iterations of the region-based registration (x-axis) and the number of iterations of the edge-based registration (y-axis) achieved on video sequence 2. Experiments on video sequences 1, 3, 4, 5 and 6 showed similar results.

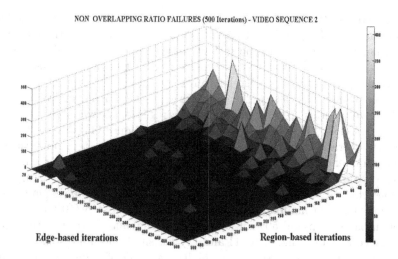

Fig. 11. Number of failures of the non-overlapping ratio (z-axis) with relation to the number of iterations of the region-based registration (x-axis) and the number of iterations of the edge-based registration (y-axis) obtained on the video sequence 2. Experiments on video sequences 1, 3, 4, 5 and 6 showed similar results.

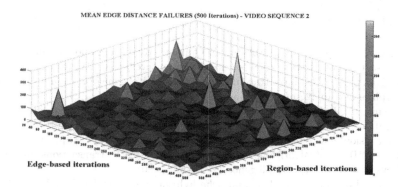

Fig. 12. Number of failures of the mean edge distance (z-axis) with relation to the number of iterations of the region-based registration (x-axis) and the number of iterations of the edge-based registration (y-axis) obtained on the video sequence 2. Experiments on video sequences 1, 3, 4, 5 and 6 showed similar results.

5 Conclusions

We have presented a 3D human motion capture algorithm by monocular vision in real-time, based on registering a 3D articulated model on color regions and then an edge distance criterion. Through experimental results (the surfaces 3D displayed above), we can understand the performance of the region-based and edge-based registration steps. From figure 9 and 10 we can see how the region-based registration step reaches to convergence faster that the edge-based registration step. The figures 11 and 12 show the instability of the edge-based registration step compared to the stability of the region-based registration step. So we need to combine the robustness and stability of the region-based registration and the precision of the edge-based registration.

In order to have the best performance in real-time for our approach, we decided to give priority to the stability of the registration when the number of iterations is below 200 (found experimentally from figure 11), thus, in this case, all the total iterations will be executed by the region based step. However, when the total number of iterations is above 200, the number of failures in region-based step registration is relatively small (figure 11), thus, we can take advantage of the precision achieved by the edge based step (figure 6 and 7) by sharing in the same proportion the total number of iterations between each step (50% of iterations for region-based step and 50% of iterations for edge-based step). Although the performance variation (fig. 9, 10, 11 and 12) was similar for all tested videos, the video sequence 6 (fig. 5) presented the highest number of failures (mistrackings) due to the ambiguity caused by the limited depth information in monocular images. Our future work aims at estimate the gradient of the edge distance in order to use a gradient descent optimization algorithm to improve the precision of our approach.

References

1. Agarwal, A., Triggs, B.: Recovering 3D human pose from monocular image. IEEE Transactions on Pattern Analysis and Machine Intelligence 28, 44–58 (2006)
2. Borgefors, G.: Distance transformations in digital images. Computer Vision, Graphics and Image processing 34, 344–371 (1986)
3. Cheung, G., Baker, S., Kanade, T.: Shape-from-silhouette for articulated objects and its use for human body kinematics estimation and motion capture. In: Computer vision and pattern recognition, Madison, Wisconsin, USA, pp. 16–22 (2003)
4. Deriche, R.: Fast algorithms for low-level vision. IEEE Transactions on Pattern Analysis and Machine Intelligence 12, 78–87 (1990)
5. Franco, J.-S., Boyer, E.: Une approche hybride pour calculer l'enveloppe visuelle d'objets complexes. In: ORASIS 2003, pp. 67–74. Gérardmer (2003)
6. Fontmarty, M., Lerasle, F., Danes, P.: Data Fusion within a modified Annealed Particle Filter dedicated to Human Motion Capture. In: IEEE / RSJ International Conference on Intelligent Robots and Systems IROS 2007, San Diego, CA, USA, October 29-November 2, pp. 3391–3396 (2007)
7. Gómez Jáuregui, D.A., Horain, P., Baroud, F.: Acquisition 3D des gestes par vision monoscopique en temps réel. In: Conférence MajecSTIC 2008, Marseille, France (2008)
8. Horain, P., Bomb, M.: 3D Model Based Gesture Acquisition Using a Single Camera. In: Proceedings of IEEE Workshop on Applications of Computer Vision WACV 2002, Orlando, Florida, December 3-4, pp. 158–162 (2002)
9. Lu, S., Huang, G., Samaras, D., Metaxas, D.: Model-based integration of visual cues for hand tracking. In: Proceedings of IEEE workshop on Motion and Video Computing, Orlando, Florida, pp. 119–124 (2002)

10. Marques Soares, J., Horain, P., Bideau, A., Nguyen, M.H.: Acquisition 3D du geste par vision monoscopique en temps réel et téléprésence. In: Actes de l'atelier Acquisition du geste humain par vision artificielle et applications, pp. 23–27. Toulouse (2004)
11. Mori, G., Malik, J.: Recovering 3D human body configurations using shape contexts. IEEE Transactions on Pattern Analysis and Machine Intelligence (PAMI) 28, 1052–1062 (2006)
12. Nelder, J.A., Mead, R.: A simplex method for function minimization. Computer Journal 7, 208–313 (1965)
13. Pang, J., Qing, L., Huang, Q., Jiang, S.: Monocular Tracking 3D People by Gaussian Process Spatio-Temporal Variable Model. In: International Conference on Image Processing, ICIP 2007, San Antonio, Texas, USA, vol. 5, pp. 41–44 (2007)
14. Poppe, R.W.: Vision-based human motion analysis: An Overview. Computer Vision and Image Understanding 108(1-2), 4–18 (2007)
15. Sminchisescu, C., Triggs, B.: Estimating Articualted Human Motion with Covariance Scaled Sampling. International Journal of Robotics Research 22, 371–393 (2003)
16. Urtasun, R., Fleet, D.J., Fua, P.: 3D people tracking with gaussian process dynamical models. In: Proceedings of the Conference on Computer Vision and Pattern Recognition CVPR 2006, New York, vol. 1, pp. 238–245 (2006)
17. Viola, P., Jones, M.: Rapid Object Detection Using a Boosted Cascade of Simple Features. IEEE Computer Vision and Pattern Recognition 1, 511 (2001)
18. Wright Jr., R.S., Lipchak, B., Haemel, N.: OpenGL SuperBible: Comprehensive Tutorial and Reference, 4th edn., pp. 127–172. Addison-Wesley Professional, Ann Arbor (2007)

Supporting Diagnostics of Coronary Artery Disease with Multi-resolution Image Parameterization and Data Mining

Matjaž Kukar and Luka Šajn

University of Ljubljana, Faculty of Computer and Information Science,
Tržaška 25, SI-1001 Ljubljana, Slovenia
{matjaz.kukar,luka.sajn}@fri.uni-lj.si

Abstract. Coronary artery disease has been described as one of the curses of the western world, as it is one of the most important causes of mortality. Therefore, clinicians seek to improve diagnostic procedures, especially those that allow them to reach reliable early diagnoses. In the clinical setting, coronary artery disease diagnostics is typically performed in a stepwise manner. The four diagnostic levels consist of evaluation of (1) signs and symptoms of the disease and ECG (electrocardiogram) at rest, (2) sequential ECG testing during the controlled exercise, (3) myocardial perfusion scintigraphy, and finally (4) coronary angiography, that is considered as the "gold standard" reference method. Our study focuses on improving diagnostic performance of the third diagnostic level. Myocardial scintigraphy is non invasive; it results in a series of medical images that are relatively inexpensively obtained. In clinical practice, these images are manually described (parameterized) by expert physicians. In the paper we present an innovative alternative to manual image evaluation – an automatic image parameterization in multiple resolutions, based on texture description with specialized association rules. Extracted image parameters are combined into more informative composite parameters by means of principle component analysis, and finally used to build automatic classifiers with machine learning methods. Our experiments with synthetic datasets show that association-rule-based multi-resolution image parameterization equals or surpasses other state-of-the-art methods for finding multiple informative resolutions. Experimental results in coronary artery disease diagnostics confirm these results as our approach significantly improves the clinical results in terms of quality of image parameters as well as diagnostic performance.

Keywords: machine learning, coronary artery disease, medical diagnostics, multi-resolution image parameterization, association rules, principal component analysis.

1 Introduction

Image parameterization is a technique for describing bitmapped images with numerical parameters – features or attributes. Traditionally popular image features are first- and second-order statistics, structural and spectral properties, and several others. Image parameterization is very useful in quality control, identification, image grouping, surveillance, image storage and retrieval, and image querying. Over the past few decades we

A. Gagalowicz and W. Philips (Eds.): MIRAGE 2009, LNCS 5496, pp. 356–367, 2009.

observe extensive use of image parameterization in medical domains where texture classification is closely related to diagnostic process [4]. This complements medical practice, where manual image parameterization (evaluation of medical images by expert physicians) frequently plays an important role in diagnostic process.

Coronary artery disease (CAD) is one of the world's most frequent cause of mortality, and there is an ongoing research for improving diagnostic procedures. The usual clinical process of coronary artery disease diagnostics is stepwise, consisting of four diagnostic levels (1) evaluation of signs and symptoms of the disease and ECG (electrocardiogram) at rest, (2) ECG testing during the controlled exercise, (3) myocardial scintigraphy, and (4) coronary angiography.

In this process, the fourth diagnostic level (coronary angiography) is considered as the "gold standard" reference method. As this diagnostic procedure is invasive and rather unpleasant, as well as relatively expensive, there is a tendency to improve diagnostic performance of earlier diagnostic levels, especially of myocardial scintigraphy [9]. State of the art approaches used for this purpose include applications of neural networks [16], expert systems [5], subgroup mining, statistical techniques [19], and rule-based approaches [11]. In our study we focus on various aspects of improving the diagnostic performance of myocardial scintigraphy.

Results of myocardial scintigraphy consist of a series of medical images that are taken both during rest and a controlled exercise. These images can be relatively cheaply obtained and the imaging procedure does not represent a threat to patients' health. In clinical practice, expert physicians use their medical knowledge and experience as well as the image processing capabilities provided by various imaging software to manually describe (parameterize) and evaluate the images.

We propose an innovative alternative to manual image evaluation – automatic multi-resolution image parameterization, based on texture description with specialized association rules, coupled with image evaluation with machine learning methods. Since this approach yields a large number of relatively low-level features (though much more informative than simple pixel intensity values), we recommend using it in conjunction with additional dimensionality reduction techniques, either by discarding some features (feature selection), or combining them into more informative, high-level features (feature extraction). Our results show that multi-resolution image parameterization equals or even outperforms the physicians in terms of the quality of image parameters. By using both manual and automatic image description parameters at the same time, diagnostic performance can be significantly improved with respect to the results of clinical practice.

2 Methods

In image processing, a lot of work has been channeled into automatic (machine) recognition, description, classification, segmentation, retrieval, and grouping of patterns [20]. These are important problems in a variety of engineering and scientific disciplines such as biology, psychology, medicine, marketing, computer vision, artificial intelligence, and remote sensing [6].

An important issue in image parameterization in general, and in our approach with association rules in particular, is to select appropriate resolution(s) for extracting most informative textural features. Structural algorithms use descriptors of some local relations between image pixels where the search perimeter is bounded to a certain size. This means that they can give different results at different resolutions. The resolution used for extracting parameters is important and depends on the properties of the observed images.

We developed the algorithm (ARes) that finds suitable resolutions at which image parameterization algorithms achieve more informative features. From our experiments with synthetic data we observe that using parameterization-produced features at several different resolutions usually improves the classification accuracy of machine learning classifiers [21]. This parameterization approach is very effective in analyzing myocardial scintigraphy images used for CAD diagnostics in the stepwise process.

The obtained high quality image parameters can be used for several purposes, among others to describe images with a relatively small number of features, and use them in machine learning process. Images corresponding to patients with known correct final diagnosis can be used as learning data that, in conjunction with the applied machine learning methods, produces reliable decision support tools (classifiers) for the diagnostic problem at hand.

2.1   Image Parameterization

Image parameterization transforms the image from the matrix form into a set of numeric or discrete features (parameters) that convey useful high-level (compared to simple pixel intensities) information for discriminating between classes. Most texture features are based on structural, statistical or spectral properties of the image. For the purpose of diagnosis from medical images it seems that structural description is most appropriate [22]. For this we use the ArTex algorithm (described in Sec. 2.3) for textural attributes which are based on spatial association rules. The association rules algorithms can be used for describing textures if an appropriate texture representation formalism is used. This representation has several good properties like invariance to global brightness and invariance to rotation.

2.2   Image Classification with Machine Learning Methods

The ultimate goal of medical image analysis is decision about the diagnosis. When images are described with informative numerical attributes, we can use various machine learning algorithms [8] for generating a classification system (classifier) that produces diagnoses of the patients, whose images are being processed. For that purpose one can choose from multitude of machine learning methods. Based on our previous experience in medical diagnostics [9], we decided to use decision trees, naive Bayesian classifiers, Bayesian networks, K-nearest neighbors, and Support Vector Machines. Our early work in the problem of diagnosing the coronary artery disease from myocardial scintigraphy images [10] indicates that the naive Bayesian classifier gives the best results. Our results conform with several others [7] who also find out that in medical diagnosis the naive Bayesian classifier frequently outperforms other, often much more complex classifiers.

2.3   The ArTex Algorithm

Most texture features are based on structural, statistical or spectral properties of the image. The most frequently used space transformations are Fourier, Laws [12], Gabor [15] and wavelet transform. However, the ArTex algorithm that was used in our experiments is based on association rules. A full and detailed explanation of the ArTex algorithm can be found in [21]. Association rules capture structural and statistical information and are very convenient to identify the structures that occur frequently, and have most discriminative characteristic. Representation of texture with spatial association rules substitutes the precise information of location and intensity of the adjacent pixels with more general information – the distance and the relative intensity of neighboring pixels. This description is rotation-invariant and is suitable for processing with general association rule algorithms. Using association rules on textures allows to extract a set of features (attributes) for a particular domain of textures. Beside the basic parameters (interestingness measure, support and confidence) other parameters were used [21].

2.4   Multi-resolution Parameterization

Why use more resolutions? Digital images are stored in the matrix form and algorithms for pattern parameterization basically use some relations between image pixels (usually first or second order statistics). By using only a single resolution, we may miss the big picture, and proverbially not see the forest because of the trees (Figure 1). Since it is too computationally complex to observe all possible relations between at least any two pixels in the image, we have to limit the search to some predefined neighborhood. This limitation makes relations vary considerably over different resolutions. This means that we may get completely different image parameterization attributes for the same image at different scales.

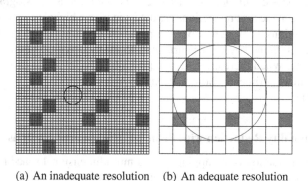

(a) An inadequate resolution   (b) An adequate resolution

Fig. 1. Detecting patterns at different scales. Taken from [21] with permission from authors.

Parameters from many resolutions. In different existing multi-resolution approaches [17], authors are using only a few fixed resolutions independently of the image contents. Usually, only two or three resolutions are used. Authors report better classification results when using more resolutions and also observe that when using more than three resolutions, the classification accuracy starts to deteriorate. In many cases authors use a set of resolutions by exponentially decreasing the resolution size ($100/2^i, i = 0 \ldots n - 1$). However, we found that in many cases [21] geometric selection of resolutions ($100i/n, i = 1 \ldots n$ where n is the number of resolutions used) gives better results. When using exponential form of resolutions, less pattern content is examined and consecutively less informative attributes are derived. Another frequently used "multi-resolution" approach is the wavelet transform [1], which describes textures with measures calculated with iterative image division. All the procedures mentioned above do not observe the image contents.

Automatic selection of a small subset of relevant resolutions. The idea for the algorithm for automatic selection of a small subset of relevant resolutions is derived from the well known SIFT algorithm [14]. SIFT is designed as a stable local feature detector which is a fundamental component of many image registration and object recognition algorithms. Since we are not interested in detecting stable image key-points but rather in detecting resolutions at which the observed image has most extremes, we devised a new algorithm ARes (Algorithm 1) for determining most informative resolutions. The algorithm was designed for the use with the ArTex parameterization algorithm.

The ARes algorithm scales down each image from 100% to some predefined lowest threshold at some fixed step while detecting appropriate resolutions. Both the lowest threshold and the resolution step are determined using the observed image dataset. At each resize step (matrix transform representing an affine scaling transformation) the *peaks* are counted. Peaks are defined as pixels which differ from their neighborhood either by the highest or the lowest intensity. This algorithm can be implemented also with DOG (Difference-Of-Gaussian) [14] method which improves the time complexity with lower number of actual resizes required to search the entire resolution space.

The detected peak counts are recorded over all resolutions as a histogram (Figure 2). From the histogram the best resolutions are detected by the highest counts of peaks. The number of resolutions we want to use in our parameterization is either predefined by the user, or can be determined by some heuristic method (such as the Minimum Description Length criterion). When there are several equal counts we chose as diverse resolutions as possible. Our experiments show that ArTex/ARes significantly outperforms other approaches [21].

2.5   Dimensionality Reduction with Principal Component Analysis

Dimensionality reduction is a mapping from a multidimensional space into a space of fewer dimensions. It is often the case that data analysis can be carried out in the reduced space more accurately than in the original space. More formally, the dimensionality reduction problem can be stated as follows: given the a-dimensional random variable $\mathbf{x} = (x_1, \ldots, x_a)$ find a lower dimensional representation of it, $\mathbf{s} = (s_1, \ldots, s_k)$ with $k < a$, that captures the content in the original data, according to some criterion.

Principal components analysis (PCA) is a linear transformation that chooses a new coordinate system for the data such that the greatest variance by any projection of the data set lies on the first axis (called the first principal component), the second greatest

Peak histogram

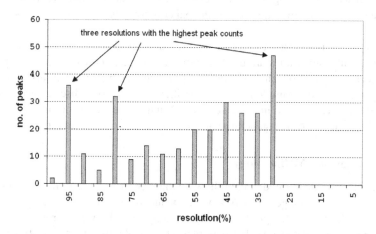

Fig. 2. The detected peak counts over all resolutions as a histogram

---

Algorithm 1. Algorithm ARes for detecting a small subset of relevant resolutions

INPUT: set of input images Θ with known classes, number of desired resolutions η, number of images to inspect in each class γ, radius ϕ which the parameterization algorithm uses later on in the process

OUTPUT: subset of resolutions Π

1:
$$W_{max} = \max_{i=1}^{|\Theta|}(\Theta_{i(width)}), H_{max} = \max_{i=1}^{|\Theta|}(\Theta_{i(height)})$$

{find the biggest image height and width}

2: extend the image sizes $\Theta_i \in \Theta$ to $W_{max} \times H_{max}$ with adding a frame of intensity equal to the average intensity of the original image Θ_i. New resized images are saved in the set Θ' {image sizes must be unified in order to be able to compare resolutions over different images}

3: $\delta = \frac{2*\phi}{3} \cdot \frac{1}{\min\{W_{max}, H_{max}\}}$ {set the resize step}

4: for each class add γ randomly selected images from the set Θ' into the set Θ_1

5: $\Omega = \{\}$

6: **for** ($\forall \theta \in \Theta_1$) **do**

7:    $\nu = 1.0$ {start with 100% resolution}

8:    **while** ($\min\{W_{max}, H_{max}\} \cdot \nu > 3 \cdot \phi$) **do**

9:       $\theta_1 = resize(\theta, \nu)$ {change the observed image's size}

10:      find local peaks in θ_1 with comparing each pixel's neighborhood inside $[3 \times 3]$ window

11:      add the pair $\{\nu, \text{number of peaks}\}$ into the set Ω

12:      $\nu = \nu - \delta$

13:    **end while**

14: **end for**

15: order the set Ω by the number of descending peaks and resolutions

16: add first η resolutions from the ordered set Ω into the final set Π

variance on the second axis, and so on. PCA can be used for reducing dimensionality in a dataset while retaining those characteristics of the dataset that contribute most to its variance by eliminating the lesser principal components (by a more or less heuristic decision).

PCA is sometimes used to extract features directly from images in matrix form, where pixel intensity values are used as primary features. Our experiments with using such a feature extraction on CAD images produced such dismal results of machine learning (on par with a simple majority classifier) that we were discouraged to further pursue in this direction. So in the case of CAD diagnostics from scintigraphic images, several thousands of ArTex-generated image features are used as an input for PCA.

2.6   Experimental Methodology

Experiments were performed in the following manner. First, 10 learning examples (images or sets of nine images for CAD) were excluded for data preprocessing and calibration of ArTex/ARes. Images from the remaining examples were parameterized; only the obtained parameters were subsequently used for evaluation. Further testing was performed in the ten-fold cross-validation setting: at each step 90% of examples were used for building a classifier, and the remaining 10% of examples for testing.

In CAD diagnostics that combines generated parameters for nine images, number of parameters (attributes) was reduced either with feature extraction – by applying PCA and retaining only the best principal components (those that together accounted for not less than 70% of data variance, amounting to the best 10 components), or with feature selection – by applying ReliefF [18] attribute quality estimation and again retaining only the best 10 (most important) ArTex/ARes generated, or physician-provided, attributes. In either case, besides the described 10 components, an equal number of the best attributes provided by physicians was used as estimated by ReliefF.

In each cross-validation step the real-valued attributes were discretized in advance by using the Fayyad-Irani [3] algorithm, if the applied method (such as the naive Bayesian classifier) required solely discrete attributes.

We applied four popular machine learning algorithms: naive Bayesian classifier, tree-augmented Bayesian network, support vector machine (SMO using RBF kernel), and J4.8 (C4.5) decision tree. We performed experiments with Weka (*www.cs.waikato.ac.nz/ml/weka*) machine learning toolkit. For CAD diagnostics, aggregated results of the coronary angiography (CAD negative/CAD positive) were used as the class variable. The results of clinical practice were validated by careful blind evaluation of images by an independent expert physician. Significance of differences to clinical results was evaluated by using the McNemar's test.

3   Materials

In our CAD study we use a dataset of 288 patients with performed clinical and laboratory examinations, exercise ECG, myocardial scintigraphy (including complete image sets) and coronary angiography because of suspected CAD. The features from the ECG an scintigraphy data were extracted manually by the clinicians. 10 patients were excluded for data pre-processing and calibration required by ArTex/ARes, so only 278 patients (66 females, 212 males, average age 60 years) were used in actual experiments. In 149 cases the disease was angiographically confirmed and in 129 cases it was excluded. The patients were selected from a population of several thousands patients who were examined at the Nuclear Medicine Department between 2001 and 2006. We selected only the patients with complete diagnostic procedures (all four levels), and for whom the imaging data was readily accessible. Some characteristics of the dataset are shown in Table 1.

Table 1. CAD data for different diagnostic levels. Of the attributes belonging to the coronary angiography diagnostic level, only the final diagnosis – the two-valued class – was used in experiments

Diagnostic level	Number of attributes		
	Nominal	Numeric	Total
1. Signs and symptoms	22	5	27
2. Exercise ECG	11	7	18
3. Myocardial scintigraphy (+9 image series)	8	2	10
4. Coronary angiography	1	6	1
Class distribution	129 (46.40%)	CAD negative	
	149 (53.60%)	CAD positive	

It must be noted that our patients represent a highly specific population, since several had already undergone cardiac surgery or dilatation of coronary vessels. This clearly reflects the situation in Central Europe with its aging population. It is therefore not surprising that both the population and the diagnostic performance are considerably different than that of our previous study, where data were collected between years 1991 and 1994 [9]. Our results are therefore not applicable to the general population, and vice versa, general findings only partially apply to our population.

Fore each patient a series of images was taken with the General Electric eNTEGRA SPECT camera, both at rest and after a controlled exercise, thus producing the total of 64 grayscale images in resolution of 64×64 8-bit pixels. Because of patients' movements and partial obscuring of the heart muscle by other internal organs, these images are not suitable for further use without heavy pre-processing. For this purpose, the EC-Toolbox workstation software [2] was used, and one of its outputs, a series of 9 polar map (bull's eye) images was taken for each patient. Polar maps were chosen because previous work in this field [13] had shown that they have useful diagnostic value.

4   Results

4.1   Results in CAD Diagnostics

As described before, out of the 288 patients, 10 were excluded for data preprocessing and calibration required by ArTex/ARes. These patients were not used in further experiments. The remaining 278 patients with 9 images each were parameterized for three resolutions in advance. ARes proposed three[1] resolutions: $0.95\times, 0.80\times$, and $0.30\times$ of the original resolution, producing together 2944 additional attributes (features, parameters). Since this number is too large for most practical purposes, it was reduced either by applying feature selection (with ReliefF) or by feature extraction (with PCA). We also performed some experiments with other image parameterization approaches such as wavelet and DFT transform (Haar and Laws), Gabor filters and SIFT features; they, however, invariably performed significantly worse than ArTex.

Experimental results were compared with diagnostic accuracy, specificity and sensitivity of expert physicians after evaluation of scintigraphic images. Results of clinical practice were validated by careful blind evaluation of images by the expert physician.

In machine learning experiments we experimented with several different settings; due to lack of space we describe here only the most advanced one, namely evaluation of the best 10 attributes (accounting for 70% of data variance) extracted by PCA from ArTex/ARres-generated attributes, used with and without the best 10 attributes, provided by physicians (as estimated by the ReliefF algorithm [18])

Evaluation of the best attributes extracted by PCA from ArTex/ARres-generated attributes. We extracted the best 10 principal components (linear combinations of original 2944 ArTex/ARes attributes) by PCA, or select the best 10 original attributes with ReliefF from the set of 2944 ArTex/ARes attributes. Additionally, we also enriched the data representation by using the same number (10) of the best physicians' attributes as evaluated by ReliefF (i.e. sex, diabetes, low HDL, LKB, exercise LCX, exercise RCA, exercise APEX, rest LCX, LUEF), and compared their results with results of machine learning.

Table 2. Experimental results of machine learning classifiers on parameterized images obtained by selecting only the best 10 attributes from ArTex/ARes (also combined with the best 10 attributes provided by physicians). Classification accurracy results that are significantly better ($p < 0.05$) than clinical results are emphasized.

	ArTex/ARes			ArTex/ARes+physicians		
	Accurracy	Specificity	Sensitivity	Accurracy	Specificity	Sensitivity
Naive Bayes	**69.4%**	58.9%	78.5%	**74.8%**	70.5%	78.5%
Bayes Net	**69.4%**	58.9%	78.5%	**74.4%**	69.8%	78.5%
SMO (RBF)	**71.9%**	65.1%	77.9%	**73.4%**	65.9%	79.9%
J4.8	**70.9%**	61.2%	79.2%	**68.0%**	63.6%	71.8%
Clinical	64.0%	71.1%	55.8%	64.0%	71.1%	55.8%

Tables 2 and 3 and Figure 3 depict the results. It is gratifying to see that — without any special tuning of learning parameters — the results are in all cases significantly better than the results of physicians in terms of classification (diagnostic) accuracy. Especially good results are that of the naive Bayesian classifier (Table 3), that improve in all three criteria: diagnostic accuracy, sensitivity and specificity. Another interesting issue is that including the best attributes provided by a physician does not necessarily

---

[1] A resolution of $0.30\times$ means $0.30 \cdot 64 \times 0.30 \cdot 64$ pixels instead of 64×64 pixels.

Table 3. Experimental results of machine learning classifiers on parameterized images obtained by selecting only the best 10 attributes from PCA on ArTex/ARes (also combined with the best 10 attributes provided by physicians). Classification accuracy results that are significantly better ($p < 0.05$) than clinical results are emphasized.

	PCA on ArTex/ARes			PCA on ArTex/ARes+physicians		
	Accuracy	Specificity	Sensitivity	Accuracy	Specificity	Sensitivity
Naive Bayes	**81.3%**	83.7%	79.2%	**79.1%**	82.9%	75.8%
Bayes Net	**71.9%**	69.0%	74.5%	**79.1%**	83.7%	75.2%
SMO (RBF)	**78.4%**	76.0%	80.1%	**76.6%**	77.5%	75.8%
J4.8	**75.2%**	78.3%	72.5%	**74.1%**	73.6%	74.5%
Clinical	64.0%	71.1%	55.8%	64.0%	71.1%	55.8%

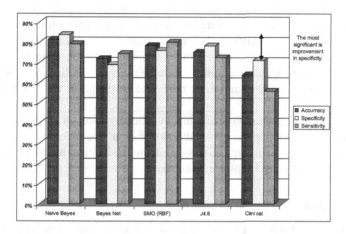

Fig. 3. Comparison of clinical results and results of machine learning classifiers on parameterized images from Table 3

improve diagnostic performance (SMO, J4.8 in Table 3). It seems that there is some level of redundancy between physicians' and principal components generated from ArTex/ARes attributes, that bothers some methods more than the others. Consequently, it seems that some of automatically generated attributes are (from the diagnostic performance point of view) at least as good as the physician-provided ones, and may therefore represent new knowledge about CAD diagnostics.

5 Discussion

In clinical practice, expert physicians use their medical knowledge and experience as well as image processing capabilities provided by various imaging software to manually describe (parameterize) and evaluated the images. We describe an innovative alternative to manual image evaluation - automatic multi-resolution image parameterization based on spatial association rules (ArTex/ARes) supplemented with feature selection or (preferably) feature extraction. Our results show that multi-resolution image parameterization equals or even betters the physicians in terms of diagnostic quality of image parameters. By using these parameters for building machine learning classifiers, diagnostic performance can be significantly improved with respect to the results of clinical practice. We also explore relations between newly generated image attributes and physicians' description of images. Our findings indicate that ArTex/ARes with PCA is likely to extract more useful information from images than the physicians do, as it significantly outperforms them in terms of diagnostic accuracy, specificity and sensitivity.

Utilizing machine learning methods can help less experienced physicians evaluate medical images and thus improve their performance (in terms of accuracy, sensitivity and specificity). From the practical use of described approaches two-fold improvements of the diagnostic procedure can be expected. Higher diagnostic accuracy (up to 17%) is by itself a very considerable gain. Due to higher specificity of tests (up to 12%), fewer patients without the disease would have to be examined with coronary angiography which is invasive and therefore dangerous method. Together with higher sensitivity this would save money and shorten the waiting times of the truly ill patients. Also, new attributes, generated by ArTex/ARes with PCA had invoked considerable interest from expert physicians, as they significantly contribute to increased diagnostic performance and may therefore convey some novel medical knowledge of the CAD diagnostics problem.

And finally, we need to emphasize again, that the results of our study are obtained on a significantly restricted population and therefore may not be generally applicable to the normal population, i.e. to all the patients coming to the Nuclear Medicine Department, University Clinical Centre Ljubljana, Slovenia.

5.1 Future Work

The presented algorithms for pattern parameterization open a new research area of multi-resolution image parameterization and enable many applications in medical, industrial and other domains where textures or texture-like surfaces are classified. The algorithm ARes can be improved with additional resolution search refinements which would be more domain oriented, and with heuristic methods for controlling selection of resolutions.

In the CAD diagnostics problem we intend to focus even more on improving the diagnostic performance of the third diagnostic level (myocardial perfusion scintigraphy), and assess different criteria for resolution quality. We will study the relations between automatically generated attributes (PCA on ArTex/ARes attributes), and attributes provided by a physician, and try to establish the correspondence between them (if existing). All of possible improvements of the parameterization and classification scheme will be used in the post-test probability estimation setting for evaluation of reliability of machine-generated diagnoses.

Acknowledgements

We thank Igor Kononenko and Ciril Grošelj for several enlightening comments on our work. Research described in this paper was supported by the Slovenian Ministry of Higher Education, Science, and Technology.

References

[1] Chui, C.K.: An Introduction to Wavelets. Academic Press, San Diego (1992)
[2] General Electric. ECToolbox Protocol Operator's Guide (2001)
[3] Fayyad, U.M.: Multi-interval discretization of continuous-valued attributes for classification learning. In: Proc. IJCAI 1993, pp. 1022–1027. Morgan Kaufmann, San Francisco (1993)
[4] Fitzpatrick, J., Sonka, M.: Handbook of Medical Imaging, Medical Image Processing and Analysis, vol. 2. SPIE, Bellingham (2000)
[5] Garcia, E.V., Cooke, C.D., Folks, R.D., Santana, C.A., Krawczynska, E.G., De Braal, L., Ezquerra, N.F.: Diagnostic performance of an expert system for the interpretation of myocardial perfusion spect studies. J. Nucl. Med. 42(8), 1185–1191 (2001)
[6] Jain, A.K., Duin, R.P.W., Mao, J.: Statistical pattern recognition: A review. IEEE Transactions on Pattern Analysis and Machine Intelligence 22(1), 4–37 (2000)
[7] Kononenko, I.: Machine learning for medical diagnosis: history, state of the art and perspective. Artificial Intelligence in Medicine 3, 89–109 (2001)
[8] Kononenko, I., Kukar, M.: Machine Learning and Data Mining: Introduction to Principles and Algorithms. Horwood publ., England (2007)
[9] Kukar, M., Kononenko, I., Grošelj, C., Kralj, K., Fettich, J.: Analysing and improving the diagnosis of ischaemic heart disease with machine learning. Artificial Intelligence in Medicine 16(1), 25–50 (1999)
[10] Kukar, M., Šajn, L., Grošelj, C., Grošelj, J.: Multi-resolution image parametrization in sequential diagnostics of coronary artery disease. In: Bellazzi, R., Abu-Hanna, A., Hunter, J. (eds.) Artificial intelligence in medicine, pp. 119–129. Springer, Heidelberg (2007)
[11] Kurgan, L.A., Cios, K.J., Tadeusiewicz, R.: Knowledge discovery approach to automated cardiac spect diagnosis. Artif. Intell. Med. 23(2), 149–169 (2001)
[12] Laws, K.I.: Textured image segmentation. Ph.D thesis, Dept. Electrical Engineering, University of Southern California (1980)
[13] Lindahl, D., Palmer, J., Pettersson, J., White, T., Lundin, A., Edenbrandt, L.: Scintigraphic diagnosis of coronary artery disease: myocardial bull's-eye images contain the important information. Clinical Physiology 6(18) (1998)
[14] Lowe, D.G.: Distinctive image features from scale-invariant keypoints. Int. J. Comput. Vision 60(2), 91–110 (2004)
[15] Manjunath, B.S., Ma, W.Y.: Texture features for browsing and retrieval of image data. IEEE Trans. Pattern Anal. Mach. Intell. 18(8), 837–842 (1996)
[16] Ohlsson, M.: WeAidU–a decision support system for myocardial perfusion images using artificial neural networks. Artificial Intelligence in Medicine 30, 49–60 (2004)
[17] Ojala, T., Pietikainen, M., Maenpaa, T.: Multiresolution gray-scale and rotation invariant texture classification with local binary patterns. IEEE Transactions on Pattern Analysis and Machine Intelligence 24(7), 971–987 (2002)
[18] Robnik-Šikonja, M., Kononenko, I.: Theoretical and empirical analysis of ReliefF and RReliefF. Machine Learning 53, 23–69 (2003)
[19] Slomka, P.J., Nishina, H., Berman, D.S., Akincioglu, C., Abidov, A., Friedman, J.D., Hayes, S.W., Germano, G.: Automated quantification of myocardial perfusion spect using simplified normal limits. J. Nucl. Cardiol. 12(1), 66–77 (2005)
[20] Vasconcelos, N.: From pixels to semantic spaces:advances in content-based image retrieval. Computer 40(7), 20–26 (2007)
[21] Šajn, L., Kononenko, I.: Multiresolution image parametrization for improving texture classification. EURASIP J. Adv. Signal Process 2008(1), 1–12 (2008)
[22] Šajn, L., Kononenko, I.: Image segmentation and parametrization for automatic diagnostics of whole-body scintigrams. In: Computational Intelligence in Medical Imaging: Techniques & Applications. CRC Press, Boca Raton (2009)

Interpreting Face Images by Fitting a Fast Illumination-Based 3D Active Appearance Model

Salvador E. Ayala-Raggi, Leopoldo Altamirano-Robles,
and Janeth Cruz-Enriquez

Instituto Nacional de Astrofísica, Óptica y Electrónica,
Coordinación de Ciencias Computacionales
Luis Enrique Erro #1, 72840 Sta Ma. Tonantzintla. Pue., México
{saraggi,robles,jcruze}@ccc.inaoep.mx

Abstract. We present a fast and robust iterative method for interpreting face images under non-uniform lighting conditions by using a fitting algorithm which utilizes an illumination-based 3D active appearance model in order to fit a face model to an input face image. Our method is based on improving the Jacobian each iteration using the parameters of lighting that have been estimated in preceding iterations. In the training stage, we precalculate a set of synthetic face images of basis reflectances and albedo generated from displacing one at the time, each one of the model parameters, and subsequently, in the fitting stage, we use all these images in combination with lighting parameters for assembling a Jacobian matrix adapted to the illumination estimated in the last iteration. In contrast to other works where an initial pose is required to begin the fit, our approach only uses a simple initialization in translation and scale. At the end of the fitting process, our algorithm obtains a compact set of parameters of albedo, 3D shape, 3D pose and illumination which describe the appearance of the input face image.

Keywords: Active appearance models, face interpreting, face alignment, 3D model fitting, face modeling.

1 Introduction

Determining the 3D shape, texture, 3D pose and illumination of a face from a single image is one of the main goals in face image interpretation. Several authors have shown that the approaches based on the "interpretation through synthesis" are effective for interpreting faces in novel images [3]. The aim is to explain novel images by generating synthetic ones that are as similar as possible. This process is also known as face alignment. 3D Morphable Models (3DMM), detailed in [2],[3], [4], and [6] are photo-realistic methods for accurate face modeling and alignment, despite their accuracy, they are slow because they handle the face surface with a dense model and require a large number of parameters

A. Gagalowicz and W. Philips (Eds.): MIRAGE 2009, LNCS 5496, pp. 368–379, 2009.

[8]. In addition, they require some anchor points manually placed over key features of the face to indicate the algorithm the rough initial pose and location of the face to fit. Concerning to illumination, the 3DMM approach handles only directed light using the Phong's reflectance model (Lambertian and specular), and is unable to model multiple illuminants and diffuse lighting. On the other hand, Active Appearance Models [7],[8], ($AAMs$) are sparse generative models for fast 2D face alignment often required in real time applications. Even though their speed, classical $AAMs$ are $2D$ and very sensitive to lighting, particulary when the lighting during the testing stage is different from illumination during the training stage. Great efforts have been done mainly on improving the quality of the alignment in 2D $AAMs$ and maintaining their speed by always using a constant Jacobian during the fitting process. Baker and Matthews et al. in [1] and [10], propose an inverse compositional optimization method for fitting active appearance models. They point out that the essentially additive method of updating the parameters in the basic framework [8] can be problematic, however $ICIA$ (Inverse Compositional Image Alignment) has a limited domain of application. It is a fitting algorithm for 2D AAM, hence cannot handle out of the image plane rotation and directed light [5]. Less attention has been given in simultaneously handling illumination, 3D pose and 3D shape in $AAMs$. Xiao et al. [9] propose a 2D+3D AAM which exploits the 2D shape and 3D shape models simultaneously. The shape instance generated by 2D AAM is varied to be consistent with a possible 3D shape. This constraint is formulated as a part of the cost function. To combine this constraint into the original cost function, a balancing weighting is added. The value of this weighting constant is determined manually. Therefore, it is not a natural and direct way for estimating 3D pose and shape. These last mentioned works do not consider the problem of illumination. In [11] and [12], authors propose methods for $2D$ face alignment under different illumination conditions by preprocessing the image to eliminate the effect of lighting before applying classical AAM fitting, however their approaches are not able to recover 3D shape and lighting after the fitting process. Kahraman et al in [13], propose an approach which integrates the classical AAM model (shape and texture) with a statistical illumination model. Their model, called AIA (Active Illumination Appearance model) is 2D and consists of two linear subspaces: one for illumination, and another for identity. In [14], it is proposed a 3D algorithm for face tracking in video sequences based on AAM models which uses a generic $3D$ human shape frame called $Candide$ developed at Linköping University and does not include the problem of illumination. In [15], Sattar et al. propose a fast face alignment method based on a 2.5D AAM model optimized by Simplex. This technique does not consider illumination. On the other hand, there are some works on lighting modeling for effective face recognition under non-uniform illumination conditions [18],[19],[20]. A widely used method on illumination modeling was proposed by Basri et al. [16], this model is known as 9D subspace model and is useful for its ability to model directed and non-directed light (multiple lights and diffuse light). According to this approach, any reflectance over a face can be approximated in 97.96% of accuracy using a linear

combination of 9 spherical harmonic reflectances, obtained from the surface normals and the albedos of the face surface. Based on the reviewed work, we notice that *AAM* models have been used for fast 2D face alignment under variable conditions of lighting but not for estimation of 3*D* pose, 3D shape, albedo and illumination under non-uniform lighting conditions, which is still a challenging problem. In contrast, some authors have proposed 3D *AAMs* for estimating 3*D* pose and shape but do not include illumination. In our work we propose to construct a 3D Active Appearance Model based on the 9D subspace illumination model. We present an efficient optimization method that matches 3D shape, 3D pose, albedo, and illumination simultaneously in each iteration, in a rapid and accurate approach. As a natural extension of 2D Active Appearance Models, our method is capable of estimating 3*D* pose, 3*D* shape, and illumination by fitting a 3*D* face model to a novel image. For modeling illumination, we use harmonic reflectances obtained from the surface normals and albedos maps of faces. In the training stage, our algorithm learns a linear model of the correlation between variation of parameters and induced residuals without considering lighting (whose parameters are not in a limited range). In the fitting stage, we use the estimated lighting parameters obtained in preceding iterations for updating the Jacobian and also the reference mean model. Instead of using a constant Jacobian as in [8], we use an adaptive one. Using the fact that a face image can be represented as a multiplication of a matrix containing basis reflectances by a vector of lighting parameters, it is possible to recalculate "on line" the Jacobian according to lighting parameters. For this purpose, we precalculate a set of matrices of basis reflectances during the training phase. For each parameter displaced in a suitable quantity from its mean state, we compute a corresponding matrix of basis reflectances. The importance of this work is its contribution with a promising and easy technique to fit a complete (3D pose and shape, albedo and illumination) parametric 3D face model to difficult face images with any kind of illumination (directed, multiple and diffuse) for interpretation purposes.

2   Modeling Lighting

In [16], Basri et al. show that any illumination over a face can be represented by a linear combination of n basis images,

$$\mathbf{I} = \mathbf{B}\mathbf{H}^T\mathbf{L} \qquad (1)$$

where $\mathbf{L}$ is a vector containing n arbitrary parameters and $\mathbf{B}$ is a matrix which columns are nine spherical harmonic images constructed by using an albedos map and a surface normals map. Columns in $\mathbf{H}$ contain samples of the harmonic functions, whereas its rows contain the spherical harmonics transform of delta functions (punctual light sources). To obtain a good approximation we should use a large set of n punctual lights uniformly distributed around the sphere. However, in [17], Lee et al., showed that it is possible to achieve good results in face recognition using only $n = 9$ punctual sources of light strategically distributed. This distribution can approximate any reflectance on a face. Thus, we can construct a matrix $\mathbf{H}$ using nine deltas in order to project them into the spherical harmonics subspace.

3 Face Synthesis Using an Illumination-Based 3D Active Appearance Model (3D-IAAM)

For constructing parametric models of 3D shape and 2D albedos, we need the 3D face surfaces of the individuals, and their corresponding 2D albedos maps. For recovering the surfaces, we used a technique based on *shape from shading*. Using 11 different images per individual, each one illuminated by a different punctual light source, it is possible to simultaneously estimate surface normals maps and albedos maps. This is accomplished by using minimum squares for solving a linear system of 11 equations, each one expressing the pixel intensity as a function of the direction of the incident light (Lambert's cosine law), for each pixel. From surface normals maps it is possible to reconstruct the surface of each face by using *shapelets* [21]. We define a *shape model* as the set of landmarks over a face surface. In order to obtain a statistical 3D shape model, first we have to place 3D landmarks over the surface of multiple faces with different identity. Then, we align the 3D shape models and apply PCA to the set in order to obtain the principal modes of variation of 3D shape. We can generate an arbitrary model using the following expression

$$\mathbf{s} = \bar{\mathbf{s}} + \mathbf{Q}_s \mathbf{c} \tag{2}$$

where $\bar{\mathbf{s}}$ is the mean shape model and $\mathbf{Q}_s$ is a matrix which contains the basis shapes (also known as *eigenshapes*) and $\mathbf{c}$ is a vector with arbitrary shape parameters. Similarly, we apply PCA to the set of shape-normalized 2D albedos maps. Before applying PCA, the albedos map of each training face must be shape-normalized (using the bidimensional projection of the mean shape frame). A triangulation is designed to warp original images to the mean shape frame. Finally, any shape-normalized albedos image can be generated with

$$\lambda = \bar{\lambda} + \mathbf{Q}_\lambda \mathbf{a} \tag{3}$$

where $\bar{\lambda}$ is the mean albedos image, $\mathbf{Q}_\lambda$ is a matrix which contains principal albedo variation modes and $\mathbf{a}$ is a vector of arbitrary parameters. Using the previous expression (eq. 3), it is possible to synthesize an arbitrary albedo and then warp it to the 2D projection of an arbitrary shape generated with eq. 2. In this way, we have a new face with an arbitrary albedo and shape. This new face is not illuminated yet. In the process of warping albedos to the new shape, it is also possible to warp a shape-normalized mean 2D map of surface normals, which was calculated during the training stage. Now, we have an albedos image and a surface normals map shaped over the new shape. Using these maps (albedos and normals), we can create 9 basis reflectance images. Any illumination can be generated by a linear combination of these basis images using eq. 1. To give a

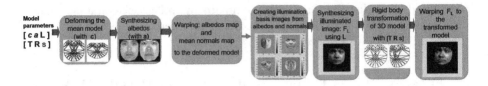

Fig. 1. Synthesizing arbitrary faces using the direct 3D-IAAM model

pose to the model, we use the 3D landmarks of the new generated 3D shape. By applying a rigid body transformation (T,R,s) to these landmarks it is possible to give any pose and size to the created face. Finally, we warp the frontal illuminated face to the 2D projection of the transformed 3D shape, as shown on Fig. 1.

4   Face Alignment Using 3D-IAAM Model

We can consider the face synthesizer as a transformation of the mean face which can result in an arbitrary face, $\mathbf{f} = T_{3D-IAAM}(\bar{\mathbf{f}})$, where $\bar{\mathbf{f}}$ is the mean face, and $\mathbf{f}$ is the resulting synthetic face with arbitrary shape, albedo, illumination and pose. Following the same notation, we should use an inverse transformation for the alignment process:

$$\mathbf{r} = T_{3D-IAAM}^{-1}(\mathbf{I}) - \bar{\mathbf{f}} \qquad (4)$$

Here, I represents a sampled region from the input image, and r, the error or residual image whose energy is a quantity to minimize by the optimization algorithm. We propose an extension of the iterative fitting algorithm in [8]. Our technique uses the $3D - IAAM$ inverse model in each iteration. During the fit, according to the parameters of the inverse model, the pixels inside a region in the image are sampled and transformed. So, the residuals image computed with (4) is a function of the model parameters $\mathbf{p}$. The first order Taylor expansion of (4) gives $\mathbf{r}(\mathbf{p} + \delta\mathbf{p}) = \mathbf{r}(\mathbf{p}) + \frac{\delta\mathbf{r}}{\delta\mathbf{p}}\delta\mathbf{p}$, here, $\mathbf{p}^T = (\mathbf{T}^T|\mathbf{R}^T|\mathbf{s}^T|\mathbf{c}^T|\mathbf{a}^T|\mathbf{L}^T)$, and the $ij - th$ element of the matrix $\frac{\delta\mathbf{r}}{\delta\mathbf{p}}$ is $\frac{\delta r_i}{\delta p_j}$. We desire to choose $\delta\mathbf{p}$ such that it minimize $|r(\mathbf{p} + \delta\mathbf{p})|^2$. Equating $\mathbf{r}(\mathbf{p} + \delta\mathbf{p})$ to zero leads to the solution

$$\delta\mathbf{p} = -\mathbf{R}\mathbf{r}(\mathbf{p}) \quad where \quad \mathbf{R} = (\frac{\delta\mathbf{r}^T}{\delta\mathbf{p}}\frac{\delta\mathbf{r}}{\delta\mathbf{p}})^{-1}\frac{\delta\mathbf{r}^T}{\delta\mathbf{p}} \qquad (5)$$

$\frac{\delta\mathbf{r}}{\delta\mathbf{p}}$ is actually a gradient matrix or simply a Jacobian changing in each iteration. Recalculating it at every step is expensive. Cootes et al. in [8], assume it to be constant since it is been computed in a normalized reference frame. This assumption is valid when we are only considering variations of texture, and lighting is ignored because it is uniform. Since texture parameters do not present a large variation between training faces, then, it is possible to compute a weighted average of the residuals images for each displaced parameter in order to obtain an average constant Jacobian. In our case, we are dealing with non-uniform illumination, therefore we propose to construct an adaptive Jacobian as is explained further.

4.1   Inverse 3D-IAAM Model

Basri [16], shows that the albedos map is actually a constant matrix which multiplies component by component to each one of the harmonic reflectance images. Thus, if we denote basis reflectances matrix as $\beta = \mathbf{BH}^T$, then eq. 1 can be expressed as

$$\mathbf{I}_{illuminated\,face} = \beta\mathbf{L} = ([\lambda..\lambda] \cdot \mathbf{\Phi})\mathbf{L} \tag{6}$$

where λ is the albedos map represented as a column vector repeated in order to form a matrix with the same dimensions as the basis reflectances matrix without albedo, represented by $\mathbf{\Phi}$. These two matrices are multiplied component by component (Hadamard product). Then, $\mathbf{I}_{illuminated\,face}$ can be rewritten as

$$\mathbf{I}_{illuminated\,face} = \lambda \cdot (\mathbf{\Phi L}) \tag{7}$$

Normalization of shape, pose and albedo. Cootes et al. [8] propose to normalize in shape a sampled region in order to compare it with a shape-normalized model in each iteration of the fitting algorithm. Because this comparison is always done in a fixed reference frame, it is possible to think that Jacobian is roughly constant during all the fitting process. Our model has a greater dimensionality because instead of only using texture and 2D shape parameters, we have included parameters of 3D pose, albedo and lighting. In particular, we know that lighting affects the appearance of a face more than identity and has an infinite number of degrees of freedom. In our case, it is not appropriate to use a constant Jacobian. Perhaps, it will be suitable if at the beginning of the fitting process, the algorithm is close to the convergence. However, when the lighting of the face is quite different from the initial illumination of the model, then it is difficult for Jacobian to remain constant. We propose to evolve the mean reference model only in lighting each iteration and use an appropriate Jacobian computed from that current lighting. Then, for comparing the sampled region with our reference model, we have to normalize the sampled region in shape, pose, and albedo.

Pose and shape normalization. Using the rigid body transformation parameters $(\mathbf{T}, \mathbf{R}, \mathbf{s})$ and the shape parameters $\mathbf{c}$, we sample a region in the image, and warp this region to the mean shape frame. This new shape-normalized image is denoted as $\mathbf{I}_{shape\,aligned}$.

Albedo normalization. If we could know the illumination of the face into the sampled region, then albedo can be estimated using eq. 7

$$\hat{\lambda} = (\mathbf{I}_{shape\,aligned}).{/}\mathbf{\Phi}\hat{\mathbf{L}} \tag{8}$$

Here, ./ denotes the element-wise division. Using this estimated albedos map, we can derive an approximated mean albedos map

$$\tilde{\lambda} \approx \hat{\lambda} - \mathbf{Q}_\lambda \mathbf{a} \tag{9}$$

Finally, the image normalized in albedo is

$$\mathbf{I}_{aligned} = (\tilde{\lambda}) \cdot (\mathbf{\Phi}\hat{\mathbf{L}}) \tag{10}$$

where $\hat{\mathbf{L}}$ is a vector containing the current estimated illumination parameters. Finally, we can rewrite eq. 10 as

$$\mathbf{I}_{aligned} = [I_{shape\,aligned} \cdot /(\mathbf{\Phi}\hat{\mathbf{L}}) - \mathbf{Q}_\lambda \mathbf{a}] \cdot (\mathbf{\Phi}\hat{\mathbf{L}}) \tag{11}$$

and the expression to minimize will be

$$\|\mathbf{r}\|^2 = \|\mathbf{I}_{aligned} - \bar{\lambda} \cdot (\mathbf{\Phi}\hat{\mathbf{L}})\|^2 \tag{12}$$

The inverse model used during alignment is shown in Fig. 2.

Fig. 2. Inverse 3D-IAAM model

4.2   Iterative Fitting Algorithm

In [8], the fitting process consists of computing a constant Jacobian matrix which is used during all the fitting process. In each iteration, a sampled region of the image is compared with a reference face image normalized in shape which is updated only in texture according to the current estimated parameters. This constant Jacobian works well in uniform lighting conditions, because texture variation is small. However, in both the Cootes approach and our method, when the lighting of the input face is harsh and very different from the lighting used in the training stage, the alignment fails if we use a constant Jacobian. The optimum procedure would be to recalculate the Jacobian each iteration by displacing parameters of albedo and illumination from their current estimated values, and displacing 3D shape and pose parameters from their mean state values, in order to synthesize the required images for computing residuals. Then, residuals and parameters displacements can be used for computing the Jacobian. That would be an expensive operation. Here, we propose to sample a region of the image and normalize it in shape and albedo. Then, we have only to *relight* the reference mean model (a model with mean shape and albedo) each iteration and *relight* a constant Jacobian using the current estimated lighting. Therefore, our reference model will be a face image with mean shape, mean albedo, mean pose and variable illumination. Updating the Jacobian with the current estimated illumination is an easy and computationally light step, because we use the fact that lighting and albedo are separated vectors and they are independent of basis reflectance images, see eq. 7. In the training phase, we construct a set of displaced images for using in the fitting phase to update the Jacobian. We know that basis reflectances $\boldsymbol{\Phi}$ (without albedo) are not affected by albedo displacements, but they can be modified by pose and shape increments. Our model uses 33 parameters: 6 for pose, 9 for 3D shape, 9 for illumination, and 9 for albedo. We construct $6 + 9 = 15$ basis reflectance matrices $\boldsymbol{\Phi}_{p_i + \Delta p_i}$ by displacing in a suitable quantity each one of the 15 parameter of pose and shape. In practice we construct 30 basis reflectance matrices because we consider 15 positive displacements and 15 negative displacements. In a similar way, by displacing each parameter with a suitable increment $p_i + \Delta p_i$ (positive and negative) we obtain 30 albedo images for positive and negative increments in pose and shape parameters, and 18 albedo images for positive and negative increments in albedo parameters. These albedo images do not have information about lighting. These 30 reflectance matrices and 48 albedo images are created during the training phase (*off-line*). During the alignment phase we can create a Jacobian *on-line* according to the current parameters of illumination $\mathbf{L}$, $\frac{\delta \mathbf{r}}{\delta \mathbf{p}} = [\frac{\delta \mathbf{r}_1}{\delta \mathbf{p}_1} \cdots \frac{\delta \mathbf{r}_{33}}{\delta \mathbf{p}_{33}}]$ where $\frac{\delta \mathbf{r}_i}{\delta \mathbf{p}_i} = [\frac{\delta \mathbf{r}_i}{\delta \mathbf{p}_{i \, (\Delta+)}} + \frac{\delta \mathbf{r}_i}{\delta \mathbf{p}_{i \, (\Delta-)}}] \times \frac{1}{2}$ and

$$\frac{\delta \mathbf{r}_i}{\delta \mathbf{p}_{i \, (\Delta+)}} = \frac{\lambda_{p_i + \Delta p_i} \cdot [\boldsymbol{\Phi}_{p_i + \Delta p_i} \mathbf{L}] - \lambda_{p_i} \cdot [\boldsymbol{\Phi}_{p_i} \mathbf{L}]}{\Delta p_i} \tag{13}$$

$$\frac{\delta \mathbf{r}_i}{\delta \mathbf{p}_{i \, (\Delta-)}} = \frac{\lambda_{p_i - \Delta p_i} \cdot [\boldsymbol{\Phi}_{p_i - \Delta p_i} \mathbf{L}] - \lambda_{p_i} \cdot [\boldsymbol{\Phi}_{p_i} \mathbf{L}]}{-\Delta p_i} \tag{14}$$

Into the Jacobian matrix, the columns corresponding to illumination parameters are maintained fixed during the fitting process and they are precalculated from a mean state of uniform lighting. One step of the iterative refinement process is as follows:

1. Project the sampled region into the mean-shape model frame using $\mathbf{I}_{aligned} = T^{-1}_{3D-IAAM}(\mathbf{I})$. This implies a pose, shape and albedo normalization using the mean lighting parameters in the first iteration.
2. Compute the residual, $\mathbf{r} = \mathbf{I}_{aligned} - \mathbf{I}_{mean-shape}$, and the current error, $E = |\mathbf{r}|^2$.
3. If it is the first iteration use a constant precalculated jacobian $\mathbf{J}$, else, assemble it by using the precomputed images of basis reflectance and albedo in combination with the estimated parameters $\mathbf{L}$ computed in last iteration, see eqs. 13 and 14.
4. Compute the predicted displacements, $\delta \mathbf{p} = -\mathbf{R}\mathbf{r}(\mathbf{p})$. Here $\mathbf{R}$ is the Moore-Penrose pseudoinverse matrix of Jacobian. In the first iteration, we use a constant Jacobian and an adaptive one in subsequent iterations.
5. Update the model parameters $\mathbf{p} \longrightarrow \mathbf{p} + k\delta \mathbf{p}$, where initially $k = 1$.
6. Using the new parameters, calculate the new face structure $\mathbf{X}$ and the new mean-shape reference model $\mathbf{I}_{mean-shape}$ by using the estimated lighting parameters $\mathbf{L}$
7. Normalize in pose, shape and albedo a region in the image:

- Pose and shape normalization: Sample the image using $\mathbf{X}$ and warp it to the mean frame
- Albedo normalization: Use eq. 11 to normalize in albedo the shape normalized image

8. Calculate a new error vector $\mathbf{r}' = \mathbf{I}'_{aligned} - \mathbf{I}'_{mean-shape}$
9. If $|\mathbf{r}'|^2 < E$, then accept the new estimate; otherwise, try at $k = 0.5, k = 0.25$, etc.

In practice, we have implemented this algorithm for two resolution levels. Using a pyramid of two different resolutions improves the convergence of the algorithm and we can situate the initial model farther from the actual face than using a single resolution.

5   Experimental Results

We utilized Yale database, which contains ten different identities with different poses and illuminations. For training the 3D shape model, we manually placed 50 landmarks over each face surface. On the other hand, all 2D surface normals maps belonging to each individual, were reshaped over the 2D projection of the mean shape in order to obtain a mean map of surface normals. This mean map was used for face synthesis during the training stage for constructing the set of basis reflectances matrices $\mathbf{\Phi}$. We evaluated our model qualitatively by synthesizing each one of the training faces in different poses and different lightings. Fig. 3 shows nine synthetic faces produced by our model and each one illuminated by a basis light source.

Fig. 3. Identity number 1 (from Yale database) illuminated by each one of the nine basis light sources

We tested the fitting algorithm over 60 real images (with a size of 320×240 pixels) taken from Yale database in the following manner: all images have the pose number 6 which presents a similar angle in azimuth to the left and elevation up. This pose has an angle of 24 degrees from the camera axis. We choose 6 different illuminations for each one of the identities, see table 1. The initial conditions of the model at the beginning of the fitting process were manually setup only in translation and scale. The rest of the parameters: rotations, 3D shape, illumination and albedo were initiated always in their mean state for all the alignments. For the 60 alignments, the translation and scale parameters were initialized using the output parameters of a manual pose detector developed by us which uses three landmarks placed by hand in both external eye corners and the tip of the nose in order to calculate translation and scale by using   3D

Table 1. Illuminations used for experiments

L1	L2	L3	L4	L5	L6
$A + 50E + 00$	$A + 35E + 15$	$A + 10E + 00$	$A - 10E + 00$	$A - 35E + 15$	$A - 50E + 00$

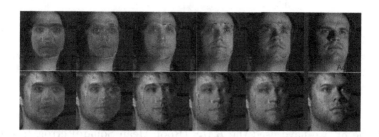

Fig. 4. Evolution of the synthetic face produced by the model during five iterations of the fitting algorithm

Fig. 5. Face alignments obtained by using pose number 6 and illumination L6

geometry. The translation computed in this way is a rough estimation with an error of ±4 pixels. Fig. 4 shows the evolution of the model during five iterations of the fitting process. Fig. 5 shows 10 alignments for each one of the identities using pose number 6 and illumination $L6$. Here, the first and fourth columns show the original face images, whereas the second and fifth columns show the synthetic faces produced by the fitting process. As a measure of similarity between the original face images and the synthetic images produced by our algorithm

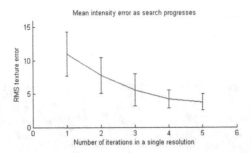

Fig. 6. Evolution of RMS error in intensity difference

during the fit, we have used the RMS (Root Mean Square) error. In order to show that our algorithm converged in all the test alignments, we obtained a mean RMS error for each one of the first five iterations of the algorithm. Fig. 6 shows a decreasing RMS error depicted by the central blue curve line, whereas perpendicular straight line segments represent the standard deviation associated to each mean RMS error. We observe a mean RMS error of 11 gray levels at the first iteration and a mean RMS error of 4 gray levels at the fifth iteration. The same thing occurs with the standard deviation from ±3 in the first iteration to ±1 in the fifth one.

6   Discussion

We presented a fast method for 3D face alignment, which is robust to non-uniform lighting conditions, and is able to fit to different identities with different albedo, shape, pose and illumination. Our model is based on learning the correlation between variation of albedo, shape, and pose parameters and the resulting residuals when the mean model is deformed by each one of the mentioned parameters. Illumination, which affects appearance more than identity is used to update an adaptive Jacobian and a mean reference model. The main contribution of this work, is a fitting algorithm which uses a novel way to normalize the albedo, according to the last estimated illumination parameters, and a novel method for recalculating the Jacobian by using the same illumination parameters and a set of precalculated albedo images and basis reflectances matrices. Our model could be used for face recognition under arbitrary pose and illumination, and it would be able of synthesizing new poses and new illuminations of an aligned face.

References

1. Baker, S., Matthews, I.: Equivalence and Efficiency of Image Alignment Algorithms. In: Proceedings IEEE Conf. on Computer Vision and Pattern Recognition (2001)
2. Blanz, V., Vetter, T.: A Morphable Model for the Synthesis of 3D Faces. In: Siggraph 1999, pp. 187–194 (1999)

3. Blanz, V., Vetter, T.: Face Recognition Based on Fitting a 3D Morphable Model. IEEE Transaction Pattern Analysis and Machine Intelligence 25, 1063–1074 (2003)
4. Romdhani, S., Blanz, V., Vetter, T.: Face Identification by Fitting a 3D Morphable Model Using Linear Shape and Texture Error Functions. In: Heyden, A., Sparr, G., Nielsen, M., Johansen, P. (eds.) ECCV 2002. LNCS, vol. 2353, pp. 3–19. Springer, Heidelberg (2002)
5. Romdhani, S., Pierrard, J.S., Vetter, T.: 3D Morphable Face Model, a Unified Approach for Analysis and Synthesis of Images. In: Face Processing: Advanced Modeling and Methods. Elsevier, Amsterdam (2005)
6. Romdhani, S., Ho, J., Vetter, T., Kriegman, D.J.: Face Recognition Using 3-D Models: Pose and Illumination. In: Proceedings of the IEEE, vol. 94, pp. 1977–1999 (2006)
7. Cootes, T.F., Edwards, G.J., Taylor, C.J.: Active Appearance Models. In: Burkhardt, H., Neumann, B. (eds.) ECCV 1998. LNCS, vol. 1407, pp. 484–498. Springer, Heidelberg (1998)
8. Cootes, T.F., Edwards, G.J., Taylor, C.J.: Active Appearance Models. IEEE Transactions on Pattern Analysis and Machine Intelligence 23, 681–685 (2001)
9. Xiao, J., Baker, S., Matthews, I., Kanade, T.: Real-Time Combined 2D+3D Active Appearance Models. In: CVPR 2004, vol. 2, pp. 535–542 (2004)
10. Matthews, I., Baker, S.: Active Appearance Models Revisited. International Journal on Computer Vision 60, 135–164 (2004)
11. Huang, Y., Lin, S., Li, S.Z., Lu, H., Shum, H.Y.: Face Alignment Under Variable Illumination. In: Proceedings of the FGR 2004, pp. 85–90 (2004)
12. Le Gallou, S., Breton, G., García, C., Séguier, R.: Distance Maps: A Robust Illumination Preprocessing for Active Appearance Models. In: VISAPP 2006, vol. 2, pp. 35–40 (2006)
13. Kahraman, F., Gökmen, M., Darkner, S., Larsen, R.: An Active Illumination and Appearance (AIA) Model for Face Alignment. In: CVPR 2007 (2007)
14. Dornaika, F., Ahlberg, J.: Fast And Reliable Active Appearance Model Search For 3d Face Tracking. In: Proceedings of Mirage 2003, pp. 10–11. INRIA Rocquencourt, France (2003)
15. Sattar, A., Aidarous, Y., Le Gallou, S., Séguier, R.: Face Alignment by 2.5D Active Appearance Model Optimized by Simplex. In: ICVS 2007, Bielefeld University, Germany (2007)
16. Basri, R., Jacobs, D.W.: Lambertian Reflectance and Linear Subspaces. IEEE Transactions on Pattern Analysis and Machine Intelligence 25, 218–233 (2003)
17. Lee, K.C., Ho, J., Kriegman, D.J.: Nine Points of Light: Acquiring Subspaces for Face Recognition under Variable Lighting. In: CVPR 2001, pp. 519–526 (2001)
18. Belhumeur, P., Kriegman, D.: What is the Set of Images of an Object Under all Possible Illumination Conditions. Int. J. Computer Vision 28, 245–260 (1998)
19. Georghiades, A.S., Belhumeur, P.N., Kriegman, D.J.: From Few to Many: Illumination Cone Models for Face Recognition Under Variable Lighting and Pose. IEEE Transactions on Pattern Analysis and Machine Intelligence 23, 643–660 (2001)
20. Ramamoorthi, R., Hanrahan, P.: An Efficient Representation for Irradiance Environment Maps. In: Proc. ACM SIGGRAPH, pp. 497–500 (2001)
21. Kovesi, P.: Shapelets Correlated with Surface Normals Produce Surfaces. In: ICCV 2005, pp. 994-1001 (2005)

EEG Data Driven Animation and Its Application

Olga Sourina, Alexei Sourin, and Vladimir Kulish

Nanyang Technological University, 50 Nanyang Ave,
Singapore 639798
{eosourina,assourin,vvkulish}@ntu.edu.sg

Abstract. Human electroencephalograph (EEG) data driven animation is often used in neurofeedback systems for concentration training in children and adults. Visualization of the time-series data could be used in neurofeedback and for the data analysis. The paper proposes a novel method of 3D mapping of EEG data and describes visualization system VisBrain that was developed for EEG data analysis. We employed a concept of a dynamic 3D volumetric shape for showing how the electrical signal changes through time. For the shape, a time-dependent solid blobby object was used. This object is defined using implicit functions. Besides just a visual comparison, we propose to apply set-theoretic ("Boolean") operations to the moving shapes to isolate activities common for both of them per time point, as well as those that are unique for either one. The advantages of the method are demonstrated with real EEG experiments examples. New emerging applications of EEG data driven animation in e-learning, games, entertainment, and medical applications are discussed.

1 Introduction

Traditionally, EEG-based technology has been applied in medical applications. Human electroencephalograph (EEG) signals are the records of electrical potential produced by the brain along with its activities. The signal is usually processed and analyzed from real-time EEG readings in frequency domain. Visualization of EEG signals is widely used in different applications. EEG is a time-series signal that can be visualized directly as a graph or 3D mapping on the model of the head/brain [1]. It can be also processed with signal processing algorithms (noise reduction, filtering and other processing) and the resulting values can be fed back to the system and depending on the application, can be used to walk through in 3D collaborative environments [2], for engaging avatars, in 3D art visualization, etc.

Currently, EEG driven animation is mostly used for visual feedback to the user of neurofeedback systems and for EEG data analysis in brain study. Neurofeedback is a process of displaying involuntary physiological processes obtained by electronic instrumentation, and then learning to voluntarily influence those processes by making changes in condition. Neurofeedback, as a therapy, treats health problems like attention deficit disorders, hyperactivity disorders and sleeping problems instead of suppressing such diseases with medication [3]. Based on visual feedback showing the user's brain activity, the user's mind could be trained to bridge new connections and

A. Gagalowicz and W. Philips (Eds.): MIRAGE 2009, LNCS 5496, pp. 380–388, 2009.

to either increase or decrease the use of specific brain functions. Intensive colors, game characters, or other visual effects can be used as visual feedback to the user.

In our works [4-7], we proposed to process EEG signals using fractal dimension model and implemented a novel algorithm to process EEG signals evoked by odor stimuli and music stimuli. In work [6], we studied brain responses to six basic olfactory stimuli given to the subjects in our experiments. In this paper, we propose a novel method of 3D modeling and mapping of EEG signals and describe the visualization systems we developed for visual analysis in brain study. The implemented system can be used as visual feedback in neurofeedback systems as well. We carried out series of experiments to study brain responses to external stimuli and analyzed the real experimental EEG data with the developed software. The results of this research open new perspectives of future application of EEG driven real time animation in human-computer interfaces that could be used to enhance the user interaction with digital media. Furthermore, new forms of human-centric and human-driven interaction with digital media that have the potential of revolutionizing entertainment, learning, and many other areas of life could be proposed.

In Section 2.1, a time dependent "blobby" model is proposed for 3D mapping of EEG signal. The set-theoretic operations that could be applied over the time-depending shapes are proposed. Visualization system VisBrain and its visualization modes are described in Section 2.2. In Section 3, experiments on brain responses to external stimuli and visual analysis of the results are described. Section 4 discusses future applications in human-computer interfaces.

2   3D Mapping and Visualization of EEG

We propose a novel method of 3D mapping of EEG data and describe visualization system VisBrain that was developed for EEG data visual analysis. We have employed a concept of a dynamic 3D volumetric "blobby" shape to visualize the electrical signal changes through time. The blob-like objects were firstly introduced in work [8] and further developed in [9]. A time-dependent "blobby" object is defined using implicit functions that allow us to propose and implement set-theoretic operations over the time changing shapes.

2.1   3D Mapping of EEG

We employed a concept of a dynamic 3D volumetric shape for visualization of the time-series data. For the shape, a time-dependent "blobby" object was used. This object is defined using so-called FRep representation [10] by the following formula:

$$f(x,y,z,t) = \sum_{i}^{24} a_i e^{-r_i \cdot b_i(t)} - g \geq 0$$

$$r_i = \sqrt{(x-x_i)^2 + (y-y_i)^2 + (z-z_i)^2}$$

(1)

where a is a scale factor, b is an exponent scale factor changing over time and g is a threshold value.

At any given point (x, y, z, t), function f can take negative, positive or zero values. The point is considered on the surface of the object if the function value is zero, inside the object if the function value is positive, and outside the object otherwise. In our case, the blobby function is built on the 24 potential functions resulting from the EEG electrodes positions on the head. The shape changes through time due to the variable values of the exponent factor b according to the signal. Its size and appearance visually reflect the brain activity. For a better visual impression, the blobby shape is superimposed on a 3D head model.

Besides just a visual comparison, we propose to apply set-theoretic ("Boolean") operations to the moving shapes to isolate activities common for both of them per time point, as well as those that are unique for either one. Furthermore, the group set-theoretic operations applied to the individual time frames of the moving shape allow us to isolate idle parts of the brain as well as to estimate an average level of the brain activity. The set-theoretic operations over two moving shapes defined with functions $f_1(x, y, z, t)$ and $f_2(x, y, z, t)$ are implemented as follows:

Union:

$$f_{union}(x, y, z, t) = \max(f_1(x, y, z, t), f_2(x, y, z, t)) \geq 0 \tag{2}$$

or group union:

$$f_{group\_union}(x, y, z) = \max_{t_1}^{t_2}(f(x, y, z, t)) \geq 0$$

Intersection:

$$f_{intersection}(x, y, z, t) = \min(f_1(x, y, z, t), f_2(x, y, z, t)) \geq 0 \tag{3}$$

or group intersection

$$f_{group\_intersection}(x, y, z, t) = \min_{t_1}^{t_2}(f(x, y, z, t)) \geq 0$$

Difference:

$$f_{difference}(x, y, z, t) = \min(f_1(x, y, z, t), -f_2(x, y, z, t)) \geq 0 \tag{4}$$

The proposed operations could be applied over one or/and over two datasets. On one data set, we can do intersection of all shapes to show constant activity, and union of all shapes to show the overall maximum activity. On two data sets, we could apply an intersection to show common activity, union to show overall maximum activity and subtraction to show activities which are characteristic to one set.

2.2 Visualization System VisBrain

The visualization system VisBrain was developed for visual analysis of EEG signal. VTK visualization toolkit [11] was used for developing the visualization software in C++ language. Three types of the visualization are implemented as follows: conventional visualization with color fields on 3D model of a head, 3D time-dependent blobby shapes based on the model described in the previous Section, and 3D moving "pins" placed on the head in the locations of electrodes. The developed software is an interactive program, which visualizes one or several signals by modeling the respective visualization type around the 3D human head. In Figure 1, graphical user interface of the system is shown. The locations of electrodes and the surfaces of the moving EEG shapes could be visualized with different colors. The parameters of the formula (1) describing blobby shapes could be tuned. Depending on the problem solved, the blobby objects can be built more isolated or more overlapped.

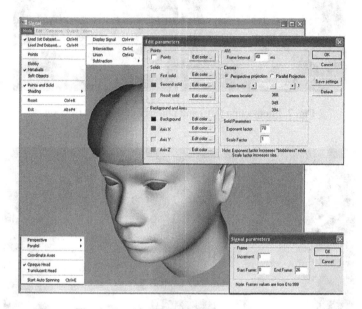

Fig. 1. Graphical User Interface of VisBrain

Two different semi-opaque moving blobby shapes corresponding to two different EEG signals can be visualized concurrently to visually analyze the difference between the respective brain activities. In fact, this method of visualization lets us notice several phenomena, which could not be possibly noticed if we used common ways of analyzing the EEG. Thus, in Figure 2, we display the minimum and maximum brain's activity by intersecting and unifying through the given time all the EEG shapes. The intersection (left image) shows those parts of the brain, which are engaged all the time. The right image (union) shows the maximum activity ever registered for the given time interval. It also shows that there are certain parts of the brain which are always idle for the given case and time interval.

In addition to "blobby" objects, "pin" objects could be used for EEG visualization. In Figure 3, an example of two signal visualization on two 3D head model is shown. Two signals can be superposed on one head model with different colors as well. Another mode of visualization, color 3D mapping is shown in Figure 4. Here the color scale could be assigned to the original signal values or for example, to fractal dimension values calculated by our fractal dimension method described in [4-5].

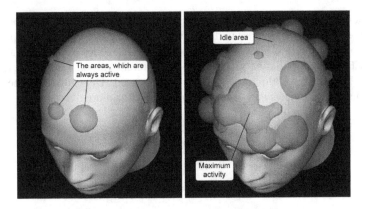

Fig. 2. Intersection (left) and union (right) of one EEG set

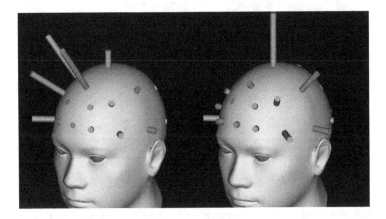

Fig. 3. "Pin" visualization of two EEG on two head model

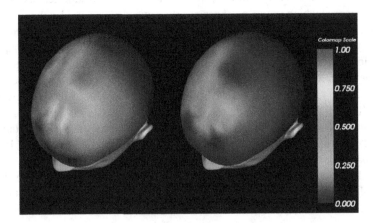

Fig. 4. Color mapping visualization of EEG. The color reflects the intensity of brain signals.

3   Experiments and Results

We proposed and carried out experiments to study brain responses to external stimuli like. We applied fractal dimension method described in [4-5] and visualization system VisBrain to analyze experiments results and validate our hypothesis. Let us describe experiments on brain responses to music stimuli, "yes/no" answers and visual analysis of the result.

3.1   Music Experiments

As human brain is processing the external information at a tremendous speed, the real time EEG signal should be sampled with a reasonably high sampling frequency. In our experiments, the equipment used for EEG signal recording, called MINDSET24, records 256 EEG samples per second for every single channel. There are 24 channels in total covering the entire scalp of the head. We proposed and conducted experiments using different music stimuli. We had series of experiments using spiritual (religious), hard rock music, classical music, and hip-hop music as external stimuli. The questionnaire was proposed to record emotional state of subjects to discover correlation between fractal dimension values and mental state induced by music. The subjects investigated in our experiments were the university students, both males and females 22-25 years old. The results of experiments were processed with the implemented dynamic fractal dimension algorithm [4-7] and were analyzed with our VisBrain system. Fractal dimension value changing over time was visualized using color mapping mode of the system. In Figure 4, dynamic fractal dimension values from two EEG signals recorded for two subjects listening one song are shown. By visual assessment, one could notice that fractal dimension values on some channels are different in two subjects. "Blobby" mode was used for validation of hypothesis that "The subjects' EEG responses do not depend on gender". By subtraction of "male" and "female" signals were shown that the same parts of the brain are responding to music in males and females and fractal dimension values on the channels depend only on music education of subjects. In Figure 3, fractal dimension values are represented by changing in heights "pins" showing which channels are active and how active they are during listening to music. The results of the processing confirmed our hypothesis that it is possible to recognize happiness and sadness emotions by computing fractal dimension values. The experiments and validated hypothesis were elaborated in our work [7]. By processing and visual analysis of the EEG it was also shown that fractal dimension model and visualization can be used to differ between concentration/relaxation states of the mind.

3.2   "Yes/No" Experiments

We carried out "yes/no" experiment and analyzed the results with dynamic fractal dimension method and visualization system VisBrain. The EEG samples were measured from 20 healthy humans – 10 females and 10 males – each of whom was to

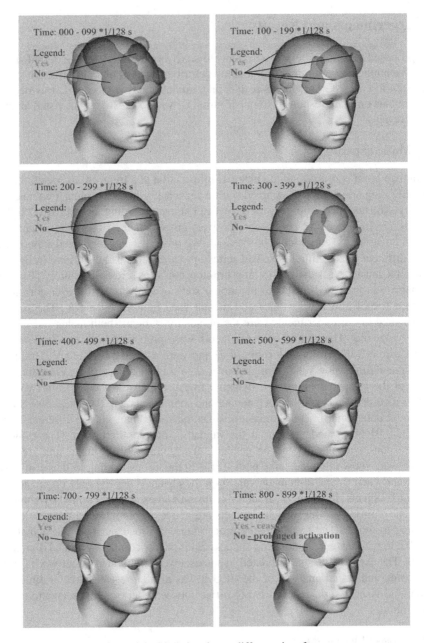

Fig. 5. "Yes/No" signals per different time frames

answer either 'yes' or 'no' to 12 questions by clicking either 'YES' or 'NO' button on a computer screen. 3D visualization of the brain activity proved to be a useful instrument while combined with the fractal dimension method. In particular, the visualization tool provides a clear picture of the asymmetric activity of the brain while performing symmetrical tasks with different emotional content. Thus, we noticed that the brain is more active while giving positive responses. Also, we observed that although it requires less mental activity in the course of responding, giving a negative response is more stressful, because it is followed by prolonged activation of the cerebral cortex and partial activation of the visual cortex of the brain. It is necessary to point out here that the latter conclusion can be drawn only thanks to the volume visualization; by means of the methods discussed in the preceding sections. In Figure 5, an example of visualization of two EEG signals corresponding to "yes" (blue) and "no" (red) answers are superposed at two time frames.

4 Conclusion and Future Work

In this paper, we proposed the model and implemented the system of 3D mapping and visualization of EEG data. An application of the VisBrain to analyze brain responses to external music stimuli and "yes/no" experiments was described. The results of our research and the implemented system VisBrain could be used in neurofeedback systems and in future development of new generation of human-computer interfaces. The music stimuli could induce human emotions and level of concentration that could be quantified and implemented as feedback in games and/or in virtual 3D spaces. Real time EEG data processing that is used in neurofeedback systems and in Brain-Computer Interfaces research can be a part of a novel human-centric and human-driven interface with digital media. In such interfaces, the content could be driven by monitoring of emotions and level of engagement/concentration, and, depending on the application (entertainment, learning, medical application, etc), different software tools should be engaged in real time.

The result of our research would contribute to the new forms of human-computer interaction leading to the next generation of interactive media. Now, new affordable electro-encephalograph cap devices with wireless data transmission are entering the market that could encourage wide spread of new applications that would bring the concentration-based and even emotion-based personalized digital experience to any user's location making such applications more mobile.

A short video about the current state of research on EEG data driven animation and brain study presented in this paper can be seen at http://intune.ntu.edu.sg/ SCE/ courses/Alexei/webpage/visualbrain.wmv

Acknowledgements

This project is supported by grant NRF2008IDM-IDM004-020 "Emotion-based personalised digital media experience in Co-Spaces" of National Research Fund of Singapore.

References

1. Jovanov, A., Kovacevic, E., Burhanpurkar, D., Samardzic, V.: Visualization of EEG evoked response potentials. In: Proc. of the 25th Annual International Conference of the IEEE on Engineering in Medicine and Biology Society, vol. 3, pp. 2099–2102 (2003)
2. Lécuyer, A., Lotte, F., Reilly, R.B., Leeb, R., Hirose, M., Slater, M.: Brain-Computer Interfaces, Virtual Reality, and Videogames. Computer 41(10), 66–72 (2008)
3. Evans, J.R., Abarbnanel, A.: An introduction to quantitative EEG and Neurofeedback. Academic Press, San Diego (1999)
4. Kulish, V., Sourin, A., Sourina, O.: Human electroencephalograms seen as fractal time series: mathematical analysis and visualization. Computers in Biology and Medicine 36(3), 291–302 (2005)
5. Kulish, V., Sourin, A., Sourina, O.: Analysis and visualization of human electroencephalograms seen as fractal time series. Journal of Mechanics in Medicine & Biology 26(2), 175–188 (2006)
6. Kulish, V., Sourin, A., Sourina, O.: Fractal spectra and visualization of the brain activity evoked by olfactory stimuli. In: Proc. of the 9th Asian Symposium on Visualization, Hong Kong, June 4-9, pp. 37-1–37-8 (2007)
7. Sourina, O., Kulish, V., Sourin, A.: Novel Tools for Quantification of Brain Responses to Music Stimuli. In: Proc. of 13th International Conf. on Biomedical Engineering. IEEE Explore (2008)
8. Blinn, J.: A Generalization of Algebraic Surface Drawing. ACM Transaction on Graphics 1(3), 235–256 (1982)
9. Wyvill, G., McPheeters, C., Wyvill, B.: Data structure for soft objects. The Visual Computer 2(4), 227–234 (1986)
10. Pasko, A.A., Adzhiev, V.D., Sourin, A.I., Savchenko, V.V.: Function Representation in Geometric Modeling: Concepts, Implementations and Applications. The Visual Computer 11(8), 429–446 (1995)
11. The Visualization Toolkit, http://www.vtk.org

Facade Structure Parameterization Based on Similarity Detection from Single Image[*]

Hong-Ping Yan[1,2], Chun Liu[2], André Gagalowicz[2],
and Cédric Guiard[2]

[1] Key Laboratory of Geo-detection, Ministry of Education, and Land Resources
Information Development Research Laboratory,
College of Information and Engineering, China University of Geosciences (Beijing),
Beijing 100083, China
hongpingyan.cn@gmail.com
www.cugb.edu.cn
[2] MIRAGES project
Batiment 23. INRIA Rocquencourt. B.P. 105 78153 Le Chesnay Cedex France
{chun.liu,andre.gagalowicz,cedric.guiard}@inria.fr
http://www.inria.fr/recherche/equipes/mirages.fr.html

Abstract. In this paper, we reverse engineer facade design from single rectified image of existing building facade by the use of similarity and hierarchy features of man-made objects. The inferred design is encoded into parametric grammar rules, named as ArchSys, which draw a compact and semantically meaningful characterization of the building structure and can be considered to support the design of other architectures. Combining with Gradient-based Mutual Information measure, we propose a rough-fine template-based similarity detection method to extract the structure patterns in a hierarchical way, which reduces computation time while increases robustness of the whole system. Our approach can be applied to various architectural typologies to detect not only symmetrical features but also similar patterns in one facade image. A feedback loop is built to refine the facade structure analysis and rule sets' parameters. Experimental results illustrate that our method is of robustness and general applications.

Keywords: template, similarity detection, grammar rules, single rectified image.

1 Introduction

Cities are of multi-dimensions, high functional and visual complexity. They are the concentration of history, culture, economy, and ecology, etc. Modeling and visualizing 3D city lies at the junction of various disciplines and is a real challenge for researchers from different domains. The modeling of 3D cities drives major

---

[*] The work reported in this paper has been performed as part of the Cap Digital Business Cluster Terra Numerica project under the support of INRIA Rocquencourt.

A. Gagalowicz and W. Philips (Eds.): MIRAGE 2009, LNCS 5496, pp. 389–400, 2009.

economical and cultural stakes and open wide market opportunities in fields ranging from the information society (3D GIS, urban planning, e-administration, navigation and nomad's services, simulation, security and defense ...) to digital content creation (tourism, museography, education, video-game, ...).

As one kind of ubiquitous and prevalent realizations in cities, buildings differ from one another in many aspects depending on their architectural styles, dimensions and appearances, while sharing regularity (symmetry, parallelism, orthogonality, etc.) and repetitive hierarchical structures inherent to man-made constructions. This observation remains at different scales from the overall facade subdivision down to the structure of inner architectural features. A typical example is windows, many of which are usually of identical and parameterizable shapes. They are commonly organized in logical hierarchies in a facade. So, all identical windows can then be represented by a single window symbol instead of a variable number of symbols, and the whole complex facade can be described with a few parametric symbols in a hierarchical and semantical unambiguous way. Buildings can be decomposed into roof and facades, and facades into floors composed of tiles (walls, windows, doors, ...). Many applications require inferring and modeling the detailed facade structure without the support of external CAD/CAM information, which is the topic of this paper. Accurate facade segmentation and description is needed to achieve an efficient visualization of large scale architectural projects, and this is also the core of numerous simulation applications (acoustic, thermic, illumination, ...) where functional roles and physical materials must be differentiated.

For modeling such kind of objects as facades with rich redundancy, procedural modeling method like L-systems [15] and shape grammars [21] is good choice. Up to now, there are many works in procedural architectural modeling [1], [13], [15], [17], [24]. Most of the works initialize the modeling process by manually defining and selecting seed rules or productions for specific buildings. Although this is a successful strategy for some applications, as we mentioned before, there are various building structures with quite different architectural styles, and therefore manual definition and selection will be time- and labor-consuming task; furthermore, regarding the existing constructions, the fidelity constraints during the interactive modeling process is difficult to be quantified. Therefore, automatic rule extraction based on existing information of the given facade is demanded. Müller, et. al [17] mimic the procedural modeling pipeline to automatically infer shape grammar rule sets from single facade image. But this method is only valid for detecting symmetrical structures and meanwhile sensitive to image noise and illumination.

Sharing a similar workflow as that in [17], our method intends to automatically reverse engineer facade design from single rectified image of existing building facade. The inferred structure will be encoded into grammar rules as a source of new designs, a way to parameterize building structure, a mean to generalize the modeling scheme as well as a pre-step to optimize the reconstructed architectural models. Here, architectural knowledge, like height of floor, dimension of tile, and other reasonable constraints on the architecture typology will be considered in order to achieve plausible results. Since the quality and complexity of the input information cannot perfectly meet our requirements (usually there are heavy image noise, measurement errors, small irregular artifacts, etc.), Gradient-weighted Mutual Information(GMI) and rough-fine template-based method are introduced to detect the similarity existing in the facade structure, which could improve the robustness, accuracy and generality of our reconstruction method. This is the main contribution of this paper. We also propose the use of an optimization process to refine the facade subdivision results and the parameters of the extracted rule sets.

The rest of the paper is structured in the following way. In Section 2, we review the related work. Section 3 details our method. In Section 4, experimental results are given. The paper is concluded in Section 5 by pointing out the space for improvements of our work.

2   Related Work

Automatic reconstruction of buildings from multiple images [6], [23], [5], [9], [11] has been extensively explored in both photogrammetry and computer vision research areas. However, it is not always possible to obtain multiple images of the architectural scene. Methods for 3D building reconstruction from a single image do exist [4], [8], [17], [10], but most of them [4] need some manual labors and are not fully automatic. While of particular interest, the method in [17] is also semi-automatic in some cases and requests an interactive offsetting of characterized architectural features.

Grammar-based modeling methods have also been tried in the field of architecture. Stiny [21] introduced the notion of shape grammars, in which production rules represent 2D/3D shape transformations. They have been successfully used by the architectural community to generate different kinds of architectures [2], [7], [24]. However, the application of rules regulated by the derivation process is not automatic since the guiding rules are mostly chosen by the user. [17] automatically extracts shape grammar rules from facade images only if there are symmetrical structures in the facade images, which is one of the limitations of this method. Our work intends to infer grammar rules from images based on similarity detection under more general hypothesis.

Image similarity detection is a mean to estimate how much information in one image is contained in another one(s). Generally, it is grounded on the visual features such as color histogram, texture, points, lines, shapes, edges, and so on. Typical intensity-based similarity measures include Cross-correlation [12], Correlation Coefficient(CC), Mutual Information(MI) [17], Normalized Mutual Information(NMI) [22], Gradient-weighted Mutual Information(GMI) [19], Regional Mutual Information(RMI) [20], and so on. CC measures the dispersion of the joint intensities along a line. This is a reasonable hypothesis in case of mono-modal image registration. MI is related to the entropy of the joint histogram and based on the marginal and joint image intensity distribution while not considering spatial information. NMI is a symmetric and normalized version of MI. GMI is a variant of MI. This measure introduces spatial information into MI by combining MI with image gradient, therefore it is more robust and accurate for similarity detection. In this paper, we will adopt GMI to find the repetitive patterns in one image.

Our method will work in four steps: 1) repetitive structure detection; 2) element parameterization; 3) rule extraction; 4) optimization of facade subdivision. The input is a single rectified image from ground-based imagery, and the output is a reconstructed 2D facade image and a set of shape grammar rules which can be reusable to thereby create a large variety of facades. In this paper, our work is centered on 2D image analysis and synthesis, and the 3D reconstruction will be done by introducing depth information into the reconstructed 2D image, which will not be covered in this paper. Furthermore, the registration of 2D image and 3D reconstructed model will be detailed in another paper.

3   Facade Structure Parameterizations Based on Similarity Detection

As mentioned above, at different scales, facade components refine in repetitive structures such as windows and doors within a tile; therefore, in order to simplify the description of facade structure, we consider template as a mean to model construction elements that share similar shapes and dimensions. In order to characterize these templates, we adopt the similarity measure, GMI, to detect the repetitive structures hidden in a facade image.

3.1   Gradient-Weighted Mutual Information

MI is an accurate measure for rigid and affine mono- and multi-modality image registration. This measure is expressed in terms of the histogram entropy of the images. Given two images I and J, their MI is defined by their marginal histogram entropies H(I) and H(J), and joint histogram entropy H(I, J) (See Equation 1).

$$MI(I, J) = H(I) + H(J) - H(I, J) \tag{1}$$

Unlike measures based on correlation of grey values or differences of grey values, MI does not assume a linear relationship between the grey values in the images. However, this measure lacks spatial information, which may lead to local maxima in some cases, for example, when the images are of low resolutions, when the images contain little information, or when there is only a small region of overlap. Conversely, image gradient is computed on a certain spatial scale. GMI thus extends MI measures to spatial information in images by integrating MI with image gradients. It builds on a combination of the mutual information cost function and a gradient-weighting function calculated from gradient magnitude and angle values from the images. For pair of corresponding pixels i and j in two images I_i and I_j, we have Equation 2 to define their spatial similarity:

$$f(i, j) = \frac{2(\Delta_i \cdot \Delta_j)(\|\Delta_i \times \Delta_j\|)}{(\|\Delta_i\|\|\Delta_j\|)^2} \tag{2}$$

Here, Δ_i and Δ_j are the gradients at pixels i and j, $\Delta_i \cdot \Delta_j$ is their scalar production, $\Delta_i \times \Delta_j$ the vector production, and $\|\Delta_i\|$ the module of Δ_i. Note that the value of $f(i, j)$ varies in the range $[0, 1]$ with the direction similarity of the two gradients Δ_i and Δ_j, and if they are of similar direction, $f(i, j)$ approaches to 1. GMI thus takes function 2 into conventional MI as a measure of spatial strength of an image in a given direction. Based on Equation 1 and 2, GMI can be expressed as following:

$$GMI(I, J) = MI(I, J)f(I, J) \qquad (3)$$

Comparing with the conventional MI, GMI combines intensity and gradient information from the images to achieve a more robust and accurate similarity matching. Therefore, in this paper, we adopt GMI to detect most similar structures/regions in one facade image separately along horizontal and vertical directions.

3.2   Rough-Fine Template-Based Similarity Detection

Pure GMI-based similarity detection is quite time-consuming. In order to speed up our method, we adopt rough-fine template-based scheme under the right-handed coordinate system originating from the left-bottom of the rectified facade image with x axis parallel to and y axis perpendicular to the ground baseline of the facade. Here, a template is such a geometric and semantic model that represents a class of architectural structures with same or similar geometric shape and same architectural function.

Fig. 1. Left: Edge profile in yellow and splitting lines in red

Considering that the ground floors of commercial buildings always lack similarity with the other floors due to wall covering and vitrified decoration, we first tackle vertical similarity detection by analyzing the similarity among different facade floors. Depicting the vertical similarity detection as illustration, we use vertical edge profile (in y axis direction) [12] to get N initial estimates of the positions of the horizontal splitting lines $y(x)$ at the mid-positions y_i of two adjacent valleys in the vertical edge profile, i.e. $y(x) = y_i$ (see Fig. 1). If there is no valley locating at the top and/or bottom margin(s) of the image, we add the top and/or bottom margin(s) as the first and/or last splitting line(s), which means that y_0 is always equal to 0 and y_N always the total height of the whole facade.

Data: N Template candidates
Result: Build template library
1 Initialize template library as empty ;
2 Initialize a temporary template *floor* as NULL;
3 Set all template candidates as *unchecked*;
4 **for** *i=N-1:0:-1* **do**
5   **if** *floor$_i$ is unchecked* **then**
6   Assign *floor$_i$* to *floor*;
7   Set *floor$_i$* as checked;
8   *j=i-1*;
9   **while** *floor$_j$ is unchecked, and j $\geqslant$ 0* **do**
10   **if** *GMI(floor, floor$_j$) $\geqslant \tau_{GMI}$* **then**
11   Set *floor$_j$* as checked;
12   Get a new template *floor$_{ij}$* by averaging *floor* and *floor$_j$*;
13   Assign *floor$_{ij}$* to *floor*;
14   **end**
15   *j=j-1*;
16   **end**
17   **end**
18   **if** *floor is not NULL* **then**
19   Put *floor* into the template library;
20   **end**
21 **end**

Algorithm 1. Rough-Fine Template-based Similarity Detection

According to the assumptions about the considered architectural typologies, the floor height h is variable in an interval [2.5m, 5.5m], so the facade can be decomposed into a set of floors with heights $h_i \in$ [2.5m, 5.5m]. Each floor is given an ID, i.e. an integer i starting from 0. In another words, 0 corresponds to the ground floor, i the *i-th* floor. In order to generalize our method and meanwhile simplify our grammar rules, we now run a pick-out process — Rough-Fine Template-based Similarity Detection as shown in Algorithm 1 to build a floor template library without repetitive floor structures inside. Each template candidate is named with the combination of the template function type and an ID number. For example, *floor$_i$* represents a floor template with ID number i. Each template is parameterized with its ID i, width w_i, and height h_i, in the form of *floor(i, w_i, h_i)*. Initially, regard all the floor structures as floor template candidates.

We exhaust all possible similar template candidates using Algorithm 1 in the vertical direction. Now, in the same way, similarity detection is performed in the horizontal direction separately on the floor templates derived from the vertical analysis. This stage successfully achieves the characterization of the basic architectural elements structuring the facade (See Fig. 2 Middle). Through this similarity detection process, a parametric floor and tile template library is generated, and each individual facade region(tile and floor. See Fig. 2 Right) is parameterized with its position, i.e., the coordinate of its top-left corner (x_{ij}, y_{ij}) in the facade, and ID number. In the next step, we will detail the tile(window, door) structure.

Fig. 2. Left: Original image with global edge profile in yellow. Middle: Tile templates. Right: Subdivision based on edge profile and GMI (splitting lines in red).

3.3   Architectural Element Analysis

Up to now, we have already subdivided facade structure into floors and further tiles, and obtained floor and tile template library in Section 3.2. At this stage, we will get the interior structures of the tiles. We first obtain the edge profile projection of the tile template (Fig. 3 Left). Based on this profile and the window/door frame width, we could detail the tile template structure by locating the frame which is centered at the valley of its edge profile projection (Fig. 3 Right). Since we have already parameterized each individual tile with its template ID number, we can match the tile with the corresponding template. Considering the influence of image noise and image distortion, we utilize the local(tile) and global(floor in horizontal direction and parcel in vertical direction) edge profile projection to slightly tune the positions of the mapped structures.

Our method is intuitive and meanwhile effective for reconstructing rectangle-shaped windows and doors. But, in fact, there are a lot of buildings with other different and complicated window and door structures other than rectangle, therefore in the cases when the geometric shapes of windows and doors are complex, the extracted outer rectangle is just the outline of the whole shape. We need further to perform edge detection in the region to get the inner structures. Here, Canny detector [3] is employed to approach the accurate shape structure.

Fig. 3. Left: Tile edge profile in yellow. Right: Extracted window structure in blue.

3.4   Rules Extraction

Since the previous procedures are hierarchical and semantic, which share the same features with shape grammar rules [21], it is easy to extract the shape grammar rules through these analyses. As mentioned above, grammar rules are very flexible and could help our method realize its potential in various applications. Therefore, our system draws a complete rule set that describes the segmented facade from the subdivision process. The extracted rule set can be applied to different dimensional facades with different styles, or as an initial state for reconstructed facade optimization.

```
#grammar rules of facade (from top to bottom):

#number of floors: 5

#number of floor types: 3

1:  facade=Subdiv(Y, 1, 3, 1){floor0|floor1|floor2}

2:  floor0=floor(0, 512, 94)Repeat(XS,  64){'tile0.obj'}

3:  floor1=floor(0, 512, 100)Repeat(XS,  64){'tile1.obj'}

4:  floor2=floor(0,512,118)Repeat(XS,64){'tile2.obj'}
```

Fig. 4. Left: Original image(512*512. This image is taken from [25]). Right: An example of our rule file extracted from the test image on the left.

Fig. 4(Right) shows an example of the grammar rules extracted from Fig. 4 (Left). Here, we named our grammar as ArchSys grammar, which is one variation of shape grammar. As shown in the example rule file, comments lines begin with the symbol #, and four rules with a rule ID number 1, 2, 3, 4 at the beginning are inferred from the test image. The first rule is the *Subdiv* operation, which means along y axis, the facade in the test image is divided into 1 floor0(one kind of floor structure) with floor height 94-pixel, 3 floor1 with floor height 100-pixel and 1 floor3 with floor height 118-pixel. Here, | is used to separate symbol between different architectural patterns/structures, and all the values are defined under the *xoy* right-handed image coordinate system originated from the left-bottom of the test image. The left 3 rules are the *Repeat* operations. For example, Rule2 tells us that floor0 is composed of tile0 which is of 64-pixel width and is repeated 8 times in x direction. The symbol between ' and ' represents a terminal shape stored in shape library. One rule is ended by a terminal symbol even though the object indicated by this symbol can be further decomposed into other components. In such a way, we can control the level of detail of the reconstructed structure.

In order to describe similar while not symmetry structures, we exploit *Translation* operator $\mathbf{T(dx, dy, dz)}$ in front of a structure, for example $\mathbf{T(dx, dy, dz)}\{tile0.obj\}$, to translate this structure from the current position(x_0, y_0,z_0) to the desired position(x_0+dx, y_0+dy,z_0+dz). Considering generalization and extension of our method, we provide *Insert* operator $\mathbf{I}(objectname)$ to insert particular object represented by the parameter *objectname* somewhere, *Scale* operator $\mathbf{S(a, b, c)}$ to scale an object with factors a, b and c separately in x, y and z direction, and *Rotate* operator $\mathbf{R}(x, y, z, \theta)$ to rotate an object with angle θ around the axis OP, $P(x, y, z)$. For now, we didn't introduce the depth information into the rules, i.e., all the z values are equal to 0, which can be done later after our subdivision results are optimized.

3.5 Optimization of the Facade Subdivision

From the extracted rules and without the support of a depth information, we can derive a 2D reconstruction of the facade image. Due to distortions in the original image and errors produced during facade subdivision, the reconstructed 2D image differs from the original image to some degree. In order to optimize the subdivision results, we project the reconstructed 2D image I_r back to the original one I_o, and get a difference image $I_\Delta = I_o - I_r$. This difference image is adopted to locally adjust the rule parameters, namely, the vertical and horizontal offset of the splitting lines and the center positions of the terminal shapes (windows, doors, etc..). This optimization loop breaks when all the position errors are less than 2 pixels.

4 Experimental Results

We implemented our system in C++ using a PC with Intel(R) Core(TM) dual-CPU, T7200, 2GHZ, and different facade images from buildings at Paris were chosen to test our method. The average running time of the whole system for a 512*512 image is about 1 minute.

For comparison with the method proposed in [17], Fig. 5 shows an original image(Left, 396*480) with tree in front of the building, and the subdivision results by using our method (Middle) and the method proposed in [17](Right). The results in Fig. 5 demonstrate that our method is insensitive to image noise, and more robust and accurate than the one in [17]. Fig. 6 shows an original image(Left) with the resolution of 512*512, the reconstructed results before optimization (Middle) and after optimization (Right). Obviously, after the optimization process, the windows are more accurately located.

Fig. 5. Left: Original image with strong noise. Middle: Subdivision result using conventional MI-based method [17]. Right: Subdivision result using our method.

Fig. 6. Left: Original image[25]. Middle: Reconstructed 2D facade image before optimization. Right: Reconstructed 2D facade image after optimization.

5   Conclusions

Our method is targeting facade reconstruction based on a single rectified image. We introduce a Rough-Fine Template-based Similarity Detection scheme making use of Gradient-weighted Mutual Information to increase the robustness and accuracy of the reconstruction process and to extend the applicability considering less regular architecture style and degraded image assumption of our method. This proposal also contributes to improve the efficiency and to speed up the whole system. By providing a feedback of the reconstructed 2D image to the original one, we build a close loop between the facade structure analysis and the parameterization of the grammar rules. This optimization process can improve the reconstruction accuracy and meanwhile lessen the risk of misregistration.

Our image-based subdivision technique results in sets of parametric grammar rules which form plausible estimates and variable constraints to consider the facade reconstruction under a model-based paradigm. Future works will also introduce depth information and consider coupled 2D-3D registration to automate the 3D facade reconstruction.

Acknowledgments. In this paper, the original image in Fig. 1, 4 and 6 is taken from the project site of Peter Wonka (`http://www.cg.tuwien.ac.at/research/vr/urbanmodels/index.html`).

References

1. Alegre, F., Dellaert, F.: A Probabilistic Approach to the Semantic Interpretation of Building Facades. In: International Workshop on Vision Technique Applied to the Rehabilitation of City Centre, pp. 1–12 (2004)
2. Bekins, D., Aliaga, D.G.: Build-by-number: Rearranging the Real World to Visualize Novel Architectural Spaces. In: Proceedings of the 16th IEEE Conference on Visualization 2005, pp. 19–27 (2005)
3. Canny, J.: A Computational Approach to Edge Detection. IEEE Transactions on Pattern Analysis and Machine Intelligence 8(6), 679–698 (1986)
4. Criminisi, A., Reid, I., Zisserman, A.: Single view metrology. In: Proceedings of the Seventh IEEE International Conference on Computer Vision, vol. 1, pp. 434–441 (1999)
5. Debevec, P.E., Camillo, J.T., Jitendra, M.: Modeling and Rendering Architecture from Photographs: a Hybrid Geometry- and Image-based Approach. In: SIGGRAPH 1996, pp. 11–20 (1996)
6. Dick, A.R., Torr, P.H.S., Ruffle, S.J., Cipolla, R.: Combining Single View Reconstruction and Multiple View Stereo for Architectural Scenes. In: Proceedings of International Conference on Computer Vision, pp. 268–274 (2001)
7. Duarte, J.: Malagueira Grammar - towards a toll for customizing Alvar Siza's mass houses at Malagueria. Ph.D Thesis, MIT School of Architecture and Planning (2002)
8. Han, F., Zhu, S.C.: Automatic Single View Building Reconstruction by Integrating Segmentation. In: Proceedings of the 2004 IEEE Computer Society Conference on Computer Vision and Pattern Recognition Workshops, pp. 53–63 (2004)
9. Kim, Z.W., Nevatia, R.: Automatic description of complex buildings from multiple images. Computer Vision and Image Understanding 96(1), 60–95 (2004)
10. Korah, T., Rasmussen, C.: Analysis of Building Textures for Reconstructing Partially Occluded Facades. In: Forsyth, D., Torr, P., Zisserman, A. (eds.) ECCV 2008, Part I. LNCS, vol. 5302, pp. 359–372. Springer, Heidelberg (2008)
11. Lafarge, F., Descombes, X., Zerubia, J., Pierrot-Deseilligny, M.: An automatic building reconstruction method: a structural approach using high resolution images. In: Proceedings of IEEE international conference on image processing (ICIP), Atlanta, USA, pp. 1205–1208 (2006)
12. Lee, S.C., Nevatia, R.: Extraction and integration of window in a 3D building model from ground view images. In: Proceedings of the 2004 IEEE Computer Society Conference on Computer Vision and Pattern Recognition, vol. 2(27), pp. 113–120. IEEE Press, Los Alamitos (2004)
13. Lipp, M., Wonka, P., Wimmer, M.: Interactive Visual Editing of Grammars for Procedural Architecture. ACM Transactions on Graphics 27(3), 1–10 (2008)
14. Maes, F., Collignon, A., Vandermeulen, D., Marchal, G., Suetens, P.: Multimodality image registration by maximization of mutual information. IEEE transactions on Medical imaging 16, 187–198 (1997)

15. Marvie, J.E., Perret, J., Bouatouch, K.: The FL-System: a Functional L-system for Procedural Geometric Modeling. The Visual Computer 21(5), 329–339 (2005)
16. Müller, P., Wonka, P., Haegler, P., Ulmer, S., Gool, L.V.: Procedural Modeling of Buildings. In: ACM SIGGRAPH 2006, vol. 25, pp. 614–623 (2006)
17. Müller, P., Zeng, G., Wonka, P., Gool, L.V.: Image-based Procedural Modeling of Facades. In: ACM SIGGRAPH 2007, pp. 181–184. IEEE Press, New York (2007)
18. Peng, J.Y., Yu, B.Z., Wang, D.K.: Image Similarity Detection Based on Directional Gradient Angular Histogram. In: Proceedings of the 16th International Conference on Pattern Recognition (ICPR 2002), vol. 1, pp. 147–153. IEEE Computer Society, Washington (2002)
19. Pluim, J.P., Maintz, J.B., Viergever, M.A.: Image Registration by Maximization of Combined Mutual Information and Gradient Information. IEEE transactions on Medical Imaging 19, 809–814 (2000)
20. Russakoff, D., Tomasi, C., Rohlfing, T., Maurer, G.: Image Similarity Using Mutual Information of Regions. In: Proceedings of the European Conference on Computer Vision, Prague, Czech Republic, pp. 596–607 (2004)
21. Stiny, G., Gips, J.: Shape Grammars and the Generative Specification of Painting and Sculpture. In: Proceedings of IFIP Congress 71, pp. 1460–1465. North-Holland, Amsterdam (1972)
22. Studholme, C., Hill, D.L.G., Hawkes, D.J.: An Overlap Invariant Entropy Measure of 3D Medical Image Alignment. Pattern Recognition 32(1), 71–86 (1999)
23. Werner, T., Zisserman, A.: New Techniques for Automated Architectural Reconstruction from Photographs. In: Heyden, A., Sparr, G., Nielsen, M., Johansen, P. (eds.) ECCV 2002, part II. LNCS, vol. 2351, pp. 541–555. Springer, Heidelberg (2002)
24. Wonka, P., Wimmer, M., Sillion, F., Ribasky, W.: Instant Architecture. ACM Transactions on Graphics 22(3), 669–677 (2003)
25. http://www.cg.tuwien.ac.at/research/vr/urbanmodels/index.html

Epipolar Angular Factorisation of Essential Matrix for Camera Pose Calibration

Władysław Skarbek[1] and Michał Tomaszewski[1,2]

[1] Warsaw University of Technology, Faculty of Electronics and Information Technology,
[2] Polish-Japanese Institute of Information Technology
W.Skarbek@ire.pw.edu.pl

Abstract. A novel epipolar angular representation for camera pose is introduced. It leads to a factorisation of the pose rotation matrix into three canonical rotations: around the dual epipole for the second camera, around the z axis, and around the dual epipole for the first camera. If the rotation around the z axis is increased by $90°$ and followed by the orthogonal projection on xy plane then the factorisation of essential matrix is produced. The proposed five parameter representation of the essential matrix is minimal. It exhibits the fast convergence in LMM optimization algorithm used for camera pose calibration. In such parametrisation the constraints based on the distance to the epipolar plane appeared slightly more accurate than constraints based on the distance to the epipolar line.

Keywords: epipolar geometry, essential matrix, angular factorisation, camera pose calibration.

1   Introduction

Essential matrix is important concept in projective geometry explored from linear algebra perspective [1,2,3,4,5]. The matrix is usually defined for two pinhole cameras using the *hat* notation for cross vector operator $E \doteq \hat{u}R$, where (R, u) is so called *camera pose*. The matrix R translates the second camera coordinates to the first one and u is the direction ($\|u\| = 1$) of translation of the second camera w.r.t. the first one. Coordinates of u are expressed in the first camera coordinates.

If both camera Cartesian systems are orthonormal, then R is an orthonormal matrix and the essential matrix E is in a sense a unique representation of camera pose (R, u). Actually there are always four camera poses having the same essential matrix, but three of them are inconsistent with image points.

The essential matrix E is a link between images x_k, y_k of the same spatial points acquired by two pinhole cameras. Mathematically this link is expressed by the bilinear form:

$$x_k^t E y_k = 0, \ k = 1, \ldots, K$$

The essential matrix is of rank two and the vector u spans its row kernel as the line which intersects the projective plane of the first camera in the so

A. Gagalowicz and W. Philips (Eds.): MIRAGE 2009, LNCS 5496, pp. 401–412, 2009.
© Springer-Verlag Berlin Heidelberg 2009

called *epipole*. It means also that the epipole of the first camera is an abstract image of the second camera centre. The image in the second camera for the first camera centre is also called the *epipole* and it spans the column kernel of E. Thus epipoles lie on projective planes on the line joining camera centres points (cf. Fig.1). This line is called the *epipole line* and any plane including this line is called the *epipolar plane* while its intersection with any projective plane is called the *the epipolar line* (cf. the plane Π_k and line L_k^i, L_k^j of Fig.1 not to be confused with epipole line $C_i C_j$ which is unique).

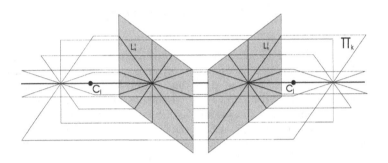

Fig. 1. Epipolar planes determined by epipole line joining pinholes C_i and C_j

Since the bilinear relation is linear w.r.t. elements of E then having $K \geq 8$ point images (x_k, y_k) makes possible to find an estimate of the matrix E. Next an estimate of the camera pose (R, u) is found using for instance SVD (Singular Value Decomposition [6]) of the essential matrix. The obtained estimate is considered as the starting point for an iterative process of a nonlinear optimization procedure. For optimization a solution space must be defined w.r.t. the admissible space. In our case the admissible space is defined as:

$$\mathcal{E} \doteq \{\hat{u}R : u \in \mathbb{R}^3,\ \|u\| = 1,\ R \in \mathbb{R}^{3 \times 3},\ R^t R = I_3,\ \det(R) = 1\}$$

There are usually two approaches considered for solution space design:

1. The admissible space is embedded into higher dimensionality solution space where it is identified by a number of constraints – in our case 12 dimensional space with 7 constraints is feasible.
2. The admissible space is parameterised using minimal number of parameters with no constraints on them and the parameter space becomes the solution space – in our case five angular parameters (two for u and three for R) define five dimensional solution space.

Probably dealing with trigonometric functions which are more complex than linear operations is the reason that in the literature we could find only recommendations for the first approach. To ensure keeping iteration on manifolds of constraints implicitly the work is performed in $12 + 7 = 19$ dimensional space what requires slower and less accurate high dimensional approaches.

In this paper we propose to consider the representation of essential matrices in angular space based on angles directly related to camera epipoles. The epipolar angles are spherical angles of epipole vectors in their camera Cartesian systems. Beside $2 + 2 = 4$ spherical angles an angle of rotation around the z axis must be added to complete characterisation. The representation is based on a factorisation of the essential matrix which separates five independent angles into three groups $(2 + 1 + 2)$. This special factorisation we call the *epipolar angular factorisation* of the essential matrix. The pose calibration procedure is described with two possible options for optimised goal function based on squared distances from epipolar line and epipolar plane. Both functions are optimised using angular factorisation of essential matrix.

In section 2 we discuss angular factorisation of any rotation matrix. While the epipolar angular factorisation can be obtained from SVD of the essential matrix $E = \hat{u}R$, in section 3 such factorisation is directly obtained from pose (R, u). In section 4 goal functions are defined for essential matrix identification and their form optimised for computational efficiency. Finally, section 5 includes experimental results.

2 Angular Factorisation of Rotation Matrix

In further discussion we will need a notation on specific rotation matrices and related formula.

Theorem 1 (Rodriguez formula[3])
The rotation matrix $\mathcal{R}_{w,\alpha}$ by the angle $\alpha \in [-\pi, \pi]$ around the axis spanned by the unit vector w, is equal to $I_3 + \sin(\alpha)\hat{w} + (1 - \cos(\alpha))\hat{w}^2$.

The following properties of spherical angles of a vector u and its related relations will be useful, too.

Lemma 1
If $u \in \mathbb{R}^3$ has spherical angles $\mu(u), \lambda(u)$ such that

$$u = [\cos \mu(u) \sin \lambda(u), \sin \mu(u) \sin \lambda(u), \cos \lambda(u)]^t$$

then

$$\lambda(u) + \lambda(-u) = \pi, \quad -\lambda(-u) = \lambda(u) - \pi \tag{1}$$

If $u^\perp \doteq e_3 \times u$ then

$$\mathcal{R}_{u^\perp,\lambda(u)}e_3 = u, \quad \mathcal{R}^t_{u^\perp,\lambda(u)}u = e_3$$
$$\mathcal{R}^t_{u^\perp,\pi} = \mathcal{R}_{u^\perp,-\pi} = \mathcal{R}_{u^\perp,\pi}, \quad \mathcal{R}_{u^\perp,\pi}e_3 = -e_3, \quad \mathcal{R}_{u^\perp,\pi}u = -u \tag{2}$$
$$(-u)^\perp = -(u^\perp) = -u^\perp, \quad \mathcal{R}_{(-u)^\perp,\lambda(-u)} = \mathcal{R}_{u^\perp,\lambda(u)-\pi} = \mathcal{R}_{u^\perp,\pi}\mathcal{R}_{u^\perp,\lambda(u)}$$

If $U \in \mathbb{R}^{3 \times 3}$ is any rotation matrix and $u \doteq Ue_3$ then this rotation can be performed in two steps:

1. rotate around e_3 (z-axis) by an angle α_0 – performed by the matrix $\mathcal{R}_{e_3,\alpha_0}$;
2. rotate around $u^\perp$ by the angle $\lambda(u)$ – performed by the matrix $\mathcal{R}_{u^\perp,\lambda(u)}$.

In order to find α_0 we compute the rotation matrix U_0 and apply the inverse Rodriguez formula: $U_0 \doteq \mathcal{R}_{u^\perp,\lambda(u)}^t U$, $\cos\alpha_0 = (\text{trace}(U_0) - 1)/2$. If the inverse Rodriguez formula returns the vector $-e_3$ as the axis then we change the sign of α_0 still keeping its value in the interval $[-\pi, +\pi]$.

The above recipe is valid if U_0 is the rotation matrix around the axis spanned by e_3. This true if and only if $U_0 e_3 = e_3 : U_0 e_3 = \mathcal{R}_{u^\perp,\lambda(u)}^t U e_3 = \mathcal{R}_{u^\perp,\lambda(u)}^t u = e_3$.

The theory [3] shows that the essential matrices have always the same singular values $1, 1, 0$. Therefore its SVD is of the form:

$$E = \pm U I_0 V^t, \quad I_0 \doteq \begin{bmatrix} 1 & 0 & 0 \\ 0 & 1 & 0 \\ 0 & 0 & 0 \end{bmatrix}, \quad U^t U = I_3, \ \det(U) = 1, \ V^t V = I_3, \ \det(V) = 1$$

The sign change of E may happen as the SVD itself does not guarantee the positive orientation of orthonormal factors.

Having SVD for $E = U I_0 V^t$ we get the angular factorisation of the essential matrix from the angular factorisations of rotation matrices U, V :

$$E = U I_0 V^t = \mathcal{R}_{u^\perp,\lambda(u)} \mathcal{R}_{e_3,\alpha_0} I_0 \mathcal{R}_{e_3,-\alpha_0'} \mathcal{R}_{v^\perp,\lambda(v)}^t = \mathcal{R}_{u^\perp,\lambda(u)} I_0 \mathcal{R}_{e_3,\alpha_0-\alpha_0'} \mathcal{R}_{v^\perp,\lambda(v)}^t$$

where $u \doteq U e_3$, $v \doteq V e_3$. Since $E^t u = 0$ and $Ev = 0$ then u spans the row kernel of E and v spans the column kernel of E. Therefore u and v are either epipolar vectors or their negations. In conclusion: singular value decomposition of the essential matrix leads to the angular factorisation of the essential matrix and indirectly produces angular representation for the camera pose. In the next section we show that the same parameters for the pose (R, u) can be obtained without using of SVD.

3   Epipolar Representation of Camera Pose

The second camera pose (R, u) is identified if we know the rotation matrix R and the directional (unit) vector u for the camera translation w.r.t. the first camera coordinate system. The columns of the matrix R describe axes of camera coordinate system. In this coordinate system the vector v describes the inverse translation if and only if $Rv = -u$. Note that unit vectors u, v point to the epipoles of the first and the second camera, respectively. Both of them determine the epipoles line but in different coordinate systems.

From the parametrisation point of view the rotation matrix R and the epipolar (unit) vector u are independent. You may move cameras apart keeping parallel cameras axes and you may rotate the camera axis by keeping its centre fixed. However, R depends on the pair of epipolar vectors since two equivalent *epipolar conditions* are true:

$$Rv = -u, \ R^t u = -v \tag{3}$$

For a convenience the vector $u^\perp$ is called the *dual epipole* vector of the epipole vector u.

Obviously, for the fixed pair of vectors u, v there are many rotation matrices R satisfying the epipolar condition. Let us denote by $\mathcal{F}_{u,v}$ the set of all such matrices:

$$\mathcal{F}_{u,v} \doteq \{R \in \mathbb{R}^{3\times 3} : R^t R = I, \ \det(R) = 1, \ Rv = -u\}$$

Assuming that u and v are not parallel to e_3, i.e. $u^\perp \neq 0$, $v^\perp \neq 0$, it appears that the matrix family $\mathcal{F}_{u,v}$ is the one dimensional set, and there exist the scalar parameter τ which is interpreted as the rotational angle around the z axis determined by the vector e_3. To this goal let us define the two helpful matrix families explicitly parameterised by τ :

$$\mathcal{F}_{u,v}^{+-} \doteq \left\{ \mathcal{R}_{u^\perp,\lambda(u)} \mathcal{R}_{e_3,\tau} \mathcal{R}_{v^\perp,\pi} \mathcal{R}_{v^\perp,\lambda(v)}^t : \tau \in [-\pi,\pi] \right\}$$

$$\mathcal{F}_{u,v}^{-+} \doteq \left\{ \mathcal{R}_{u^\perp,\lambda(u)} \mathcal{R}_{u^\perp,\pi} \mathcal{R}_{e_3,\tau} \mathcal{R}_{v^\perp,\lambda(v)}^t : \tau \in [-\pi,\pi] \right\}$$

The upper indices $\pm$ were fixed according the signs which appear in the following relations:

$$\mathcal{R}_{u^\perp,\lambda(u)} \mathcal{R}_{e_3,\tau} \mathcal{R}_{v^\perp,\pi} \mathcal{R}_{v^\perp,\lambda(v)}^t = \mathcal{R}_{u^\perp,\lambda(u)} \mathcal{R}_{e_3,\tau} \mathcal{R}_{(-v)^\perp,\lambda(-v)}^t$$

$$\mathcal{R}_{u^\perp,\lambda(u)} \mathcal{R}_{u^\perp,\pi} \mathcal{R}_{e_3,\tau} \mathcal{R}_{v^\perp,\lambda(v)}^t = \mathcal{R}_{(-u)^\perp,\lambda(-u)} \mathcal{R}_{e_3,\tau} \mathcal{R}_{v^\perp,\lambda(v)}^t$$

The above equalities follow directly from the properties collected in the lemma 1. The following theorem shows the parametrisation of the matrices related by the pair of epipolar unit vectors (u, v).

Theorem 2 (on parameterisation of epipolar matrices)
Let $\mathcal{F}^t$ denotes the family of the transposed matrices of $\mathcal{F}$. Then the characterisation of the matrix families $\mathcal{F}_{u,v}$ and $\mathcal{F}_{u,v}^t$ follows:

$$\mathcal{F}_{u,v}^{+-} = \mathcal{F}_{u,v} = \mathcal{F}_{u,v}^{-+}$$

$$\mathcal{F}_{u,v}^t = \mathcal{F}_{v,u}^{-+} = \mathcal{F}_{v,u}$$

$$\mathcal{F}_{v,u}^t = \mathcal{F}_{u,v}^{+-} = \mathcal{F}_{u,v}$$

Again the lemma 1 helps to prove the theorem. Just to illustrate the approach let us prove the inclusion $\mathcal{F}_{u,v} \subset \mathcal{F}_{u,v}^{+-}$. Let $R_0 \doteq \mathcal{R}_{u^\perp,\lambda(u)}^t R \mathcal{R}_{(-v)^\perp,\lambda(-v)}$:

$$\mathcal{R}_{(-v)^\perp,\lambda(-v)} e_3 = -v, \ Rv = -u \longrightarrow R(-v) = u \longrightarrow R_0 e_3 = \mathcal{R}_{u^\perp,\lambda(u)}^t u = e_3$$

Since $R_0 e_3 = e_3$ then e_3 spans the rotation axis for R_0. Therefore there exists $\tau_0 \in [-\pi,\pi]$ such that $R_0 = \mathcal{R}_{e_3,\tau_0} \longrightarrow R = \mathcal{R}_{u^\perp,\lambda(u)}^t R_0 \mathcal{R}_{(-v)^\perp,\lambda(-v)}^t \in \mathcal{F}_{u,v}^{+-}$. The unknown parameter τ can be easily found from the inverse Rodriguez formula.

Let us observe that the proposed parametrisation is actually the factorisation of the rotation matrix R into three rotation matrices:

1. the rotation around $u^\perp$ by the angle $\lambda(u)$;
2. the rotation around e_3 by the angle τ_0;
3. the rotation around $-v$ by the angle $-\lambda(-v)$.

This type of factorisation is called *epipolar angular factorisation*. Having the angular factorisation for the rotation matrix we may attempt to build the factorisation for the essential matrix $E = \hat{u}R$. Firstly, the factorisation of the *hat* matrix is performed based on the following lemma.

Lemma 2 (on factorisation of hat matrix)
Let $I_0 = [e_1, e_2, 0_3]$ be the diagonal matrix created from I_3 by replacing the last column by zeros. Then

$$\hat{e}_3 = I_0 \mathcal{R}_{e_3, \pi/2}$$

and

$$\hat{u} = \mathcal{R}_{u^\perp, \lambda(u)} \hat{e}_3 \mathcal{R}^t_{u^\perp, \lambda(u)}$$

The factorisation of $\hat{e}_3$ is obvious. However, the proof for $\hat{u}$ follows directly from the deep formula frequently exploited in the epipolar geometry:

$$A \in \mathbb{R}^{3 \times 3}, \ \det(A) \neq 0 \longrightarrow A^t \widehat{Au} A = \det(A) \hat{u}$$

by substituting $A \doteq \mathcal{R}_{u^\perp, \lambda(u)}$ and using equality $u = Ae_3$.

Combining factorisations for $\hat{u}$ and for $R \in \mathcal{F}^{+-}_{uv}$ we obtain:

$$E = \hat{u}R = \mathcal{R}_{u^\perp, \lambda(u)} \hat{e}_3 \mathcal{R}^t_{u^\perp, \lambda(u)} \mathcal{R}_{u^\perp, \lambda(u)} \mathcal{R}_{e_3, \tau} \mathcal{R}^t_{(-v)^\perp, \lambda(-v)}$$
$$= \mathcal{R}_{u^\perp, \lambda(u)} \hat{e}_3 \mathcal{R}_{e_3, \tau} \mathcal{R}^t_{(-v)^\perp, \lambda(-v)} = \mathcal{R}_{u^\perp, \lambda(u)} I_0 \mathcal{R}_{e_3, \tau + \pi/2} \mathcal{R}^t_{(-v)^\perp, \lambda(-v)}$$
$$= \mathcal{R}_{u^\perp, \lambda(u)} \mathcal{R}_{e_3, \tau + \pi/2} I_0 \mathcal{R}^t_{(-v)^\perp, \lambda(-v)}$$

The last equality follows from commutativity of I_0 with the rotation matrix around the z axis.

We conclude that the factorisation of the essential matrix differs slightly from the factorisation of the rotation matrix – there is factor I_0 which nullifies the z component and there is the rotation angle around the z axis increased by $90°$. Namely, the following steps are identified in factorisation of the essential matrix E :

1. the rotation around $-v$ by the angle $-\lambda(-v)$;
2. the rotation around e_3 by the angle $\tau_0 + \pi/2$;
3. the nullifying of z component – $I_0[x, y, z]^t = [x, y, 0]^t$
4. the rotation around $u^\perp$ by the angle $\lambda(u)$.

Note that steps two and three can be interchanged. The above can be concluded in the theorem.

Theorem 3 (on factorisation of essential matrix)
Let $u^\perp \neq 0$, $v^\perp \neq 0$. We consider all essential matrices for all camera poses (R, u) such that $Rv = -u$:

$$\mathcal{E}_{u,v} \doteq \{E \in \mathbb{R}^{3 \times 3} : E = \hat{u}R, \ R^t R = I, \ \det(R) = 1, \ Rv = -u\}$$

Let us define the parameterised family of the essential matrices:

$$\mathcal{E}^{+-}_{u,v} \doteq \{\hat{u}R : R \in \mathcal{F}^{+-}_{u,v}\}$$

Then the following characterisation of essential matrices from $\mathcal{E}_{u,v}$ is valid:

$$\mathcal{E}_{u,v} = \mathcal{E}_{u,v}^{+-} = \left\{ \mathcal{R}_{u^\perp,\lambda(u)} I_0 \mathcal{R}_{e_3,\tau+\pi/2} \mathcal{R}_{v^\perp,\pi} \mathcal{R}_{v^\perp,\lambda(v)}^t : \tau \in [-\pi,\pi] \right\}$$

Let us observe that the above representation of the camera pose (R, u) and its essential matrix $E = \hat{u}R$ is the implicit angular representation:

$$(R, u) \mapsto E \mapsto [\alpha_0, \ldots, \alpha_4]^t$$

where $\alpha_1 \doteq \mu(u)$, $\alpha_2 \doteq \lambda(u)$, $\alpha_3 \doteq \mu(-v)$, $\alpha_4 \doteq \lambda(-v)$ and α_0 is obtained by the inverse Rodriguez formula for the matrix R_0 :

$$R_0 \doteq \mathcal{R}_{u^\perp,\lambda(u)}^t R \mathcal{R}_{(-v)^\perp,\lambda(-v)}$$

To conclude this section let us justify and emphasise its title by the following theorem.

Theorem 4 (on single point correspondence for pose identification)
Let us consider two cameras with arbitrary relative pose. If both epipoles u, v are known and two images x, y of an unknown spatial point are measured in the local camera coordinates then the pose (R, u) is identified by finding the missing angle α_0 in the epipolar angular factorisation of R :

1. *Rotate x : $[x_1, x_2, x_3]^t \leftarrow \mathcal{R}_{u^\perp,\lambda(u)}^t x$;*
2. *Rotate y : $[y_1, y_2, y_3]^t \leftarrow \mathcal{R}_{(-v)^\perp,\lambda(-v)}^t y$;*
3. *If $a_1 = x_1 y_1 - x_2 y_2$, $a_2 = x_2 y_1 + x_1 y_2$ then $\alpha_0 = \arctan_2(\pm a_1, \mp a_2)$.*

The change of sign in α_0 formula exhibits the epipolar ambiguity of rotations R and $R' \doteq \mathcal{R}_{u,\pi} R$.

4   Application for Essential Matrix Identification

Essential matrix identification is performed by fitting its parameters to K points x_k in the image from the first camera and the corresponding points in the image of the second camera, $k = 1, \ldots, K$. The error term contributed by point image pair (x, y) is based on the distance from the measured point either to the epipolar line or to the epipolar plane (Fig. 2):

1. The error ϵ_{xy} of bilinear form $x^t E y$: $\epsilon_{xy} \doteq x^t E y$.
2. The distance δ_{xy} from the point x to the epipolar line determined by the vector $E y$ and the distance δ_{yx} from the point y to the epipolar line determined by the vector $E^t x$:

$$\delta_{xy} \doteq \frac{|\epsilon_{xy}|}{\|Ey\|}, \quad \delta_{yx} \doteq \frac{|\epsilon_{xy}|}{\|E^t x\|}$$

3. The distance Δ_{xy} from the point x to the epipolar plane y belongs to, and the distance Δ_{yx} from the point y to the epipolar plane x belongs to:

$$\Delta_{xx'} \doteq \frac{|\epsilon_{xx'}|}{\|x'\|}, \quad \Delta_{x'x} \doteq \frac{|\epsilon_{xx'}|}{\|x\|}$$

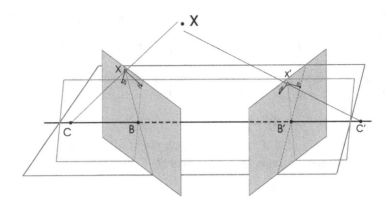

Fig. 2. Shows distances from the point image x (x') to the epipolar line and to the epipolar plane of its counterpart image

Since each image point pair (x, y) contributes two distances, the K point pairs makes total of $2K$ error terms for LMM (Levenberg Marquardt Method [7]). However, in case of plane distance since the nominators are equal and the denominators are parameter free, the two terms can be combined into one term.

1. For the distance to the epipolar line, two components of error function ϵ_{2k-1} and ϵ_{2k} are evaluated for $k = 1, \ldots, K$:

$$\epsilon_{2k-1} \leftarrow \frac{x_k^t E x_k'}{\|E x_k'\|}; \quad \epsilon_{2k} \leftarrow \frac{x_k^t E x_k'}{\|E^t x_k\|}$$

2. For the distance to the epipolar plane, one component of error function ϵ_k is evaluated for $k = 1, \ldots, K$:

$$\epsilon_k \leftarrow x_k^t E x_k' \sqrt{\frac{1}{\|x_k\|^2} + \frac{1}{\|x_k'\|^2}}$$

The epipolar angular factorisation of the essential matrix gives us an opportunity to optimize the design of the algorithm.

Let $E = R_1 I_0 R_0 R_2^t$ be the epipolar angular factorisation of the essential matrix E :

$$R_1 \doteq \mathcal{R}_{u^\perp, \lambda(u)}, \quad R_2 \doteq \mathcal{R}_{(-v)^\perp, \lambda(-v)}$$

The error term for a least square method which brings a matched pair (x, y) of point images can be separated as follows:

$$x^t E y = (R_1^t x)^t (I_0 R_0)(R_2^t y) = (Q_1 x)^t (Q_2^* y)$$

where $Q_i \in \mathbb{R}^{2 \times 3}$ is R_i^t with the third row dropped $i = 1, 2$:

$$Q_2^* \doteq R_0^* Q_2, \quad R_0^* \doteq \begin{bmatrix} c & -s \\ s & c \end{bmatrix}, \quad c = \cos \alpha_0, \quad s = \sin \alpha_0$$

The special form of rotation matrix Q_i enables saving a number of algebraic operations:

1. Evaluate the epipolar vector $u = [u_1, u_2, u_3]^t$ from its spherical coordinates:

$$u_1 \leftarrow \cos\alpha_1 \sin\alpha_2, \quad u_2 \leftarrow \sin\alpha_1 \sin\alpha_2, \quad u_3 = \cos\alpha_2$$

2. Compute Q_1 :

$$Q_1 \leftarrow \left[I_2, -[u_1, u_2]^t \right] - \left[\frac{[u_1, u_2]^t [u_1, u_2]}{1 + u_3}, [0, 0]^t \right].$$

3. Evaluate the negated epipolar vector $-v = [v_1, v_2, v_3]^t$ from its spherical coordinates:

$$v_1 = \cos\alpha_3 \sin\alpha_4, \quad v_2 = \sin\alpha_3 \sin\alpha_4, \quad v_3 = \cos\alpha_4$$

4. Compute Q_2 :

$$Q_2 \leftarrow \left[I_2, -[v_1, v_2]^t \right] - \left[\frac{[v_1, v_2]^t [v_1, v_2]}{1 + v_3}, [0, 0]^t \right]$$

Further improvements we seek in norm computations for the vectors $Ey, E^t x$:

1. The norm of Ey is computed w.r.t. the epipolar vector v :

$$\|Ey\|^2 = y^t E^t E y = y^t (R_2 I_0 R_0^t R_1^t)(R_1 R_0 I_0 R_2^t)y = $$
$$y^t (R_2 I_0 R_2^t)y = y^t [I_3 - vv^t]y = \|y\|^2 - (v^t y)^2$$

2. The norm of Ey is computed w.r.t. the epipolar vector u :

$$\|E^t x\|^2 = x^t E E^t x = x^t (R_1 I_0 R_0 R_2^t)(R_2 R_0^t I_0 R_1^t)x = $$
$$x^t (R_1 I_0 R_1^t)x = x^t [I_3 - uu^t]x = \|x\|^2 - (u^t x)^2$$

Finally the optimized algorithm for camera pose identification has the form:

1. *Input:* K projections (x_k, y_k'), $k = 1, \ldots, K$, $K \geq 8$.
2. For the distance to
   (a) the epipolar line two components of error function ϵ_{2k-1} and ϵ_{2k} are evaluated for $k = 1, \ldots, K$:

   $$\epsilon_{2k-1} \leftarrow \frac{(Q_1 x_k)^t Q_2^* y_k}{\sqrt{\|y_k\|^2 - (v^t y_k)^2}}; \quad \epsilon_{2k} \leftarrow \frac{(Q_1 x_k)^t Q_2^* y_k}{\sqrt{\|x_k\|^2 - (u^t x_k)^2}}$$

   (b) the epipolar plane one component of error function ϵ_k is evaluated for $k = 1, \ldots, K$:

   $$\epsilon_k \leftarrow (Q_1 x_k)^t Q_2^* y_k \sqrt{\frac{1}{\|y_k\|^2} + \frac{1}{\|x_k\|^2}}$$

3. *Output:* result of the LMM procedure (Levenberg Marquardt Method) applied to find the five angular parameters by nonlinear optimization (init LMM by the angular parameters obtained from the output of the eight-point algorithm).

5 Experimental Results

To verify the correctness of the pose identification procedure the following simulation procedure has been implemented.

Random camera poses were generated $M = 1600$ times. For each camera pose $K = 9$ random spatial points were generated and their perfect projections onto the image planes were applied.

The imperfect pixel measurements are simulated by normal random noise clamped to exponentially decreasing ranges. The geometric sequence of ranges is initialised to $[-0.01, +0.01]$ w.r.t. the image size, e.g. for the resolution 1024×756 the first noise range is large – about ± 10 pixels. The next range is half of the previous one. Such 20 ranges cover the interval $[10^{-7}, 10^{-2}]$ and therefore include the pixel resolutions for nowadays and future digital cameras.

The following averaged measures were analysed for the above goal functions and for the eight-point algorithm.

dE : the Frobenius norm for the difference of original and reconstructed essential matrices and its relative error w.r.t. the eight-point algorithm $ddE = (1 - \frac{dE}{dE_8}) \cdot 100$.

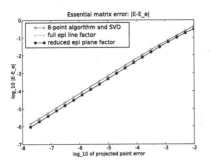
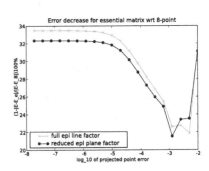

dR : the Frobenius norm for the difference of original and reconstructed rotation matrices and its relative error w.r.t. the eight-point algorithm $ddR = (1 - \frac{dR}{dR_8}) \cdot 100$.

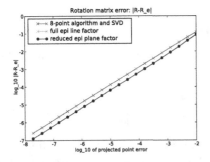
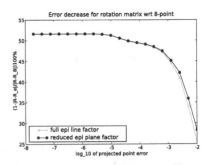

du : the Frobenius norm for the difference of original and reconstructed epipolar vector and its relative error w.r.t. the eight-point algorithm $ddu = (1 - \frac{du}{du_8}) \cdot 100$.

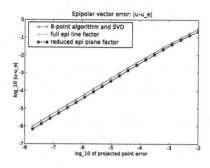
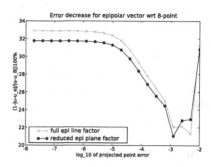

We see that in the logarithmic range of errors in $[-4, -2]$ corresponding to the current practice, the distance to the epipolar plane is slightly better than the distance to the epipolar line.

Additionally two graphs has been generated: the number of iterations (error function calls) in LMM and the actual relative decrease of goal functions if we switch from the result of the eight-point algorithm to the one produced by LMM.

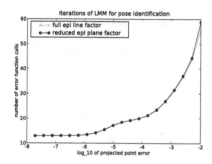
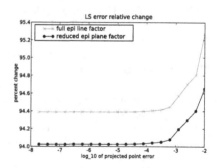

Both techniques in time complexity measured by the number of error function calls in the LMM algorithm are equivalent. Note also that about 95% of the initial error produced by the eight-point algorithm, i.e. the value of the goal function is reduced by LMM. It means that the nonlinear optimization is very efficient – takes almost all overhead introduced by the eight-point method. Why this level of improvement is not transferred to the pose parameters for which the improvements are three times lower? The reason is in the continuity of essential matrix space and the continuity of the corresponding camera pose manifold – randomly disturbed parameters correspond to feasible poses which differ from perfect poses. The convergence of pose to true value occurs when the measurement error tends to zero. However, the speed of convergence for pose, counted per one component, is about one order of magnitude slower.

6   Conclusion

The epipolar angular representation for camera pose leads to a factorisation of the pose rotation matrix into three canonical rotations: around the dual epipole for the second camera, around the z axis, and around the dual epipole for the first camera. If the rotation around the z axis is increased by 90° and followed by the orthogonal projection on xy plane then the factorisation of essential matrix is produced.

The proposed five parameter representation of the essential matrix is minimal. It exhibits the fast convergence in LMM optimization algorithm used for camera pose calibration.

In such a parametrisation the constraints based on the distance to the epipolar plane appeared to be slightly more accurate than constraints based on the distance to the epipolar line.

Acknowledgment. The work presented was developed within VISNET 2, a European Network of Excellence (http://www.visnet-noe.org), funded under the European Commission IST FP6 Programme.

References

1. Hartley, R., Zisserman, A.: Multiple View Geometry in Computer Vision. Cambridge University Press, Cambridge (2000)
2. Faugeras, O., Luang, Q.T.: The Geometry of Multiple Images. The MIT Press, Cambridge (2001)
3. Ma, Y., Soatto, S., Kosecka, J., Sastry, S.: An Invitation to 3-D Vision. The MIT Press, Cambridge (2004)
4. Luong, Q., Faugeras, O.: On the determination of epipoles using cross-ratios. Computer Vision and Image Understanding 71(1), 1–18 (1998)
5. Xu, G., Zhang, Z.: Epipolar Geometry in Stereo, Motion, and Object Recognition. Kluwer Academic Publishers, Dordrecht (1996)
6. Golub, G., Loan, C.: Matrix Computations. The Johns Hopkins University Press, Baltimore (1989)
7. Press, W., Teukolsky, S., Vetterling, W., Flannery, B.: Numerical Recipes in C. In: The Art of Scientific Computing. Cambridge University Press, Cambridge (2006)

Integrated Noise Modeling for Image Sensor Using Bayer Domain Images

Yeul-Min Baek[1], Joong-Geun Kim[1], Dong-Chan Cho[1], Jin-Aeon Lee[2], and Whoi-Yul Kim[1]

[1] Dept. Electronics and Computer Engineering, Hanynag University
Haengdang-dong, Seongdong-gu, Seoul, Korea 133-791
{ymbaek,jkkim,dccho}@vision.hanyang.ac.kr,
wykim@hanyang.ac.kr
[2] Samsung Electronics
Giheung-eup, Yongin-si, Gyeonggi-do, Korea 449-712
jalee@samsung.com

Abstract. Most of image processing algorithms assume that an image has an additive white Gaussian noise (AWGN). However, since the real noise is not AWGN, such algorithms are not effective with real images acquired by image sensors for digital camera. In this paper, we present an integrated noise model for image sensors that can handle shot noise, dark-current noise and fixed-pattern noise together. In addition, unlike most noise modeling methods, parameters for the model do not need to be re-configured depending on input images once it is made. Thus the proposed noise model is best suitable for various imaging devices. We introduce two applications of our noise model: edge detection and noise reduction in image sensors. The experimental results show how effective our noise model is for both applications.

1 Introduction

Many image processing algorithms assume that noise is an additive white Gaussian noise (AWGN) with some constant standard deviation. However, the noise of real images acquired by image sensors for digital camera is not AWGN. In fact, the noise of real images has some spatial correlation (not white), the dependency of intensity values (not constant standard deviation) and non-Gaussian distribution. Therefore, many image processing algorithms using AWGN assumption are not effective and need to adjust parameters manually with real images acquired by such image sensors.

We characterized image sensor noise model from Bayer domain images. Since we focus on dark-current noise, shot noise and fixed-pattern noise together, the precise noise level can be estimated using the model without under-estimation problem.

The paper is organized as follows. We review related works in section 2. In section 3, we describe how to build the integrated image sensor noise model using Bayer domain images. We provide experimental results in section 4 and apply our integrated image sensor noise model to two image processing algorithms in section 5. We conclude in section 6.

A. Gagalowicz and W. Philips (Eds.): MIRAGE 2009, LNCS 5496, pp. 413–424, 2009.

2 Related Work

Many image denoising methods have been proposed using various techniques, such as wavelet [1], anisotropic diffusion [2] and bilateral filtering [3], etc. With these methods it is assumed that the noise is AWGN with constant standard deviation for varying intensity values. Thus, they are not efficient to remove real noise and need to adjust parameters manually [5]. On the other hand, the study of noise modeling is very limited. Most image noise modeling methods use multiple images [4]. However, since noise estimation using multiple images suffers from the over-constrained problem [6], noise estimation methods using a single image have been studied [4][6][7]. In [4], Hwang et al. modeled shot noise, the dominant noise of image sensor, as the Skellam distribution from a single image. In general, shot noise is often characterized by the Poisson distribution. However, in many noise estimation methods shot noise is modeled as Gaussian distribution on the assumption that photon arrival rate is high enough. Thus, when the intensity value of image is low, treating the shot noise as Gaussian distribution may not be proper. Skellam distribution, the discrete probability distribution of the difference between two random variables having Poisson distributions, was proposed by Hwang et al. as the shot noise model for intensity difference. However, as mentioned in [7][9][10], since noise characteristics tends to be distorted due to the post-processing of camera pipeline such as demosaicing, gamma correction, and white balancing, it was difficult to find a proper noise model in terms of *Intensity-Skellam line*. To avoid such distortions of noise characteristics, Liu et al. [6] suggested the noise model, called *noise level function* based on the *piecewise smooth image prior*. They built *the space of noise level functions* and use the Bayesian MAP inference to infer the *noise level function* from a single image. However, in Liu et al.'s method parameters for the model had to be re-estimated for each input image. For that reason, that method was difficult to be used for consumer imaging devices such as digital camera, phone camera, etc. In [7], Yoo et al. proposed *Gaussian and impulsive noise reduction method*. They estimated the parameters for the noise model in Bayer domain to avoid the distortion of noise characteristics, and then adaptively determined the filter coefficients using their model. However, due to modeling shot noise only, Yoo et al.'s method becomes imprecise when intensity is high.

3 Integrated Noise Modeling

In this section, we model the image sensor noise characteristic using Bayer domain images.

As mentioned in [8], there are mainly five different types of noise sources, namely fixed-pattern noise, dark-current noise, shot noise, amplifier noise and quantization noise. Among these different noise sources, we ignore amplifier noise and quantization noise because these two types of noise are not only very small but also can usually be considerably reduced by built-in image processors [4]. Most of noise modeling methods deals with shot noise only. However, when the image intensity is high or low, fixed-pattern or dark-current noise becomes a dominant source, respectively [9]. Therefore, we propose an integrated noise modeling that can cope with dark-current noise, shot noise and fixed-pattern noise simultaneously to build the precise noise model of image sensors for digital cameras.

3.1 Noise Modeling in Bayer Domain

Generally, noise modeling means the process of modeling the noise level as a function of image intensity. In many noise modeling methods, noise level samples are collected from homogeneous patches within an input image. Fig. 1 shows such noise samples in our experiment as described in [4], where a line equation was fitted to the samples for building noise model. However, as Fig. 1 shows, it is hard to see any linear relationship between noise samples. The reason for this non-linearity is caused by the post processing of camera pipeline as discussed above [6][9][10][13]. Fig. 2 shows a typical camera pipeline structure. As Fig. 2 illustrates, an image in image domain from an imaging device is generated by a series of post-processing of the image sensor output. In Fig. 2, the camera response function (CRF) including gamma correction and white balance makes that the noise level bears non-linear characteristics [6][9]. In addition, demosaicing process for Bayer pattern makes the noise to have the correlation among color channels, as well as a spatial correlation [10]. Therefore, it is difficult to build the precise noise model from an image in image domain.

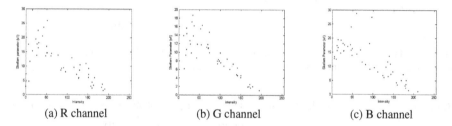

(a) R channel   (b) G channel   (c) B channel

Fig. 1. Non-linear characteristics of noise in image domain

To build the precise yet simple noise model, we propose the following form of integrated noise model for image sensors using Bayer domain images.

$$I = f\left(L + n_{\text{int egrated}}\right) \quad (1)$$

Here, L and I are the intensity in Bayer domain and image domain, respectively, and $n_{\text{integrated}}$ is noise estimated by our integrated noise model. f is a CRF that transforms the intensity in Bayer domain into the intensity in image domain. In general, photon arrival obeys Poisson distribution. However, when photon arrival rate is high, Poisson distribution can be approximated to Gaussian distribution [7]. Therefore, we can assume that integrated noise term has zero mean additive Gaussian noise given by Eq. (2).

$$n_{\text{int egrated}} \sim N\left(0, \sigma^2_{\text{int egrated}}\left(\sigma^2_D(t), \sigma^2_S(L), \sigma^2_F(L), d_{bias}\right)\right)$$

In Eq. (2), $\sigma^2_D(t)$ is the variance of the dark-current noise as a function of exposure time(t), $\sigma^2_S(L)$ is the variance of the shot noise as a function of intensity (L) in

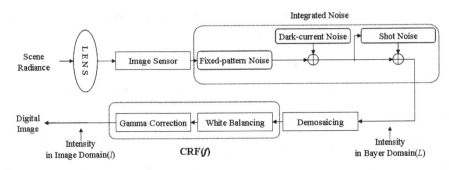

Fig. 2. Camera Pipeline

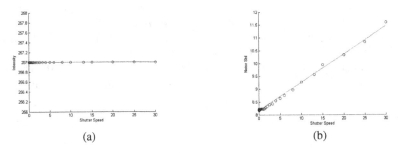

(a)   (b)

Fig. 3. (a) is the intensity value of dark frames as a function of exposure time and (b) is the noise variance of dark frames as a function of exposure time

Bayer domain, $\sigma_F^2(L)$ is the variance of the fixed-pattern noise as a function of intensity in Bayer domain, and d_{bias} denotes dark-current bias intensity.

3.2 Dark-Current Noise Modeling

When there is no light projected on the sensor, the output signal from image sensor should be zero. However, the actual output signal from image sensor is more than zero and has dark-current noise because of dark current electrons.

To model dark-current noise, we captured dark frame images with respect to exposure time. Dark frame images were captured by closing the lens cap of camera in the dark room. We then computed global mean intensity values from dark frame images. As Fig. 3(a) shows, global mean intensity values of dark frame images are almost the same ($\sigma = 0.1$) even though the exposure time (t_n) is changed. However the standard deviation (σ_{Dn}) of the dark current noise has linear relationship with exposure time (t_n) as shown in Fig. 3(b). Therefore, we modeled the dark-current noise $\sigma_D(t)$ by fitting a line to the dark-current noise samples (t_n, σ_{Dn}) using the least square regression.

3.3 Shot Noise Modeling

Shot noise or photon shot noise is caused by the inherent natural variation of incident photon flux. To model the shot noise, we acquired the single color checker image as shown in Fig. 4(a). Then we extracted 24 homogeneous color patches from the color checker image and collected the shot noise samples (L_n, σ^2_{Sn}) by computing mean (L_n) and variance (σ^2_{Sn}) of each patch. As Fig. 4(b) indicates, we could find the linear relationship of the shot noise samples (L_n, σ^2_{Sn}). Therefore, we build the shot noise model $\sigma^2_s(L)$ again by fitting a line to the shot noise samples (L_n, σ^2_{Sn}) using the same method.

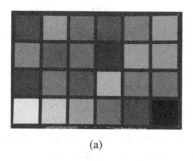

(a)

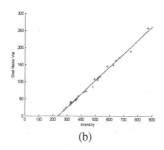

(b)

Fig. 4. (a) is GretagMacbeth^{TM} color checker chart for modeling shot noise and (b) is the shot noise model obtained by fitting a line to the shot noise samples marked "+" (L_n, σ^2_{Sn})

3.4 Fixed-Pattern Noise Modeling

Fixed-pattern noise (FPN) is caused by the difference of pixel response. Therefore, FPN is usually a dominant source when the intensity is high. According to the noise measurement method specified by ISO 15739 standard [14], FPN standard deviation of some intensity was obtained by averaging a sequence of n images to reduce random noise as Eq. (3).

$$\sigma_{fp} = \sqrt{\sigma^2_{ave} - \frac{1}{n-1}\sigma^2_{diff}} \qquad (3)$$

Here, σ^2_{ave} is the variance of mean image computed by averaging an image sequence, σ^2_{diff} is the average value of variance computed from the difference images between the mean image and each one in the sequence. In our experiment, eight homogeneous images were acquired by taking pictures of white paper, and the FPN standard deviation (σ_{fp}) and mean intensity value (L_{fp}) were computed. Since FPN is caused by the difference of pixel response, we checked five different pixel responses which are selected randomly. Since the pixel response increased linearly as shown in Fig. 5(a), we assumed that the FPN standard deviation increased linearly with respect to intensity. Therefore, as Fig. 5(b) shows, we used a line equation that passes through the origin and (L_{fp}, σ_{fp}) as the fixed-pattern noise model, $\sigma_F(L)$.

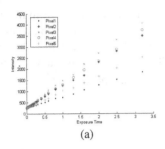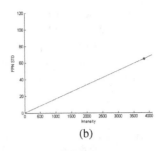

(a)   (b)

Fig. 5. (a) is 5 randomly selected pixels' response and (b) is the FPN standard deviation model $\sigma_F(L)$

3.5 Integrated Noise Terms

Now, we integrate the dark current noise model, shot noise model and fixed-pattern noise model discussed above altogether. The standard deviation of the integrated noise model as function of intensity and exposure time is given by Eq. (4).

$$
\begin{aligned}
if \quad & (L < d_{bias}) \quad \sigma_{int\,egrated}(L,t) = \sigma_D(t) \\
else \quad & \sigma_{int\,egrated}(L,t) = \sqrt{\sigma_S^2(L) + \sigma_F^2(L)}
\end{aligned}
\tag{4}
$$

In Eq. (4), when intensity is lower than d_{bias}, image is only corrupted by dark-current noise. On the other hand, when intensity is higher than d_{bias}, the integrated noise should consist of dark current noise, shot noise and fixed-pattern noise. However, since shot noise is associated with the portion of dark current [15], only shot noise term and fixed-pattern noise term are added when intensity is higher than d_{bias}.

4   Experimental Results and Discussion

To verify our noise model, we performed experiments on both real and synthetic noise images, and tested how our noise model behaves depending on the camera setting. All images for the experiment were acquired by Canon™ EOS 400D. Since our noise modeling method uses Bayer domain images, we acquired RAW images and used a software *dcraw* to extract Bayer domain images from RAW files. *dcraw* is the open source program which is able to read numerous raw image formats [11].

4.1   Robustness of Integrated Noise Model

We ran some experiments to see how our noise model works due to the changes in camera parameter settings such as aperture, shutter speed, and ISO. At first, we fixed ISO and changed aperture and shutter speed. The result is shown Fig. 6 and Table 1. As Fig. 6 indicates, although the aperture and shutter speed is changed, our noise model has very little variation. Table 1 shows the RMSE value between the maximum and minimum noise standard deviation with respect to intensity values, respectively. In Table 1, intensity is normalized between 0 and 1.

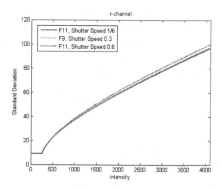

Fig. 6. The integrated noise model for R-channel. The integrated noise model is not changed even at different camera settings, The red solid line is for the setting with f11 and 1/6s, the green dash line indicates the setting with f9 and 0.3s, and the blue dot line corresponds to the setting with f11 and 0.6s.

Table 1. The variation of integrated noise model corresponding to aperture and shutter speed change

	R channel	GR channel	GB channel	B channel
RMSE	0.00055	0.000281	0.000438	0.000235

In the same manner, the result of our noise model variation with different ISO's is shown in Fig. 7. As Fig. 7 indicates, our noise model was only changed by image sensor gain, i.e. ISO. This property means that our noise model does not need to be reconfigured with different input images if it is done once. Therefore, our compact noise model according to each ISO setting can be applicable to consumer imaging device in forms of LUT.

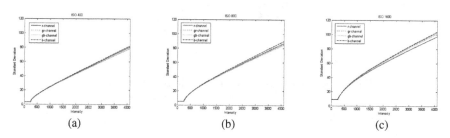

Fig. 7. The integrated noise model at different ISO settings, (a) is ISO 400, (b) is ISO 800 and (c) is ISO 1600

4.2 Comparison with Ground-Truth Noise Sample

To collect ground-truth noise samples, we acquired 52 color checker images with various exposure times and computed mean and standard deviation of 24 patches from each color checker image. Then, we compared our noise model with the ground-truth noise samples. The result is shown in Fig. 8. As anticipated, our noise model estimated upper-boundary of ground-truth noise samples with success as shown in Fig. 8. The figure shows that the ground-truth noise samples are scattered when the intensity is high. This is caused by FPN, i.e. the upper-boundary noise samples are corrupted by FPN. On the other hand, the lower boundary noise samples that have only shot noise were bounded by the proposed shot noise model (blue dash line).

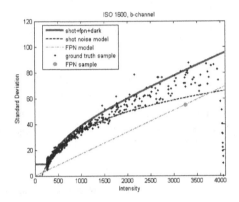

Fig. 8. The ground-truth noise samples and our noise model

4.3   Comparison with Synthetic Noise Model

To generate the synthetic noise images as shown in Fig. 9, we specified the synthetic integrated noise model. To build FPN model, we generated eight synthetic images.

Fig. 9. The synthetic image with pre-defined integrated noise

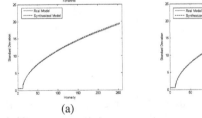
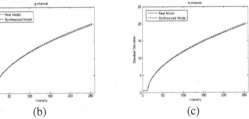

Fig. 10. Comparison between our noise model and the synthetic noise model. Red solid line is our estimated noise model and blue dash line is the synthetic noise model.

Table 2. The RMSE error between our estimated noise model and the synthetic noise model

	R channel	G channel	B channel
RMSE	0.00814	0.000887	0.000897

The result, as shown Fig. 10, our noise model is very close to the defined synthetic noise model. Table 2 shows the RMSE value between our noise model and synthetic noise model. In table 2, intensity values are normalized from 0 to 1.

5 Image Processing Applications

Many image processing algorithms are noise-dependent. Therefore, our noise model allows them robust against noise. In this section, we introduce two typical applications of our noise model: a noise removal and edge detection.

5.1 Adaptive Bilateral Filter

Bilateral filtering [3] is a well-known algorithm for edge-preserving image smoothing. Bilateral filter expressed in Eq. (5) is defined as the product of domain filter and range filter.

$$B_s = \frac{\sum_{p \in \Omega} w_d(p,s) w_r(L_p, L_s) L_s}{\sum_{p \in \Omega} w_d(p,s) w_r(L_p, L_s)} \tag{5}$$

Here, Ω is the neighborhood set of bilateral filter kernel, s is the center pixel position and p is the neighborhood pixel position in bilateral filter kernel.

The coefficient of domain filter, w_d is computed based on the spatial distance as Eq. (6). Thus, the domain filter performs the image smoothing. On the other hand, the coefficient of range filter, w_r is computed based on the intensity difference as Eq. (7) to preserve edges.

$$w_d(p,s) = \exp\left(-\frac{1}{2}\left(\frac{\|p-s\|}{\sigma_d}\right)^2\right) \tag{6}$$

$$w_r(p,s) = \exp\left(-\frac{1}{2}\left(\frac{\|L_p-L_s\|}{\sigma_r}\right)^2\right) \tag{7}$$

As mentioned above, since the real noise is not AWGN with a constant standard deviation, the original bilateral filter with a fixed σ_r in Eq. (7) is not efficient for real images.

To apply our noise model to bilateral filter, we change σ_r to $\sigma_{integrated}(L_s, t)$. Since our noise model and adaptive bilateral filter are defined in Bayer domain, we used bilinear interpolation as demosaicing process and estimated the CRF of Canon^{TM} EOD 400D to change the result image in Bayer domain into the image in image domain. The result of an adaptive bilateral filtering is shown in Fig. 11(c). For comparison, the result of original bilateral filter with constant parameter setting $\sigma_d = 3$ and $\sigma_r = 10$ is shown in Fig. 11(b). As Fig. 11 illustrates, bilateral filter with our noise model showed better performance of removing noise while preserving details.

5.2 Canny Edge Detection

The basic idea of Canny edge detection [12] is to find an optimal filter so that the most salient edges can be found in the presence of noise. The optimal filter is designed for theoretically independent of noise, but thresholds are noise dependent [6].

Generally, noise model can give us the probability that the magnitude of gradient is caused by whether noise or edge. In this section, we simply apply our noise model for the threshold of Canny edge detector. The high threshold was set to $6\,\bar{\sigma}_{int\,egrated}$, where $\bar{\sigma}_{int\,egrated}$ is the average of integrated noise standard deviation. The lower threshold is set to be 0.4 of higher threshold. The result is shown in Fig. 12(c). For comparison, the result of Canny edge detection with an automatic parameter setting in MAT-LAB^{TM} is shown in Fig. 12(b). The same low-pass filters are used for both methods. As Fig. 12 indicates, the performance of Canny edge detector was enhanced by our noise model despite of heavy noise.

(a)   (b)   (c)

Fig. 11. (a) Is the original image, (b) Is the result of bilateral filter and (c) Is the result of our adaptive bilateral filter. The bottom row shows enlarged patches for the top row images.

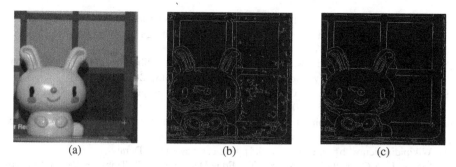

(a)   (b)   (c)

Fig. 12. (a) Is original image, (b) Is the result of Canny edge detector in MATLAB^{TM} and (c) Is the result of Canny edge detector with adaptive thresholds incorporated with our noise model

6 Conclusion

In this paper, we proposed the integrated noise model for digital camera sensors using Bayer domain images. To build the accurate noise model, our noise model defined in Bayer domain incorporated dark-current noise, shot noise and fixed-pattern noise together. In the experimental results, we showed that our noise model was changing only by image sensor gain. It means that our noise model does not need to be reconfigured if it is done once, and is best suited for consumer imaging devices such as digital camera, phone camera, etc. Also, we showed that our noise model estimated real noise accurately by comparing with ground-truth noise samples and synthetic noise model. In addition, we showed that the performance of some image processing algorithms were enhanced using our noise model.

References

1. Portilla, J., Strela, V., Wainwright, M.J., Simoncelli, E.P.: Image denoising using scale mixture of Gaussian in the wavelet domain. IEEE Trans. Image Processing 12(11), 1338–1351 (2003)
2. Perona, P., Malik, J.: Scale-space and edge detection using anisotropic diffusion. IEEE Trans. Pattern Analysis and Machine Intelligence 12(7), 629–639 (1990)
3. Tomasi, C., Manduchi, R.: Bilateral filtering for gray and color images. In: Proc. Sixth Int'l Conf. Computer Vision, pp. 839–846 (1998)
4. Hwang, Y., Kim, J., Kweon, J.-S.: Sensor noise modeling using the Skellam distribution: Application to color edge detection. In: Proc. IEEE Conf. Computer Vision and Pattern Recognition, pp. 1–8 (2007)
5. Weeratunga, S.K., Kamath, C.: Comparison of PDE-based non-linear anisotrofic diffusion techniques for image denoising. In: Proc. SPIE-IS&T Electronic Imaging, vol. 5014, pp. 201–212 (2003)
6. Liu, C., Szeliski, R., Kang, S.B., Zitnick, C.L., Freeman, W.T.: Automatic estimation and removal of noise from a single image. IEEE Trans. Pattern Analysis and Machine Intelligence 30(2), 299–314 (2008)

7. Yoo, Y., Lee, S., Choe, W., Kim, C.-Y.: CMOS image sensor noise reduction method for image signal processor in digital cameras and camera phones. In: Proc. SPIE-IS&T Electronic Imaging, vol. 6502, p. 65020 (2007)
8. Healey, G.E., Kondepudy, R.: Radiometric CCD camera calibration and noise estimation. IEEE Trans. Patter Analysis and Machine Intelligence 16(3), 267–276 (1994)
9. Faraji, H., MacLean, W.J.: CCD noise removal in digital images. IEEE Trans. Image Processing 15(9), 2676–2685 (2006)
10. Lim, S.: Characterization, of noise in digital photographs for image processing. In: Proc. SPIE-IS&T Electronic Imaging, vol. 6069, p. 60690O (2006)
11. Wikipidia. dcraw, http://en.wikipedia.org/wiki/Dcraw
12. Canny, J.: A Computational Approach to Edge Detection. IEEE Trans. on Patten Analysis and Machine Intelligence 8(6), 679–698 (1986)
13. Baek, Y.-M., Cho, D.-C., Lee, J.-A., Kim, W.-Y.: Noise Reduction for Image Signal Processor in Digital Cameras. In: Proc. Int'l Conf. Convergence and Hybrid Information Technology, pp. 474–471 (2008)
14. ISO 15739: Photography – Electronic still-picture cameras
15. Hytti, H.T.: Characterization of Digital Image Noise Properties based on RAW Data. In: Proc. SPIE, vol. 6059, p. 60590A (2006)

Searching High-Dimensional Neighbours: CPU-Based Tailored Data-Structures Versus GPU-Based Brute-Force Method

Vincent Garcia and Frank Nielsen

Ecole Polytechnique, Palaiseau, France
{garciav,nielsen}@lix.polytechnique.fr
http://www.lix.polytechnique.fr/~nielsen/

Abstract. Many image processing algorithms rely on nearest neighbor (NN) or on the k nearest neighbor (kNN) search problem. Several methods have been proposed to reduce the computation time, for instance using space partitionning. However, these methods are very slow in high dimensional space. In this paper, we propose a fast implementation of the brute-force algorithm using GPU (Graphics Processing Units) programming. We show that our implementation is up to 150 times faster than the classical approaches on synthetic data, and up to 75 times faster on real image processing algorithms (finding similar patches in images and texture synthesis).

Keywords: kNN, GPU programming, NVIDIA CUDA, image processing, finding similar patches, texture synthesis.

1 Introduction

Many image processing algorithms rely on nearest neighbor (NN) or on the k nearest neighbor (kNN) search problem. Typical applications are for instance finding similar patches in images [12], texture synthesis [8], object tracking [5], content based image indexing [15], deblurring [3,2], image filtering, etc.

The simplest way to solve the kNN search problem is the brute-force algorithm, also known as *exhaustive search*. However, the main issue of this algorithm is its huge complexity. Several methods have been proposed to reduce the computation time. For instance, a kd-tree [4] creates a partition of the point sets using a tree structure. The kNN search problem can take advantage of this structure by computing the distances between a given query point and a subset of the reference points. Another famous approach, named LSH (for *Locality Sensitive Hashing*) [11,9,6,1], uses hash functions to compute the distances between a given query point and a subset of the reference points. However, both of these approaches are inefficient (in terms of computation time) in many image processing algorithms because they still are very slow in high-dimensional space.

In this paper, we propose a fast implementation of the brute-force algorithm using GPU (Graphics Processing Units) programming. We show first that our

A. Gagalowicz and W. Philips (Eds.): MIRAGE 2009, LNCS 5496, pp. 425–436, 2009.
© Springer-Verlag Berlin Heidelberg 2009

implementation is up to 150 times faster than classical approaches (tree based) on synthetic data. Second, we apply our GPU implementation to two different image processing algorithms: finding similar patches in images and texture synthesis. These algorithms both use points in high-dimensional spaces (respectively up to 1323 and 660). In comparison to classical approaches (tree based), our GPU implementation is up to 75 times faster for finding similar patches in images, and up to 50 times faster for texture synthesis.

2   K Nearest Neighbor Search

2.1   Problem Definition

Let $\mathcal{R} = \{r_1, r_2, \cdots, r_m\}$ be a set of m reference points with values in $\mathbb{R}^d$, and let $\mathcal{Q} = \{q_1, q_2, \cdots, q_n\}$ be a set of n query points in the same space. The kNN search problem consists in finding the k nearest neighbors of each query point $q_i \in \mathcal{Q}$ in the reference set $\mathcal{R}$ given a specific distance. Commonly, the Euclidean or the Manhattan distance is used but any other distance can be used instead such as the Chebyshev norm or the Mahalanobis distance. Figure 1 illustrates the kNN problem with $k = 3$ and for a point set with values in $\mathbb{R}^2$.

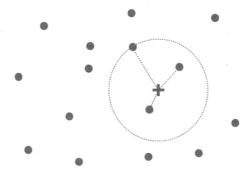

Fig. 1. Illustration of the kNN search problem for $k = 3$. The blue points correspond to the reference points and the red cross corresponds to the query point. The circle gives the distance between the query point and the third closest reference point.

2.2   Classical Approaches

Brute force The kNN search problem can be solved using the basic *brute force* algorithm (noted BF) and also called *exhaustive search*. Basically, for a given query point q_i, this algorithm consists in computing all the distances between q_i and the reference points and to select the k reference points providing the smallest distances. To be more precise, the BF algorithm is the following:

1. Compute all the distances between q_i and r_j, $\forall j \in [1, m]$.
2. Sort the computed distances.
3. Select the k reference points corresponding to the k smallest distances.

The main issue of this algorithm is its huge complexity: $O(nmd)$ for the nm distances computed (approximately $2nmd$ additions/subtractions and nmd multiplications) and $O(nm \log m)$ for the n sorts performed (mean number of comparisons).

Space partitionning Several kNN algorithms have been proposed in order to reduce the computation time. Generally, the idea is to reduce the number of distances computed [13]. A kd-tree [4] is a partition of the point sets using a tree structure. The kNN search problem can take advantage of this structure by computing the distances between a given query point and a subset of the reference points: only distances within nearby volumes are computed. Mount and Arya propose [14] a highly optimized implementation (written in C++) of the kNN search using a kd-tree structure. Their library, nammed ANN (for Approximate Nearest Neighbor) supports both exact and approximate nearest neighbor searching in spaces of various dimensions. ANN is currently one of the fastest kNN search using space partionning.

Locality-Sensitive Hashing (LSH) For methods based on space partitionning (e.g. using kd-tree), it has been shown [16] that the kNN search in a high dimensional space was comparable to the BF algorithm. Andoni *et al.* have proposed [11,9,6,1] a kNN search method, nammed LSH (for *Locality Sensitive Hashing*), very efficient for such a dimension. The basic idea is the following: two closed points are hashed in the same *bucket* (collision) with high probability. Basically, the authors propose to use a set of hash functions to compute the buckets related to the reference points. Then, the hash functions are applied for each query point. A simple hash table allows to find quickly the reference points closed to the considered query point. Finally, the distances are computed only between the current query point and the selected reference points. LSH is known to be faster than ANN in high dimensional space. Indeed, the computation of hash function value is very fast. However, the construction of the buckets should be preprocessed due to its slowness. However, in many image processing applications, buckets cannot be preprocessed.

3   GPU Programming and Application to kNN Search

Through the C-based API CUDA (Compute Unified Device Architecture), NVIDIA[1] recently brought the power of parallel computing on Graphics Processing Units (GPU) to general-purpose algorithmic [7,10]. This opportunity represents a promising alternative to solve the kNN problem in reasonable time. In this paper, we propose a CUDA implementation for solving the brute force kNN search problem. We compared its performances to several CPU-based implementations. Besides being faster by up to two orders of magnitude, we noticed

---

[1] http://www.nvidia.com/page/home.html
   http://www.nvidia.com/object/cuda_home.html

that the dimension of the sample points has only a small impact on the computation time with the proposed CUDA implementation, contrary to the C-based implementations.

The BF method is by nature highly-parallelizable. This property makes the BF method perfectly suitable for a GPU implementation. Let us remind that the BF method has two steps: the distance computation and the sorting. For simplicity, let us assume here that the reference and query sets both contain n points.

The computation of the n^2 distances can be fully parallelized since the distances between pairs of points are independent. Two kinds of memory are used: global memory and texture memory. The global memory has a huge bandwith but the performances decrease if the memory accesses are non-coalesced. In such a case, the texture memory is a good option because there are less penalties for non-coalesced readings. As a consequence, we use global memory for storing the query set (coalesced readings), and texture memory for the reference set (non-coalesced readings). Therefore, we obtain better performances than when using global memory and shared memory[2] as proposed in the matrix multiplication example provided in the CUDA SDK.

The n sortings can also be parallelized while the operations performed during a given sorting of n values are clearly not independent of each other. Each thread sorts all the distances computed for a given query point. The sorting consists in comparing and exchanging many distances in a non-predictable order. Therefore, the memory accesses are not coalesced, indicating that the texture memory could be appropriate. However, it is a read-only memory. Only the global memory allows readings and writings. This penalizes the sorting performance.

The Quicksort is a popular algorithm because it is one of the fastest algorithms. However, it is recursive and CUDA does not allow recursive functions. As a consequence, it cannot be used in our implementation. The comb sort complexity is $O(n \log n)$ both in the worst and average cases. It is also among the fastest algorithms and simple to implement. Nevertheless, keeping in mind that we are only interested in the k smallest elements, k being usually very small compared to n, we consider an insertion sort variant which only outputs the k smallest elements. As illustrated in figure 2, this algorithm is faster than the comb sort for small values of parameter k. For this experiment, $n = 4800$ points (both reference and query sets) drawn uniformly in a 64 dimensional space were used. Using the comb sort, the computation time is constant whatever the value k because all the distances are sorted. On the contrary, using the insertion sort, the computation time linearly increases with k. We define k_0 as follow: the comb sort and the insertion sort are performed in the same computation time for $k = k_0$. k_0 is the abscissa value of the intersection of the two straight lines shown in figure 2. For $k < k_0$, the insertion sort is faster than comb sort. Beyond k_0, the comb sort is the fastest. Figure 3 shows the value of k_0 as a function of the

---

[2] Memory shared by a set of threads with high bandwidth and no penalties for random memory accesses.

size of sets. k_0 approximately increases linearly. According to our experiments, the affine function approximating this increase, computed by linear regression, is given by:

$$k_0(n) = 0.0247n + 1.3404 \qquad (1)$$

where n is the size of the reference and query sets. The judicious choice of the sorting algorithm used depends both on the size of sets and on the parameter k. In our experiments, we used the insertion sort because it provided the smallest computation time due to the value of k and the size of point sets used.

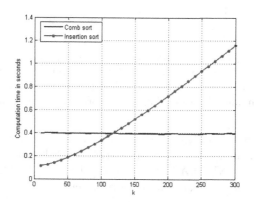

Fig. 2. Evolution of the computation time for comb sort (blue line) and insertion sort (red line) algorithms as a function of parameter k. For this experiment, 4800 points (reference and query sets) are used in a 64 dimensional space. The computation time is constant for the comb sort and linearly increases for the insertion sort.

4   Experimental Results

In this section, we consider two sets of n points (reference and query points) in a d dimensional space. These points are drawn uniformly in $[0, 1]^d$. The values n and d are specified bellow.

The initial goal of our work was to speed up the kNN search process in a Matlab program. In order to speed up computations, Matlab allows to use external C functions (Mex functions). Likewise, a recent Matlab plug-in allows to use external CUDA functions. In this section, we show, through a computation time comparison, that CUDA greatly accelerates the kNN search process. We compare three different implementations of the BF method and one method based on kd-tree (ANN). The methods compared are:

- BF method implemented in Matlab (noted BF-Matlab)
- BF method implemented in C (noted BF-C)
- BF method implemented in CUDA (noted BF-CUDA)
- ANN C++ library (noted ANN-C++)

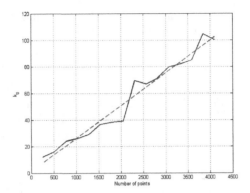

Fig. 3. Evolution of k_0 as a function of the size of sets in a 64 dimensional space. The red dashed line is the linear approximation of the experimental curve (blue solid line) computed by linear regression. Bellow this line, the insertion sort is faster than the comb sort algorithm, and above this line, comb sort is the fastest algorithm.

The computer used to do these experimentations is a Pentium 4 3.4 GHz with 2GB of DDR2 memory PC2-5300 (4×512MB dual-channel memory). The graphic card used is a NVIDIA GeForce 8800 GTX with 768MB of DDR3 memory and 16 multiprocessors (128 processors) interfaced with a PCI-express 1.1 port.

The table 1 presents the computation time of the kNN search process for each method and implementation listed before. This time depends on the size of the point sets (reference and query sets), on the space dimension, and on the parameter k. In this paper, k was set to 20. The computation time, given in seconds, corresponds respectively to the methods BF-Matlab, BF-C, ANN-C++, and BF-CUDA. The chosen values for n and d are typical values that can be found in papers using the kNN search.

The main result of this paper is that CUDA allows to greatly reduce the time needed to resolve the kNN search problem. According to the table 1, BF-CUDA is up to 407 times faster than BF-Matlab, 295 times faster than BF-C, and 148 times faster than ANN-C++. For instance, with 38400 reference and query points in a 96 dimensional space, the computation time is 57 minutes for BF-Matlab, 44 minutes for BF-C, 22 minutes for the ANN-C++, and less than 10 seconds for the BF-CUDA. The considerable speed up we obtain comes from the highly-parallelizable property of the BF method.

The figure 4 shows the evolution of the computation time as a function of the dimension d for sets of $n = 4800$ points. The dimension d influences only the duration of the distance computation process. The computation time seems to increase linearly with the dimension of the points. The major difference between these methods is the slope of the increase. For sets of 4800 points, the slope is 0.54 for BF-Matlab method, 0.45 for BF-C method, 0.20 for ANN-C++ method, and quasi-null (actually 0.001) for BF-CUDA method. In other words,

Table 1. Comparison of the computation time, given in seconds, of the methods BF-Matlab, BF-C, ANN-C++, and BF-CUDA. BF-CUDA is up to 407 times faster than BF-Matlab, 295 times faster than BF-C, and 148 times faster than ANN-C++.

	Methods	n=1200	n=2400	n=4800	n=9600	n=19200	n=38400
d=8	BF-Matlab	0.51	1.69	7.84	35.08	148.01	629.90
	BF-C	0.13	0.49	1.90	7.53	29.21	127.16
	ANN-C++	0.13	0.33	0.81	2.43	6.82	18.38
	BF-CUDA	**0.01**	**0.02**	**0.04**	**0.13**	**0.43**	**1.89**
d=16	BF-Matlab	0.74	2.98	12.60	51.64	210.90	893.61
	BF-C	0.22	0.87	3.45	13.82	56.29	233.88
	ANN-C++	0.26	1.06	5.04	23.97	91.33	319.01
	BF-CUDA	**0.01**	**0.02**	**0.06**	**0.17**	**0.60**	**2.51**
d=32	BF-Matlab	1.03	5.00	21.00	84.33	323.47	1400.61
	BF-C	0.45	1.79	7.51	30.23	116.35	568.53
	ANN-C++	0.39	1.78	9.21	39.37	166.98	688.55
	BF-CUDA	**0.01**	**0.03**	**0.08**	**0.24**	**0.94**	**3.89**
d=64	BF-Matlab	2.24	9.37	38.16	149.76	606.71	2353.40
	BF-C	1.71	7.28	26.11	111.91	455.49	1680.37
	ANN-C++	0.78	3.56	14.66	59.28	242.98	1008.84
	BF-CUDA	**0.02**	**0.04**	**0.11**	**0.40**	**1.57**	**6.65**
d=80	BF-Matlab	2.35	11.53	47.11	188.10	729.52	2852.68
	BF-C	2.13	8.43	33.40	145.07	530.44	2127.08
	ANN-C++	0.98	4.29	17.22	73.22	302.44	1176.39
	BF-CUDA	**0.02**	**0.04**	**0.13**	**0.48**	**1.98**	**8.17**
d=96	BF-Matlab	3.30	13.89	55.77	231.69	901.38	3390.45
	BF-C	2.54	10.56	39.26	168.58	674.88	2649.24
	ANN-C++	1.20	4.96	19.68	82.45	339.81	1334.35
	BF-CUDA	**0.02**	**0.05**	**0.15**	**0.57**	**2.29**	**9.61**

all the methods are sensitive to the space dimension in term of computation time. However, regarding to the tested methods, the impact of the dimension on the performances is quasi-negligible for the method BF-CUDA. This characteristic is particularly useful for applications using high dimensional space.

The figure 5 shows the evolution of the computation time as a function of the number of points n in a $d = 32$ dimensional space. The number of points n influences the duration of both the distance computation process and the sorting process. The computation time increases polynomially with n. Indeed, n^2 distances are computed. However, the impact of n is one more time quasi-negligible for the method BF-CUDA in comparison to other tested methods.

The figure 6 shows the evolution of the computation time as a function of the parameter k for sets of $n = 4800$ points in a $d = 32$ dimensional space. The parameter k influences the duration of the sorting process. The computation time increases linearly with k. BF-CUDA is less sensitive to k than other tested methods.

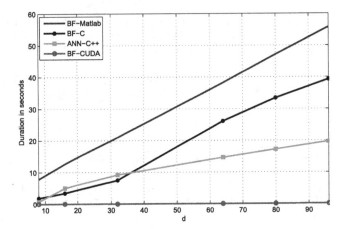

Fig. 4. Evolution of the computation time as a function of the point dimension for methods BF-Matlab, BF-C, BF-CUDA, and ANN-C++ for a set of 4800 points, $k = 20$. The computation time linearly increases with the dimension of the points whatever the method used. However, the increase is quasi-null with the BF-CUDA.

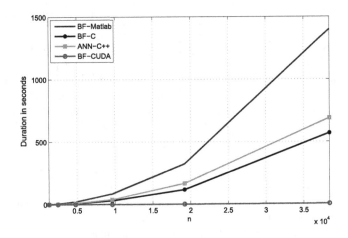

Fig. 5. Evolution of the computation time as a function of the number of points for methods BF-Matlab, BF-C, BF-CUDA, and ANN-C++ for a set in a $d = 32$ dimensional space, $k = 20$. The computation time polynomially increases with n whatever the method used. However, the increase is negligible with the BF-CUDA in comparison to other tested methods.

5   Application to Image Processing Problems

5.1   Finding Similar Patches in Images

The search of similar patches in images is a crucial problem in many computer vision algoriths. Given an initial hand edited patch in a image, the problem consists in finding the k most similar patches in the considered image. By treating each image patch as a point in a high-dimensional space, we can use a kNN algorithm to find the k most similar patches. In this context, Kumar *et al.* [12] study many different kNN search algorithms. They conclude that the tree based algorithms (vantage point trees [17]) have the best overall construction and search performance.

In this part, we compare our CUDA implementation of the BF method to ANN-C++. The initial patch size is 21×21 and the image size is 128×128. So, the problem of finding similar patches consists in finding the k nearest neighbors among 16384 points in a 441 dimensional space for gray level images and in a 1323 dimensional space for color images. Note that k is set at $k = 10$. For gray level images (dimension=441), the kNN search process takes 3.55 seconds with ANN-C++ and 0.06 seconds with BF-CUDA. In this case, BF-CUDA is 60 times faster than ANN-C++. For color images (dimension=1323), the kNN search process takes 11.03 seconds with ANN-C++ and 0.15 seconds with BF-CUDA. In this case, BF-CUDA is 75 times faster than ANN-C++.

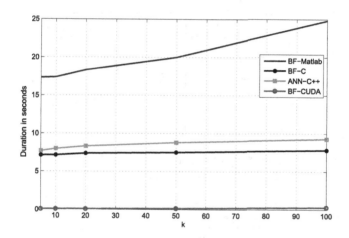

Fig. 6. Evolution of the computation time as a function of the parameter k for methods BF-Matlab, BF-C, BF-CUDA, and ANN-C++ for a set of 4800 points in a $d = 32$ dimensional space. The computation time linearly increases with k whatever the method used. However, BF-CUDA is less sensitive to k than other tested methods.

5.2   Texture Synthesis

Efros and Leung proposed in [8] a very simple but very efficient texture synthesis algorithm. Consider the problem of synthesizing a large picture I_t (of size $w_t \times h_t$) given a small texture sample I_s (of size $w_s \times h_s$). The synthesis algorithm first starts by filling the target image by random-colored pixels, and then synthesis the target image I_t by (re)assigning pixel colors, pixel by pixel, following the horizontal scaline order. For a given pixel position $(x, y) \in I_t$, we consider a square window centered at (x, y) of side $2s + 1$ where s denotes an integer parameter defining the neighborhood size and related to texture synthesis quality. In that window, note that $2s^2 + 2s$ pixels have already been synthesized. These pixels form an L-shape. For assigning the color of the pixel (x, y) in the I_t, we search for the best match of the current L-shape in the image I_s (see figure 7). The best match is defined as the matching position in I_s that minimizes the sum of square differences. Each L-shape can be map into a high dimensional vector where $d = 2s^2 + 2s$ (see figure 8).

We consider small texture samples (I_s) of size 64×64 pixels and we want to create a large picture (I_t) of size 128×128 pixels. The window size used is 21×21 pixels ($s = 10$). For gray level images (dimension=220), the kNN search process takes 0.72 seconds with ANN-C++ and 0.018 seconds with BF-CUDA. In this case, BF-CUDA is 40 faster than ANN-C++. For color images (dimension=660), the kNN search process takes 2.00 seconds with ANN-C++ and 0.04 seconds with BF-CUDA. In this case, BF-CUDA is 50 faster than ANN-C++.

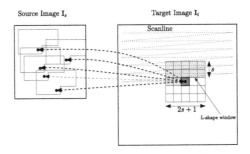

Fig. 7. Synthesis of a 2D texture image

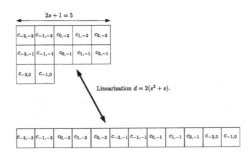

Fig. 8. Linearization of a L-shape into a high dimensional vector

6   Conclusion

In this paper, we have proposed a fast, parallel k nearest neighbor (kNN) search implementation using a graphics processing units (GPU). We have shown that the use of the NVIDIA CUDA API accelerates the kNN search by up to a factor of 150 compared to a classical tree-based approach on synthetic data. Likewise, our implementation is up to 75 times faster on real image processing algorithms (finding similar patches in images and texture synthesis).

References

1. Andoni, A., Indyk, P.: Near-optimal hashing algorithms for approximate nearest neighbor in high dimensions. In: IEEE Symposium on Foundations of Computer Science, vol. 51(1), pp. 459–468 (2006)
2. Angelino, C.V., Debreuve, E., Barlaud, M.: Image restoration using a knn-variant of the mean-shift. In: IEEE International Conference on Image Processing, San Diego, California, USA (October 2008)
3. Angelino, C.V., Debreuve, E., Barlaud, M.: A nonparametric minimum entropy image deblurring algorithm. In: IEEE International Conference on Acoustics, Speech, and Signal Processing, Las Vegas, Nevada, USA (April 2008)
4. Arya, S., Mount, D.M., Netanyahu, N.S., Silverman, R., Wu, A.Y.: An optimal algorithm for approximate nearest neighbor searching fixed dimensions. Journal of the ACM 45(6), 891–923 (1998)
5. Boltz, S., Debreuve, E., Barlaud, M.: High-dimensional statistical distance for region-of-interest tracking: Application to combining a soft geometric constraint with radiometry. In: IEEE International Conference on Computer Vision and Pattern Recognition, Minneapolis, USA (2007)
6. Datar, M., Immorlica, N., Indyk, P., Mirrokni, V.: Locality-sensitive hashing scheme based on p-stable distributions. In: Symposium on Computational Geometry, pp. 253–262. ACM Press, New York (2004)
7. Dudek, R., Cuenca, C., Quintana, F.: Accelerating space variant gaussian filtering on graphics processing unit. In: Moreno Díaz, R., Pichler, F., Quesada Arencibia, A. (eds.) EUROCAST 2007. LNCS, vol. 4739, pp. 984–991. Springer, Heidelberg (2007)
8. Efros, A.A., Leung, T.K.: Texture synthesis by non-parametric sampling. In: IEEE International Conference on Image Processing, Corfu, Greece, September 1999, pp. 1033–1038 (1999)
9. Gionis, A., Indyk, P., Motwani, R.: Similarity search in high dimensions via hashing. In: International Conference on Very Large Data Bases, pp. 518–529 (1999)
10. Heymann, S., Muller, K., Smolic, A., Frohlich, B., Wiegand, T.: Sift implementation and optimization for general-purpose gpu. In: 15th International Conference in Central Europe on Computer Graphics, Visualization and Computer Vision, WSCG 2007 (2007)
11. Indyk, P., Motwani, R.: Approximate nearest neighbors: Towards removing the curse of dimensionality. In: Symposium on Theory of Computing, pp. 604–613 (1998)
12. Kumar, N., Zhang, L., Nayar, S.K.: What is a Good Nearest Neighbors Algorithm for Finding Similar Patches in Images? In: Forsyth, D., Torr, P., Zisserman, A. (eds.) ECCV 2008, Part I. LNCS, vol. 5302, pp. 364–378. Springer, Heidelberg (2008)
13. Lv, Q., Josephson, W., Wang, Z., Charikar, M., Li, K.: Multi-probe lsh: efficient indexing for high-dimensional similarity search. In: VLDB 2007: Proceedings of the 33rd international conference on Very large data bases, pp. 950–961. VLDB Endowment (2007)

14. Mount, D.M., Arya, S.: Ann: A library for approximate nearest neighbor searching, http://www.cs.umd.edu/~mount/ANN/

15. Piro, P., Anthoine, S., Debreuve, E., Barlaud, M.: Image retrieval via kullback-leibler divergence of patches of multiscale coefficients in the knn framework. In: IEEE International Workshop on Content-Based Multimedia Indexing, London, UK. IEEE Computer Society, Los Alamitos (2008)

16. Weber, R., Blott, S.: A quantitative analysis and performance study for similarity-search methods in high-dimensional spaces. In: International Conference on Very Large Data Bases, pp. 194–205 (1998)

17. Yianilos, P.N.: Data structures and algorithms for nearest neighbor search in general metric spaces. In: Proceedings of the Fifth Annual ACM-SIAM Symposium on Discrete Algorithms, SODA (1993)

Author Index